Advances in 21st Century Human Settlements

Indexed by SCOPUS

This Series focuses on the entire spectrum of human settlements – from rural to urban, in different regions of the world, with questions such as: What factors cause and guide the process of change in human settlements from rural to urban in character, from hamlets and villages to towns, cities and megacities? Is this process different across time and space, how and why? Is there a future for rural life? Is it possible or not to have industrial development in rural settlements, and how? Why does 'urban shrinkage' occur? Are the rural areas urbanizing or is that urban areas are undergoing 'ruralisation' (in form of underserviced slums)? What are the challenges faced by 'mega urban regions', and how they can be/are being addressed? What drives economic dynamism in human settlements? Is the urban-based economic growth paradigm the only answer to the quest for sustainable development, or is there an urgent need to balance between economic growth on one hand and ecosystem restoration and conservation on the other – for the future sustainability of human habitats? How and what new technology is helping to achieve sustainable development in human settlements? What sort of changes in the current planning, management and governance of human settlements are needed to face the changing environment including the climate and increasing disaster risks? What is the uniqueness of the new 'socio-cultural spaces' that emerge in human settlements, and how they change over time? As rural settlements become urban, are the new 'urban spaces' resulting in the loss of rural life and 'socio-cultural spaces'? What is leading the preservation of rural 'socio-cultural spaces' within the urbanizing world, and how? What is the emerging nature of the rural-urban interface, and what factors influence it? What are the emerging perspectives that help understand the human-environment-culture complex through the study of human settlements and the related ecosystems, and how do they transform our understanding of cultural landscapes and 'waterscapes' in the 21st Century? What else is and/or likely to be new vis-à-vis human settlements – now and in the future? The Series, therefore, welcomes contributions with fresh cognitive perspectives to understand the new and emerging realities of the 21st Century human settlements. Such perspectives will include a multidisciplinary analysis, constituting of the demographic, spatio-economic, environmental, technological, and planning, management and governance lenses.

If you are interested in submitting a proposal for this series, please contact the Series Editor, or the Publishing Editor:

Bharat Dahiya (bharatdahiya@gmail.com) or

Loyola D'Silva (loyola.dsilva@springer.com)

Vien Thuc Ha · Hieu Ngoc Nguyen ·
Hans-Joachim Linke
Editors

Proceedings of the 8th International Conference on Sustainable Urban Development

ICSUD 2022, Binh Duong - Ho Chi Minh City, Vietnam

Springer

Editors
Vien Thuc Ha
Vietnamese-German University
Binh Duong - HCMC, Vietnam

Hieu Ngoc Nguyen
Vietnamese-German University
Binh Duong - HCMC, Vietnam

Hans-Joachim Linke
FB Bau-u. Umweltingenieurwissenschaften
TU Darmstadt
Darmstadt, Hessen, Germany

ISSN 2198-2546 ISSN 2198-2554 (electronic)
Advances in 21st Century Human Settlements
ISBN 978-981-99-8005-5 ISBN 978-981-99-8003-1 (eBook)
https://doi.org/10.1007/978-981-99-8003-1

This Springer imprint is published by the registered company Springer Nature Singapore Pte Ltd.
The registered company address is: 152 Beach Road, #21-01/04 Gateway East, Singapore 189721, Singapore

Preface and Acknowledgements

It is with great pleasure to present the proceedings of the 8th International Conference on Sustainable Urban Development (ICSUD). As an annual event organized by the Sustainable Urban Development Study Program at Vietnamese-German University, ICSUD serves as a dynamic platform for fostering knowledge exchange and collaborative networking among academics, researchers, policymakers, and representatives of non-governmental organizations in the field of urban development. The 2022 conference marked 10th anniversary of the Sustainable Urban Development Study Program (SUD) that took place at the newly established campus of Vietnamese-German University in Binh Duong Province, Vietnam.

The theme "*Innovative and Inclusive Urban Growth Models*" was chosen to resonate with the current scholar discussions and the National Urban Agenda on finding new growth models to sustain the growth after 30 years of rapid urbanization. Innovations should not be centered on technology alone; it should include new governing and business models to enable and sustain the growth among different interest groups, within and crossing disciplines such as design, planning, or mobility. Better growth models not only ensure environmental integrity but also promote social cohesion and economic prosperity.

The conference gathered distinguished speakers, scholars, and participants from various backgrounds in urban studies-, urban planning-, and urban development-related fields. Participants shared their new findings and experiences over three key cross-disciplinary themes including "governance", "climate change resilience", and "mobility". Of the 22 papers submitted, 16 papers were selected to publish in this Springer Nature Conference Proceedings under 3 chapters including Governance Based Transforming Growth Models, Climate Change Resilience Based Transforming Growth Models, and Mobility Based Transforming Growth Models.

We were privileged to have esteemed support from organizing teams. We specially thank for the diligent efforts of the Organizing Committee from Vietnamese-German University led by Dr. Vien Thuc Ha, Dr. Hieu Ngoc Nguyen, and Dr. Dao Thi Bich Van. We deeply express gratitude for the great work and support of Prof. Hans Joachim Linke and his team from the University of Darmstadt. We show gratitude for the financial support from Fredrick Neuman Foundation and the personal contribution of Prof. Andreas Stoffers. We acknowledged the contribution of the scientific board members who had contributed numerous works for the conference paper and the later proceedings. Their commitments were pivotal in creating a vibrant and intellectually stimulating conference environment. We extend our heartfelt gratitude to all committee members for their untiring work and support.

We would also like to express our sincere appreciation to all individuals and organizations who have contributed to the success of the conference. The participants, with their diverse backgrounds and expertise in urban studies-, urban planning-, and urban development-related fields, played a crucial role in enriching the academic discourse and laying the foundation for future collaborations and advancements in the field.

We would like to acknowledge the assistance provided by the conference staff, volunteers, and all those who contributed their time and effort to ensure the smooth operation of the event. Their dedication and professionalism were invaluable in creating a welcoming and stimulating conference environment.

Lastly, we express our appreciation to all the authors who contributed their research papers to this proceeding's publication. Their contributions have made this proceeding a valuable resource for further research, innovation, and policymaking in sustainable urban development. May the insights, ideas, and recommendations shared during ICSUD contribute to the advancement of *inclusive and resilient urban growth models* that will shape our cities in the future.

Binh Duong - HCMC Vietnam — Vien Thuc Ha
Binh Duong - HCMC Vietnam — Hieu Ngoc Nguyen
Darmstadt, Germany — Hans-Joachim Linke

Contents

Climate Change Resilience Based Transforming Growth Models

Mobility Based Transforming Growth Models

Governance Based Transforming Growth Models

20-Minute to Health-Oriented City: The Case of Ho Chi Minh City

Hieu Ngoc Nguyen, Anh Thai Pham, and Son Nam Pham

Abstract Building a 20-minute city is a desire but a big challenge for any city in the car-booming era. When the city became a metropolis, commuting choices narrowed when travel distance increased while alternative means of public transport were limited. The cost of pollution and congestion due to car transformation urge city authority to act; however, it seems that different vectors of growth such as mobility choices and property development could hardly match the planning schemes, resource for infrastructure development, and governing capacity. This article discusses alternative solutions to build a 20-minute city in the existing development pattern in Ho Chi Minh City in the context of resource limitation. Using an analytical framework to explore possibility to avoid car in the existing urban form, population and service density, land use changes and urban growth pattern, the authors argue how E-bike could increase the accessibility to jobs, housing, and service as well as reduce pollution and expenses for both households and city authorities. The discussion implicated a strategy to enable health-oriented city by developing service-oriented zones where green mobility, integrated planning, and new urban design to be supported by an innovative and inclusive governing model.

Keywords 20-minute city · Active transportation · Time-based strategy · Smart growth models · Integrated urban planning

1 Introduction

Ho Chi Minh City is transforming into an urban region with concern on the mobility. Sprawling and shifting to cars from scooters lifted-up the congestion level, especially in the arterial roads where suburban commuters have less choices to reach the city centers. The increasing commuting distance to jobs and basic services, extension

H. N. Nguyen (✉) · A. T. Pham · S. N. Pham
Vietnamese German University, Binh Duong, Vietnam
e-mail: hieu.nn@vgu.edu.vn

V. T. Ha et al. (eds.), *Proceedings of the 8th International Conference on Sustainable Urban Development*, Advances in 21st Century Human Settlements,
https://doi.org/10.1007/978-981-99-8003-1_1

of highways, and higher car affordability add up to the car-dependent region from transformation.

Public health became a critical issue during this transformation. The shift to cars only helped affordable groups living in the periphery to avoid pollution. Higher income commuters still use scooters for jobs and services due to congestion in the city center.

There are alternative solutions to a health-oriented society nowadays. Different approaches to combat regional growth from Transit-Oriented Development to smart growth to improve accessibility to jobs and services [1, 2]. Most solutions require to integrate transport and urban planning and attack the higher volume of traffic from demand management instead of supply. In general, regional transformation in the car-booming era requires a structural change to promote active transport.

Of the discussions nowadays, time-based strategies became a fashion in many cities; however, this is a highly challenging issue. The 15- or 20-minute city was chosen to call for integrated actions from public transport to last mile solutions [3, 4]. They shared lessons from car dependency and sprawling suburban problems. They also shared the concerns of integrating efforts from planning to implementation, from decision-makers to enterprises and societies.

The following article discusses opportunities to adopt time-based strategies to transform the city and region toward health-oriented growth in Ho Chi Minh City. The solution proposed to synergize development efforts to transform cities to service-centric zones to cluster the growth at shorter distance to the main destinations within 20 minutes of commuting. The proposal pivoted on the three alternatives to shift transport modes, guiding proper urban renewal target the public health from 20 minute approachability using greener and active transportation vehicle such as E-assisted bicycle. The reduction of emission was calculated based on the projection of new average distance travel, new modal shares, and the new population distribution in the expanding urban space.

As zone solution became the main approach, the division of zones was carefully chosen to ensure the effectiveness of measures and their practicality. Each zone was chosen with a combination of accessibility to equal population size, rather than the population size or area themselves. The zone also reflected a changing distribution of population and travel demand by 2040 from the 2020 perspective and the potential growth from the inertia projected from 2010.

2 Literature Review

The 15-minute city is a new concept but it had its roots in the tradition and development framework concentrate on the proximity to develop city center and enhance the local products for local needs. In this model, transport mode plays a central role to promote ensuring people may cycle, walk, or use other green vehicles for their routines. This aligns to the paradigm shift from Jane Jacobs to revitalize city centers from free-wheel suburbia toward net-zero mobility when changing of means and commuting time [5]. Nowadays, the specific time slogan: 15-, 20-, or even 30-minute city reflected a more flexible and tailored-made objectives to transform city's mobility and structure according to their development contexts (population size, urban form, or public transit and infrastructure) [3].

Most of these time-based strategies relied upon four transforming dimensions including density, proximity, diversity, and digitalization. Density means not only a higher number of populations over space, but also maximized supply of urban services for higher demand. Proximity is aimed at reducing travel distance to satisfy citizens' basic needs with better social connection, more efficient use of resources, and lower impacts to urban environment which is detrimental to public health. Diversity implies the mixtures of land use and building types to nurture local economy and housing as ecosystems for all income groups. In return, density and proximity reinforce city attractiveness from its diversity.

Recently, new enablers such as technologies, climate change, and the recent COVID pandemic also support faster transformation. Digitalization-enabled multiple solutions to organically enhance services for better people participation, increase road safety for non-motorized transport (NMT) with tracking and tracing vehicles over the "cloud". The saving of greener mobility will reduce greenhouse gas to support mainstream policies of climate adaptation [3]. NMT or similar means of vehicle will become the right choice for pro-environment groups that significantly contribute to the survival of the earth and human beings [6]. The COVID reinforced the need for health-oriented cities as walking or biking could increase city competitiveness [7]. In not too long distance, active transport is better than public transportation due to lower cost, better safety, easy route connectivity, and timing [6].

From the perspective of implementation, there are diverse strategies to implement the concept to match their development contexts. Paris center is a well-known case to exploit well-established physical infrastructure to promote active mobility [8]. Similar approaches were applied in Europe (Milan, Barcelona) to revisit its historical use of city centers. The spread of active mobility adopted in Bogota (Colombia) helped a chain of cities to share an integrate plan [9, 10]. At larger scale, Shanghai metro area adopted its Master Plan from 2017 to 2035 to target "*15-minute community life circle*". The city of Portland (USA) did not have propaganda in terms of minutes, but their "*Complete neighborhood*" scheme was to increase accessibility to services within 20 min using walking and cycling [4]. In short, each city found their own chance to implement the concept matching their development context (see Table 1).

Application of the concept requires better integration of efforts. In the case of Milan, 15-minute cities adopted multiple strategies to regenerate the railway and surrounding open spaces as well as commercial space. The surrounding areas were integrated with downtown services (public square, restaurant-open cooking, office, vineyards-green terraces, and event spaces) to make a pleasant place. The station zone was integrated with larger neighborhoods and up to the city scale via the circle line of green buffer where residents can reach the TOD zone from within 15-min walk to neighborhoods (Fig. 1).

In the case of Singapore, multiple strategies were introduced to enable time-based strategies. The efforts to change the workplace-home patterns to more time efficient

Table 1 Highlighted cases to adopt time-based urban models

Cities	Continents	Strategies	Details	Outcomes
Paris, France	Europe	Conversion	Place: school playgrounds to be transformed into parks; bike lanes add-ins Time: free to access after school hours	–
Milan, Italy	Europe	Conversion		
Singapore	Asia	Goals within city Master plan	20-minute towns 45-minute city	On-going to improve existing infrastructure, which enhances transportation service
Lagos, Nigeria	Africa	Conversion	Schools à food markets in pandemic	Commuting time decrease
Shanghai, China	Asia	Goals within city Master plan	15-minute community life circles	–
Bogota, Columbia	South America	Conversion	Bike lanes (84 km)	–
Portland	North America	Plan	Complete neighborhoods	–
Melbourne, Australia	Oceania	Goals within city Master plan	20-minute neighborhoods	–

Source [3, 4, 9–15]

(reaching destinations within 20 min in the town scale and 45 min in city scale) were designed and implemented with integrated, collaborative, and participatory plans. Through the multiple solutions of addressing first-last mile problem and making use of mass transit system, authorities and people worked together on multiple strategies such as Autonomous system, Dynamically Routed Services (DRS), rail and biking expansion, Transit Priority Corridors (TPCs), and point-to-point mobility (P2P) schemes. The recent striking strategy was to expand the cycling network from 400 to 700 km before 2040 and to ensure new employment centers enabling people to commute less [13].

However, lessons learnt from the cases may not reflect barriers for changes. What we learnt in each case was what works in their context; however, we could hardly learn why some were not chosen. Change is not easy when people abandon their comforts. Besides, change should overcome efficiency, the lifestyle, or the habit. When cities were adapted to the extent that scooter riders could sit on their moped to buy groceries and take their children from their door to school, people tended to excuse their inactive mobility due to economic and habitual reasons[1] [17, 18]. We should accept this to design a proper framework for changes.

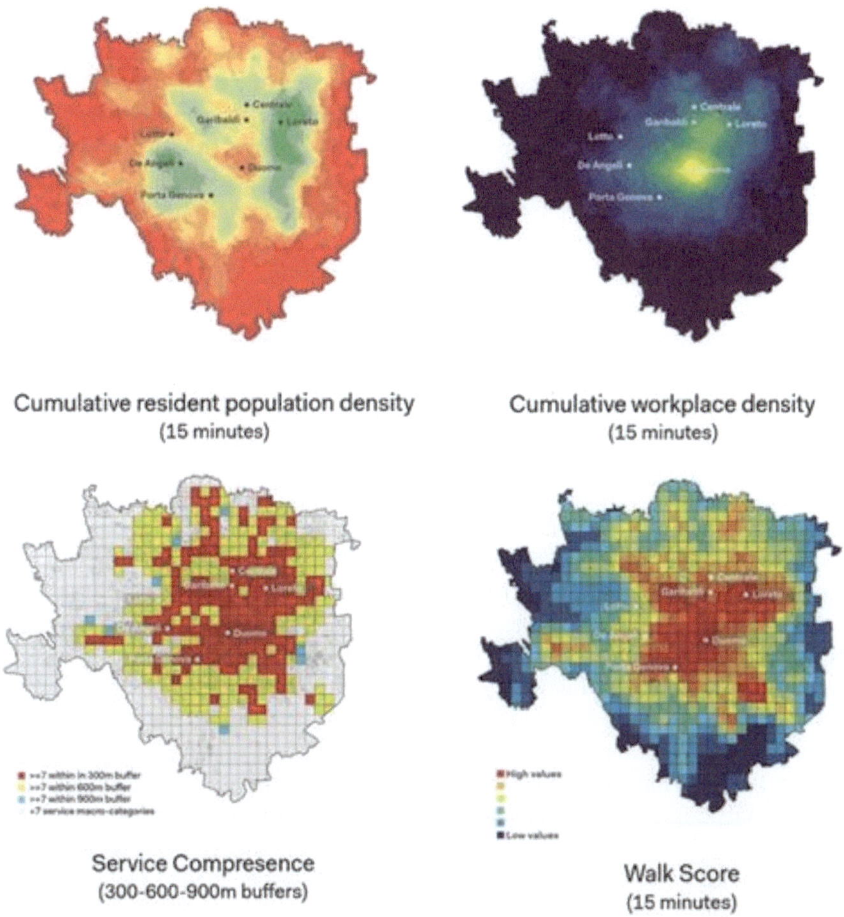

Fig. 1 Milan 15-minute city model from various aspects. *Source* [16]. *Notes* From left to right: 15-minute access model to: cumulative resident population density, cumulative workplace density, service compresence 300–600–900 m buffers, and the walk score

3 Ho Chi Minh City Transport Development Situation

Over the past three decades, the Ho Chi Minh City transport system fell behind the development needs. The land dedicated for traffic is well below the demand in both Central and Suburban Districts [19]. Near saturated level of service occurs during the day and over-saturated situations are common during the rush hours in both city center and arterial roads in the periphery. Note that given the two-digit growth of cars over the past two decades, the car ownership level is still the early transformation from scooter to car (about 7% of households using cars at the year 2021) [20]. We imagine how severe the transportation situation would be if more people switched to cars because of longer travel distance, severe pollution, and higher level of income.

The threat of car transformation was compelling over the existing poor transport infrastructure. Low road density in the densifying new districts[2] and narrow road system unfit for cars[3] explained critical congested city mobility problems. Increasing private vehicles and the low share of public transport exaggerated the situation. The number of motorbikes keeps increasing from the saturated level (7.44 million–nearly one scooter per person) and higher rate of car increase (1.9 times over past 5 years) pushed harder to the already overburden city transport infrastructure [21].

The reliance on regional infrastructure also faced with barriers in resource mobilization and other reasons. Over past three decades, the city footprint increased nearly four times that spill over seven provinces surround [22]. However, the regional infrastructure has not developed to match the multi-nodal regional model (as planned in the regional and transport development plan) [23]. No ring roads were completed with either suspended status or under patchy development over the past 10 years. Satellite cities and suburban towns have poor links to the large urban core as no rapid transit or expressways were developed from the adopted the master transport plan in 2013 [24]. The first metro line delayed more than 10 years with uncertain date of completion at the year 2024. The demand for financial resource was compelling; but barriers came from other areas such as site clearance, outdated planning, and incentive policies [25]. In short, the transit system and regional infrastructure could not be improved quickly (Fig. 2).

Given that we could develop the metro system on time, the desire of mass transit system may not function as good as expected due to the existing urban form. Besides the fact that city has high informal economy, Ho Chi Minh City was built upon years of adapting to scooter dependency with over 91% travel needs over 2 km [28]. Private vehicles are preferred because roads and lanes are made for them, not for walking. Apart from that, the accessibility to station plazas may reduce the ridership of the urban rail network. A preliminary analysis showed that only 17% of current building apartments are within a walking radius of 8 planned metro lines, and less than 25% population within the walking range, which is well below this figure in Tokyo or Shanghai of 80% [29]. Although the suburban area has potential to build new high-rise building apartments; however, the walking radius of current 27% (new districts) indicates that metro system alone does not significantly improve the situation, especially during the next decade [21] (Fig. 3).

Meanwhile, the city's targeted higher goals for sustainable development. In 2021, the Prime Minister adopted a task to revise the Ho Chi Minh City Master Plan to cope with the fundamental changes [30]. The new plan asked the city to sustain its development from a lower profile of overloaded transport and basic infrastructure, pending socio-economic problems, and higher threats from climate change and geopolitical uncertainties. Those who accept the plan should live with low commitment on large-scale infrastructure projects. Therefore, we should look for solutions that fit to availability of resources and capacity.

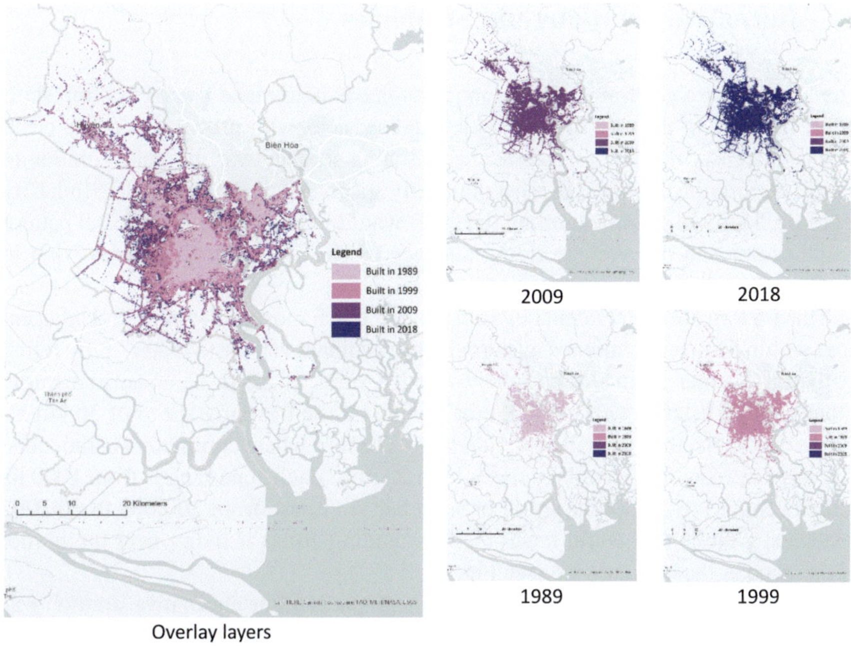

Fig. 2 Land conversion in Ho Chi Minh City from 1987 to 2018. *Source* [26, 27]

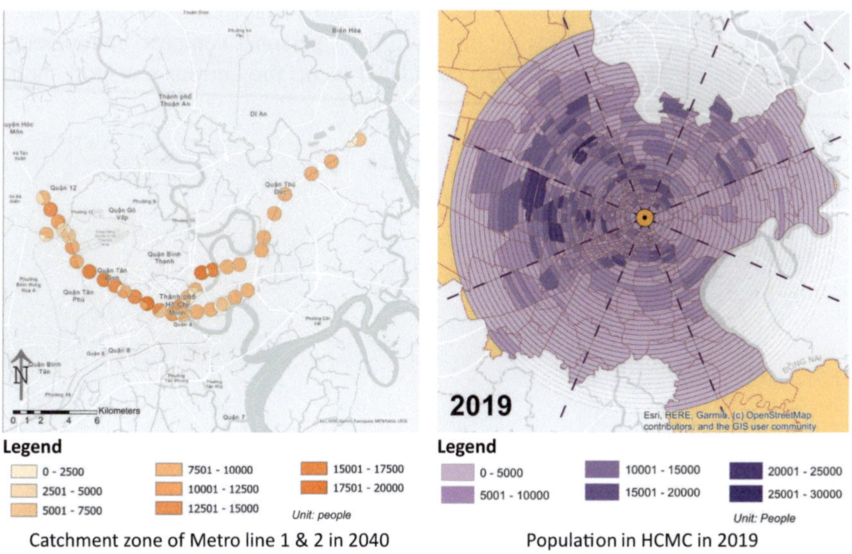

Fig. 3 Limited metro line's catchment in Ho Chi Minh City space. *Source* Authors

4 Alternative Solutions and Strategies

The basic solution to transform the mobility needs is to make use of existing short travel distance in the city proper to jobs and services with proper green and clean means of transport. The "*service-centric*" Zoning strategy was the main solution to reduce commuting distance in most of the city space. Applying to Ho Chi Minh City region, a 5 km radius zone circling the CBD would fit both the current travel pattern to the center and the average riding distance of Ho Chi Minh City (over 7 km in 2017).

This pattern allows residents in each zone to reach most of the zone and adjacent ones within 5 km or 20 minutes by the existing dominant vehicle (scooter). If traveling to the city center from the periphery or vice versa at off-peak or normal traffic situation, the distance will be 20 km or 40 min (at the traveling speed of 30 km/h). On the city scale, the city will be divided into 27 zones where commuters in the urban districts can reach its central service within 20 min. Each zone ranges from 1000 ha (urban centers) to 10,000 ha (rural or suburban) to house from 200,000 to 500,000 residents. Commuters could reach the surrounding zones if living near the border of the zone at the similar time (see Fig. 4).

The reduction of emissions to compare the impacts of the alternative solution was calculated following the Bosserhoff method [31]. The emission from individuals was a product of emission from vehicle per kilometer,[4] the average trips generated by individuals in Ho Chi Minh City (3.5 trips per person daily following recent research about vehicle occupancy in 2019), and the average commuting distances (~7.4 km) [21, 32, 33]. Besides the emission from personal vehicles, commercial/ public vehicle's emissions were also calculated using the current transport mode

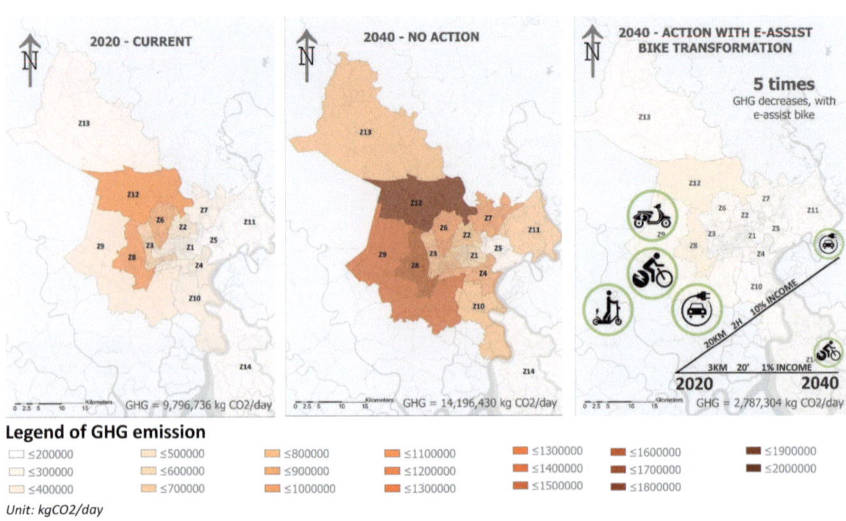

Fig. 4 Emission reduction with E-assist bike transformation at city scale. *Source* Authors

share. The emission in 2040 was calculated using a changing structure of modal share and an assumption that personal average trips per day and commuting distance will not change significantly (see Table 2).

As cities are transforming rapidly to urban regions, the biggest threat is to satisfy the mobility or accessibility needs in a larger spatial scale, with longer and higher cost, under slowly improved transport infrastructure during the car-booming era. If a time-based strategy is chosen, then there are multiple choices to be made from mode choice to exploitation city infrastructure and urban form to affordability of the transforming measures.

This zone strategy is expected to improve health through the reduction of emission together with the mode share transformation. The localized jobs and services within the zone enable commuters to restrain using cars at the long run. Shorter travel distance enables commuters to use active transport modes such as bike or E-assist bike. A scenario of new mode share by 2040 (see above) illustrated that the emission level may reduce 5 times (2700 Tons of GHG versus14,000 per day by 2040) if E-assist bike was the first choice in the mixed mode to commute within the zone. This choice may also reduce 10 times individual traveling costs (from about 10% of household income to 1% if shifting from the car to E-bike). The travel time may also reduce significantly from 2 h per day (if using car to cover 20 km distance) to 20 min (using E-assist bike or E-scooter to cover 5–10 km distance).

The zoning solution is a part of strategic transformation to more active transport modal share. To make the gain, a multi-modal transformation strategy shall be adopted to match the affordability of the stakeholders in the economy. A new structure of mode share chosen for the city in 2040 consists of UMRT, E-car, E-scooter, and E-assist bike. Each mode of transport has its advantages on certain travel distance and function; however, from the perspective of health-induced benefits for individuals and society, we can rank them by enablers to use them in the back-to-back format.

Table 2 Transport mode share 2040 scenario and 2020 mode share

Mode share	2040_Action		2040_No action		2020	
	General %	Detail %	General %	Detail %	General %	Detail %
Public	30		30		10	
Rail	30	9.0	30	9.00		0
Bus	60	18.00	60	18.00		
Taxi	10	3.00	10	3.00		
E-private	60		60		80	
E-bike	70	42.00	90	54.00	90	72
E-scooter	20	12.00	0	0.00	0	0
E-car	10	6.00	10	6.00	10	8
Others	10	10	10	10	10	
		100.00				

Source Authors

Table 3 shows that E-assist bike could be the best solution to individual health and societal environment.[5] E-assist bike could be strategic choice to form the backbone of transport mode for individuals from teenagers to retired ones commuting daily to shops, schools (bringing one kid along), and jobs within 20 minutes.

The choice of E-assist bike also fits the urban zones where the chance to widen the streets is small. In fact, the urban form of built-up area in Ho Chi Minh City grew to support scooters, so E-assist bike also fits well to the existing conditions. Analysis at Zone 2 showed that a rider at the center of this zone could reach a catchment of 160,000 person within 5 minutes riding and up to over 400,000 persons by 20 minutes riding. Within that time limits, the rider can also access the two richest targets of jobs and services (CBD and the airport) (Fig. 5).

However, E-assist bikers may be concerned about their safety in using slower mode, pollution impedes their health during riding time, and hot weather conditions may discourage office workers. To address these concerns, the following integrated approaches are proposed:

1. Better accessibility to destinations (jobs, home, and service network) via the land use and building approach.
2. Better commuting on roads (separated lanes, speed limits, managing at intersections, setting up biking corridors, and parking convenience) via transport management approach.
3. Better health environment (improve shading and reduce pollution level from controlling the health hazard emission level from vehicles and industries at specific zones and the city centers) via environmental approach (Fig. 6).

Table 3 Strategic modal choices and enablers

Factor	UMRT	E-car	E-scooter	E-assist bike
Land use	+1	−2	+2	+1
Access jobs	+1	−1	+3	+3
Access services	+1	−3	+2	+3
Building form				
Commuting	+1	−3	+3	+2
Time	+2	−3	+2	+3
Cost to home	−2	−3	+2	+3
Cost to city	+2	−3	+2	+1
Easy riding	−2	−3	+2	+3
Transfer/park				
Others	+1	−1	+2	+3
Emission	+2	+3	−1	−2
Safety	+3	+3	−1	−2
Weather and air pollution	+2	−3	−2	+3
Exercise	−2	+3	+2	−1
Carrying passengers	−2	+3	+2	+1
Carrying goods				
Total	7	−15	20	21

Source Authors

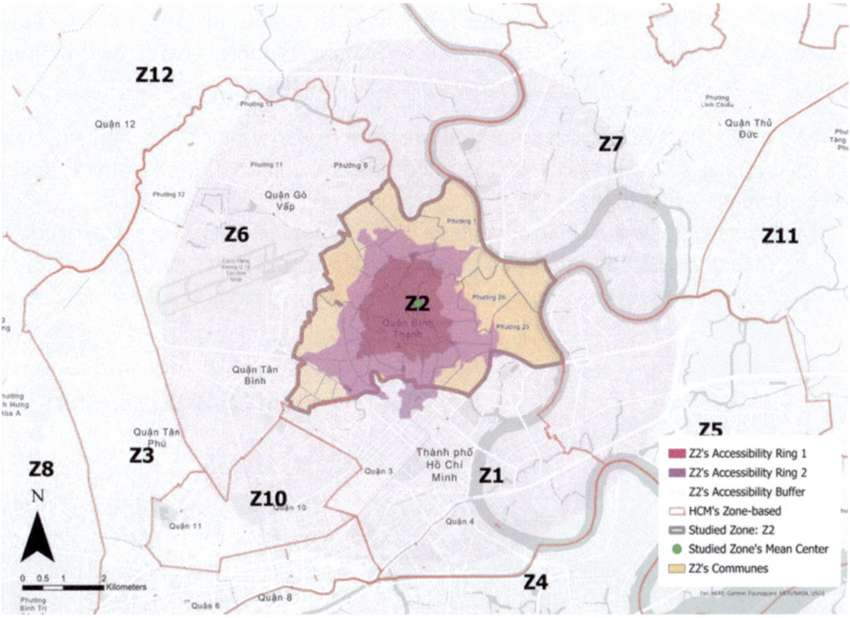

Fig. 5 Time-based accessibility analysis at Zone 2 using E-assist bike. *Notes* Accessible ring 2: 10 minutes by E-motorbike equals 20 minutes by E-assist bike. Accessible ring 1: 5 minutes by E-motorbike equals 10 minutes by walking. *Source* Authors

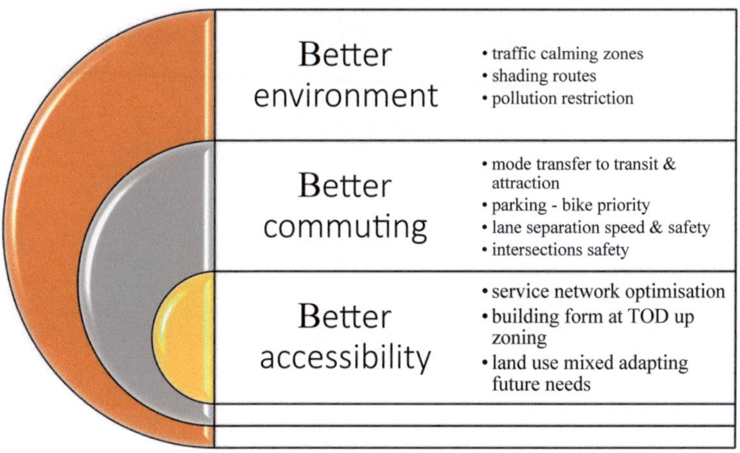

Fig. 6 Strategic approaches to health transformation. *Source* Authors

These approaches should be integrated in each zone and city space to help stakeholders navigating a successful transformation. Figure 7 highlighted actions to health-oriented society coming from both sides, including:

1. Make the most out of existing land use pattern and urban form, and improve shading on street network and service network distributions to streets favor commuters using bikes.
2. Enable stakeholders to take advantage of technological changes (electrification, automated vehicle, and renewable energy), digital and new trending economies (circular, home working, minimalism, eco-friendly lifestyle), and investment on

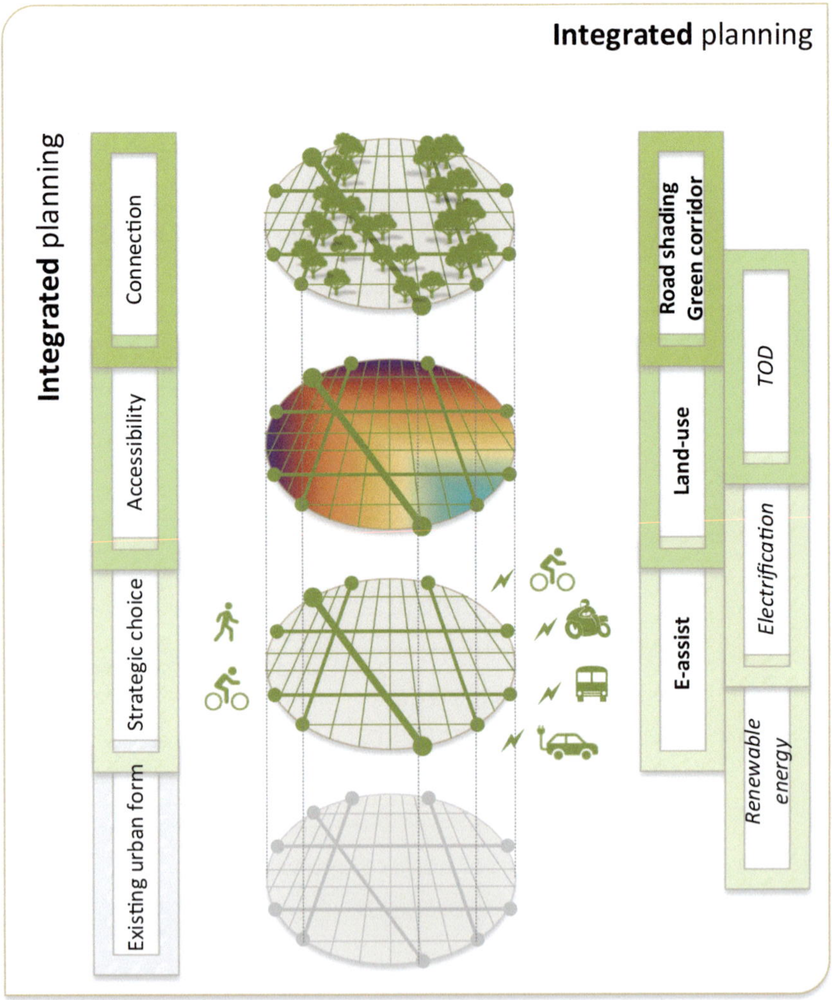

Fig. 7 Integrated solutions to transform the time-based strategy. *Source* Authors

transit system at both city and regional scale (Transit Oriented Development–TOD) (see below).

The integration shall be implemented flexibly to maximize the local features. In the city center, E-assist bike support could be the priority to take advantage of the downtown and TOD zones. To encourage bikers to ride without "*E-assist*", the authority shall reinforce innovative measures from shading (trees and any means to improve shading and reduce heat, noise, pollution, and dust) to favorable parking solutions and discouraging the abused of motorized vehicles, especially cars (combined measures on lanes separation, speed limits, and parking high charges and restriction).

However, in the newly established urban districts or rural ones, E-scooter or E-car may take a higher mode share to match the local conditions. Similarly, in the better planned zone with better access to future transit infrastructure development, walking and biking shall be favored with the above-mentioned measures; however, in poorer planned areas, we may have to adopt a higher share of E-scooters. The rural districts with low connectivity to transit system shall adopt a larger share of motorized vehicles (including E-cars) to maximize the economic opportunities and long-distance commuting demand.

5 Conclusion

The paper supplied evidence for planners to guide the city transformation toward healthier and competitive and sustainable society. The time-based strategies were developed upon practical zoning and flexible modal share solutions to meet the 20-minute target. The importance of affordability and practicality was considered, given the slower than expected mass transit system development. The innovative strategy of multi-modal oriented (MOD) instead of TOD reflected city was driven by scooter for more than half a century. The active transport should make use of the existing structure and technologies (electrification, digitalization, circularization, etc.) to "*intercept*" the compelling demand during the car-booming era in the fast-growing urban region.

Health-oriented cities were incrementally and collaboratively improved through innovative and integrated efforts. Better health was improved not only from more active transport, but also lower pollution due to shorter commuting to destinations with efficient and nature-friendly transport modes. More E-assist biker transformation will save huge resources for future growth and spare more space for unpredicted uncertainties from political change or climate change.

The proposed solutions of integration in planning and implementation are demanding; however, they were built from reality. The multi-level approaches to integrate actions asked people to change their behavior with huge benefit of conveniency and affordability. The green cover and shading solutions are cheaper than road widening or building tunnel solutions. The flexible and zoning solutions could spread the opportunities and cost to local businesses and individuals. This would create profound impacts on public finance and structurally reinforce society to become resilient and inclusive.

Assumptions in calculations may limit the claims on benefit or value, given that there are complicated and changing factors to predict the future. However, this article illustrated a bigger picture for framing priority and actions toward healthier society from the current situation. The choice of 20 minutes or 40 minutes to the city center from the periphery could become an icon for united actions against harmful threats of excessive use of cars and fragmented growth. Anyway, Ho Chi Minh City should create its opportunities from their undeniable problems and threats.

Let's do it together and do it now.

Notes

1. Average daily steps in Vietnam are only 3600 compared with 6880 in Hong Kong.
2. 2.01 km/km^2 in the new urban districts in 2017, only 8.8% of land dedicated for transport.
3. 50% of the 4000 km road length is less than 7 m.
4. Average emissions of selected vehicles are 45 g/CO_2/Pax-km for motorcycle, 9.2 g/CO_2/Pax-km for E-scooter, 135 g/CO_2/Pax-km for diesel car, and 34 g/CO_2/Pax-km for E-car.
5. This analysis assumes higher middle-income commuters will use cars more often if they must travel longer distance without better alternatives, while the lower middle-income group will shift to cars when they could afford in the next decade(s). Also, the highest income group and some of the lowest income commuters (10% + 10%) keep their current choice.

References

1. Suzuki H, Cervero R, Iuchi K (2013) Transforming cities with transit: transit and land-use integration for sustainable urban development. Urban Development. World Bank Publications
2. Cervero R (2001) Integration of urban transport and urban planning, in The Challenge of Urban Government. In: Freire SR (ed). World Bank Institute, Washington, pp 407–427
3. Introducing the "15-Minute City": Sustainability, Resilience and Place Identity in Future Post-Pandemic Cities. Smart Cities (2021)
4. 15-Minute City: Decomposing the New Urban Planning Eutopia. Sustainability (2021)
5. Deloitte, Urban Future With a Purpose: 12 trends shaping the future of cities by 2030
6. Green Mobility and Well-Being (2021)
7. Active mobility in historical centres: towards an accessible and competitive city. Transp Res Proc (2022)
8. Moro A (2022) Co-design of public spaces for pedestrian use and soft-mobility in the perspective of communities reappropriation and activation. Transp Res Proc
9. Bloomberg, Bogotá Is Building its Future Around Bikes
10. Akinfenwa G (2020) Covid-19: Lagos creates makeshift food markets in schools, in The Guardian
11. Analysis and optimization of 15-minute community life circle based on supply and demand matching: a case study of Shanghai (2021)
12. Administ SUPALR (2017) Shanghai Master Plan 2017–2035—Striving for The Excellent Global City
13. Authority SsLT (2019) Land Transport Master Plan 2040

14. Government VS. 20-minute neighborhood
15. Government VS. Plan Melbourne 2017–2050
16. Transport T (2022) Walkability and the 15-minute City Model: An Integrated Approach for the City of Milan. In: Livable Streets
17. Quang BT (2019) Vietnamese are the worst lazy for physical exercise ' Nguoi Viet luoi tap the duc nhat the gioi. https://vnexpress.net/nguoi-viet-luoi-tap-the-duc-nhat-the-gioi-3887015.html
18. Van NTC (2018) Strategies for Integrated Transport and Urban Development in Asian Developing Countries. In: Transport. Technische Universitat Darmstadt: Darmstadt, Germany
19. Prime_Minister (2013) Decision 568—Approval of Ho Chi Minh City transport master plan towards 2020 and beyond. In: 568/TTg, Prime_Minister, Editor. 2013: Hanoi
20. Hieu NN, Nguyen HD, Thong NM (2020) Mobility rights approach and some alternatives to control private vehicles in Ho Chi Minh City centre. Urban Rural Plan 2020(107+108):52–59
21. Hai PM (2018) Ho Chi Minh City Sustainable Urban Transport Index (SUTI)
22. Group WB (2020) Vietnam's urbanization at the crossroad: embarking on an Efficient, Inclusive, and Resilient Pathway. Washington, DC
23. Minister VP (2017) Decision No. 2076/QĐ-TTg on Approving the HCMC's Regional Construction Planning to 2030, Vision to 2050
24. Prime_Minister (2013) Phe duyet dieu chinh quy hoach giao thong thanh pho Ho Chi Minhden 2020—Decision 586. In: 568/TTg, Prime_Minister, Editor. 2013: Hanoi
25. Minh G (2021) Ho Chi Minh City ring road 3 was suspended for years (Duong vanh dai 3 thanh pho Ho Chi Minh dang do nhieu nam). from: https://vnexpress.net/3-duong-vanh-dai-o-tp-hcm-dang-do-nhieu-nam-4274972.html
26. enCity D (2019) Sasaki, Thu Duc Concept Master City Plan, transforming cities through innovation. Ho Chi Minh City, enCity
27. USAID (nd) Land Cover Portal
28. Van NTC, Tuan VA (2015) Strategies for integrated urban and transport development in motorcycle dependent cities. In: Urban Transport
29. Son NVN. Sustainable spatial strategy towards climate change resilience - Chien luoc phat trien khong gian vung theo huong ben vung va thich ung voi bien doi khi hau. in Ho Chi Minh City Master Concept Plan Revision towards 2030 vision 2050 - Y tuong dieu chinh quy hoach xay dung vung thanh pho Ho Chi Minh den nam 2030 tam nhin 2050
30. Prime_Minister (2021) General Contruction Planning Tasks of Ho Chi Minh City to 2040 and the Vision to 2060. Hanoi
31. Bosserhoff D (2000) Integration von Verkehrsplanung und raeumlicher Planung - Teil 1: Grundsaetze und Umsetzung; Teil 2: Abschaetzung der Verkehrserzeugung
32. TDSI (2017) Strengthening the management of individual motorized vehicles to reduce traffic
33. MCPT. Statistical data of Ho Chi Minh city management center of public transport

Public–Private Partnerships (PPPs) for Smart Transport Infrastructure Development: A Promising Approach and Potential Risks

Indra Gunawan, Tiep Nguyen, and Leonie Hallo

Abstract There is a growing body of literature examining the benefits and downsides of smart transport infrastructure development. Particularly, the integration of advanced technology regarding the Internet of Things (IoT), Big Data, and Artificial Intelligence (AI) within smart transport infrastructure systems provides promising solutions in coping with significant problems of urban cities, including rapid population growth, environmental pollution, and traffic jam. However, the lack of expertise and huge financial funding for smart transport projects may create potential risks during the implementation and operation stage in practice. Even though recent studies have provided some suggestions for the use of hybrid Public–Private Partnership (PPP) models for financial rising and allocation, there still have been some "unknown" aspects of smart transport infrastructure projects that are not completely captured with high consideration and detailed evaluation. Thus, this study aims to investigate the potential risks of applying PPP models for smart infrastructure development through reviewing articles indexed by Scopus, Web of Science, PubMed, and ScienceDirect from 2010 until the present. The initial results released that 19 critical risks were categorized into five main groups (political-legal, technological-technical, economic-financial, social, and environmental) that highly contribute to the success of smart transport infrastructure implementation.

Keywords Public–private partnership · Smart transport infrastructure · Risk management

---

I. Gunawan · L. Hallo
The University of Adelaide, Adelaide, South Australia, Australia

T. Nguyen (✉)
International University, Ho Chi Minh City, Vietnam
e-mail: nvtiep@hcmiu.edu.vn

Vietnam National University, Ho Chi Minh City, Vietnam

V. T. Ha et al. (eds.), *Proceedings of the 8th International Conference on Sustainable Urban Development*, Advances in 21st Century Human Settlements,
https://doi.org/10.1007/978-981-99-8003-1_2

1 Background

1.1 The Need for Smart Transport Infrastructure Development

The proportion of the world's population residing in cities by 2030 is predicted to grow to over 60% [1]. This degree of urbanization will create challenges across the board, including transportation. Cities around the world are already noticing the impacts of this shift in both developed and developing countries [2]. The concept of smart cities has developed over the last two decades and has been the subject of much academic research. A smart city generally aims at increasing operational efficiency, increasing information sharing with all stakeholders and improving the well-being of citizens using ICTs [3]. Technologies are used to improve the performance of many aspects of the smart city, including transportation, which is a major difficulty and source of much public complaint.

Public transport systems are playing an increasing role in large cities, and increasing the capacity of a system is expensive and takes time. As noted by Yang et al. [4], current policy strategies for sustainable urban growth are inadequate. As they say, "...most 'smart city' projects seek to find cost effectiveness within existing structures, which have not been built in keeping with the sustainability goals" [4]. As the demand for public transportation grows exponentially in the world's largest cities, along with growth in population density and lack of availability of land, public transportation networks have become increasingly congested, especially during peak hours. Thus, traveling times become longer, resulting in inefficiencies and reduced productivity. Rapid urbanization, as well as rapid motorization, brings challenges. Exponential growth in the number of private vehicles has escalated the requirement to come up with smart public transport systems in order to address congestion issues [5]. Many major cities around the world are experiencing challenges in managing transportation networks which are huge, expensive, and complex. Sustainable development must be supported by smart transportation infrastructure, with increased participation of the public and user-friendly interfaces giving real-time urban data.

One method proposed to reduce congestion is that of transportation demand management (TDM), and automated fare collection data can be helpful in the design phase of management systems for public transport demand. Travel demand management can help increase the efficiency and effectiveness of public transportation, including introducing price discrimination to reduce travel during peak periods and providing free transport in off-peak times. Peak surcharges have been implemented in many large cities worldwide, including Washington, London, Tokyo, and Sydney. Reward programs have also encouraged people to travel in nonpeak periods [6].

Of course, credible public transportation requires planning. It also requires effective governance. Factors involved in exacerbating difficulties include population growth, cost of various alternatives, lack of resourcing, geographic issues, affordability, constraints regarding land, and difficulties in acquiring land, often from private ownership [3]. Public–private partnerships (PPPs) is one method by which the transportation sector can be expanded [7]. PPPs are our long-term arrangements between a government and a private agency generally with the private capital financing the project up front and then taking revenue during the PPP contract [8]. Such ventures can promote better quality infrastructure with optimal risk and cost allocation [9]. While there may be different specific arrangements, common features include and a long-term commitment resulting in specific goods or services [10]. As noted by Calderini et al. [11], PPPs are appropriate ways to overcome the shortages in public finances often experienced. Various forms of PPP have been suggested for use in smart city projects, but drawbacks in high transaction costs, the lengthy time between conception and completion, political risk, and lack of finances have limited the application of this approach. In its favor, the exchange of information from private actors to public actors (who generally have less experience) is a very positive feature of such approaches.

1.2 Debates on PPPs Selection for Transport Infrastructure Development

PPPs have received increasing academic attention in relation to transportation infrastructure. In their meta-analysis, Chen et al. [7] examined studies in this field in terms of performance, contracts, risk, value for money, and institutional factors. They make the point that a comprehensive understanding of the performance of PPPs is not easily obtained because of limited data due to commercial confidentiality reasons. In their consideration of the risks involved, they highlight the appropriate allocation to private and public partners. A study by Sinha and Jha [12] reports on the use of PPPs in the National Highway development project of India in 1998 and subsequent use of PPP procurement throughout the transportation sector there. They draw attention to disputes over land acquisition and other approvals from government and the need therefore for dispute resolution clauses within agreements. They comment that PPPs are complex in their nature and can suffer from many unexpected circumstances with frequent disputes. Some studies have used a case study approach to look at the various risks. Thus, Grimsey and Lewis [13] and Chung et al. [14] described the risk scenarios existing within specific case studies. Other studies have been based on surveys of the professionals' perceptions of PPP undertakings.

Studies of investors have also been carried out, and a study by Demirag et al. [15] investigated the perceptions of investors in the finance industry and concluded that risks are a common concern that may vary; their respondents focused on the need to control these risks through ensuring or allocating risks to others. They also noted that investors prefer to partner with firms with whom they have some background. In addition, a study by Chen and Hubbard [16] investigated the market for PPPs in China and looked at differences in risk perception between the public and private sectors and between practitioners and academics. They found that PPP project risks commonly held to be important included government intervention, poor decision-making by public agencies, and corruption. The question of how risks should be managed has also been investigated, generally using surveys. Thus, Bing et al. [17] surveyed experts on preferred risk allocation in the UK and found that both macro- and micro-level risks should be kept by the public sector or at least shared with their private partners, according to the respondents. On the other hand, meso-level risks should be borne by the private sector.

There are many risks when PPPs are selected and undertaken in developed and developing countries. This situation is mirrored when we consider smart transport infrastructure PPPs.

2 Research Method

The literature review was undertaken in three phases. Firstly, a comprehensive literature review identified potential risks during metro railway project implementation. The Scopus, Web of Science, PubMed, and ScienceDirect databases were selected, and the following strings of TITLE-ABS-KEY were used, including 'Smart Transport Infrastructure'; AND 'Project Development' to provide a holistic picture of smart infrastructure project development over the world. The initial search retrieved 279 articles for subsequent refinement. The scope was then narrowed with the following string searched: 'Smart Transport Infrastructure'; AND 'Project Development'; AND 'Investment Models' within the specific time from 2010 to 2022 with popular and well-known publications. This period was selected because of the significant contributions of scholars to smart infrastructure literature. The second phase of the literature review refined the search. Abstracts and introduction sections were analyzed, and articles of a specific technical nature were not considered. This second phase produced a sample of 126 research outputs. In the third phase of the literature review, papers selected in the second phase were read, and those relevant to the 'smart' factor and associated investment models, mainly focusing on public–private partnership models. Next, a qualitative content analysis was undertaken to identify the advantages and potential risks of PPP model selection for smart transport infrastructure projects. The final database consisted of 57 papers published in international journals, books, conference proceedings, or other research outputs.

3 Findings and Discussions

3.1 The Real Meaning of 'Smart' in Transport Infrastructure Development

This literature search produced several views about a 'smart' factor that was reflected through many aspects of smart infrastructure projects, including technical-technological, financial-economic, legal, social, and environmental aspects.

Technological-Technical Aspect

Many studies concern the use of technology to enhance transportation efficiency and usability. Smartcard data was often mentioned in studies in this suite. A study by Kim et al. [18] investigated how to gain data about public transportation infrastructure and systems and their usage. Smartcard data can be used to provide information about passengers' timings but does not give information about trip purposes, which is a downfall. Donnici et al. [19] also investigated sustainable mobility to reduce congestion, pollution, and energy consumption. They suggest that integrating the various transport systems and encouraging the use of electric vehicles will achieve these aims. These authors focus on standards that help manufacturers to produce low-emission cars, encouragement for investment in alternative fuels, developing long-distance bus connections in Europe, and developing batteries.

Interestingly, an integrated development strategy is the focus of the paper by Shichiyakh et al. [20] that reports on the development of smart cities within Russia. Given that transport is one of the eight major components of a smart city, intelligent transport systems are needed, and these could allow riders to track the movement of buses, trains, and underground trains in real time and therefore be able to plan the trip more accurately. People can also see where there are traffic jams. The study reports on progress in several cities around the world along these lines. In addition, optimizing public transport is the topic of an article by Tirachini and Hensher [21] in which the authors include two key decision variables which had not hitherto been much researched: the best choice of a fare collection system and investment in dedicated infrastructure for buses in relation to their optimal speed. While previous literature has considered frequency, capacity, and spacing between stations, the running speed of buses and the fare payment method have not been investigated. The authors conclude that an off board payment system is the most efficient method during high demand periods, and that optimal speed for the bus grows in line with demand.

Moreu et al. [22] investigate problems within North America's railroad infrastructure, particularly the hundred thousand bridges that need to be regularly inspected. Knowing which bridges need to be replaced or repaired is critical. This study suggests using autonomous wireless smart sensors to measure the responses of bridges and report the result of using this technique in the US and Canada. According to these authors, previous methods of monitoring bridges' health are inadequate. The authors conclude that this technique suits harsh railroad environments and conditions. This method could be elaborated on and data collected remotely, which would be a safe and inexpensive approach. Stipanovic et al. [23] investigated transportation infrastructure failure due to several factors, including aging infrastructure, increased numbers of extreme weather events, and increased loading of traffic. The cost of maintenance is therefore increasing. This study looked at predicting the fatigue life of steel bridges. It used the case study of a bridge in Rotterdam in Europe, suffering cracking due to fatigue, concluding that their whole life cycle cost model considering both direct and indirect costs was able to demonstrate various investment and maintenance strategies and their differing costs.

The safety of pedestrians via a warning signal to prevent road accidents was investigated by Ess et al. [24], who noted that pedestrian deaths had remained high despite the rollout of various automatic braking systems and cooperative intelligent transport systems. The authors looked at uncontrolled crossings and user modeling software for traffic accident investigation. They conclude that such intelligence warning systems have great potential in preventing vehicle-pedestrian collisions if an effective warning signal can predict the trajectories and acceleration of road users and the stopping distance of vehicles. There may be false positive alarms, but this is a low price. To be useful, smart pedestrian crossings need to calculate stopping distances and provide a suitable warning to the driver. They also need to predict road users' behavior: many drivers take no action at all, but 1% accelerate. Pedestrians slow down as a whole. High collision risks occur when the driver and the pedestrian accelerate or do nothing.

Financial-Economic Aspect

A study by Borozan et al. [25] investigated risks involved in the electrification of transport systems and linked the difficulties with climate change uncertainties. They comment upon the need for cost-effective integration of electrified transport using smart charging concepts and investigate various investment and operation models of grid-to-vehicle, vehicle-to-grid, and vehicle-to-building charging options. Demand for EV charging will be shaped by the vehicle and the owner's driving patterns. This is a complicated connection, with planning uncertainties around timing, and energy consumption during trips, which will depend upon the initial amount of charge. The authors conclude that the economic value of smart charging depends upon reductions in the costs they enable. The risks they mention include the uncertainties involved in long-term developments within power provision and the risk of stranded investments. It is challenging to advise investors without greater certainty in the system. Incorporating smart charging was demonstrated in this study to reduce option values significantly and save money over a 40-year time horizon. Smart charging concepts, in other words, are very worthwhile if properly planned.

A study by Katrenko et al. [26] examined the relationship between the effectiveness of transportation systems and the level of economic development. The main very good point that highly effective transport systems have is that they provide many opportunities and many positive outcomes, whereas insufficient transport systems have significant economic costs, including quality of life. This paper is concerned with mathematical methods of managing traffic flows and developing a software product for this purpose. They comment that no general theory has yet emerged that is able to adequately describe real road traffic conditions. There are many models, but in fact, this is a complex system that must include drivers' reactions under different circumstances. Most models are based upon average speeds measured at checkpoints over a period. They suggest a software tool and apply it to the central part of the city of Lviv in Ukraine. There is a clear need for additional parking with infrastructure organized so that people can leave their car at the entrance to the parking center and then easily access public transport. Other cities also use this concept of parking the car and go. We can only assume that this research has been suspended at this time.

Smartcard data was also the focus of attention in an article by Siripanich et al. [27], who looked at various information derived from transit smartcard data in Australia. The article demonstrated that properties in well-connected and well-serviced locations generally have higher property values. Value adding occurs from public investment in transportation infrastructure. Residential and other properties near public transport options will encourage further development and increase residential land values. People want to be close to transport infrastructures such as roadways and public transport stations. Real estate around the most accessible stations for public transport will have a greater premium. Smartcard ticketing systems enable convenience and integration, but they also provide the opportunity for advanced research in transport. User data is collected at a lower cost; however, as noted in a previous study, it does not provide information about the journey's purpose. Sometimes smartcard data is not easily accessible due to privacy issues. This study examined the connection between the pricing of real estate and public transport usage from smartcard data examined through spatial regression models. A positive correlation between peak trips and real estate was not found for properties around the CBD.

Legal Aspect

The introduction of smart characteristics into an existing infrastructure brings unique legal risks. Many of these risks concern legalities around data rights, protection of personal data and privacy. A study by Losavio et al. [28] investigated legal challenges arising within smart cities, such as risks concerning the massive amount of data collected and analyzed. According to these authors, there is a great opportunity for a collision between public policy and personal autonomy when smart characteristics are introduced into a city.

Given the novelty of the situation, legalities have not necessarily kept up with such developments, and 'legal digital investigation-whether public or private in the age of the IoT invokes elements of information security, public security, and the legal regulation of public and private life. These may be unprecedentedly not yet directly addressed by law or tradition' Losavio et al. [28]. The situation parallels the enormous transformation caused by technology within health care at the lag in the domain of ethics in keeping up with that technology. The introduction of smart poles within smart cities, for example. Abosaq [29] also considered privacy issues which arise when smart infrastructure is deployed and the paramount need to consider the secrecy of private data and enact measures to ensure this. Protecting user data from exploitation will require strict regulations.

Moving to transport infrastructure, a study by Ge et al. [30] looked at legal issues around smart transportation in China and the US and included the downsides of legal costs associated with highway accidents. They point out that 'it is important to …. Form a rule base of administrator policies and judicial regulations in the fields related to intelligent transportation systems'. The future large-scale acceptance of smart transportation systems requires more detailed regulatory plans than the currently existing one. These authors suggest that China has a more systematic approach than the US on this topic. Regarding privacy protection, there needs to be strict protection for users' sensitive data and its control. The types of information that need protection will include data from the vehicle sensors and cameras and personal information from what drivers and passengers are doing in the car, such as their website browsing, text messaging, phone calls, etc. As they say, the wide application of new technology brings problems alongside the many opportunities.

Along with the significant social and economic benefits, there should be safety gains, especially from autonomous vehicles. There is a potentially huge economic plus to this development. Mobility will be transformed by these changes, enabling groups that currently cannot easily access good transportation options, such as the elderly and disabled, to move about much more easily.

Another study of the legal and policy issues connected with smart transportation infrastructure is by Glancy [31]. Connected vehicles could interact with the legal system of countries in various ways, including product liability, law enforcement access to connected vehicles, and privacy issues which this author asserts relate to autonomy, personal information, and surveillance. Each of these privacy interests needs addressing. Dewi and Putra [32] pick up the issue of law enforcement on highways and the lack of suitability of current methods in dealing with smart systems for legal enforcement against road traffic violations. Another relevant study here is that of Selim et al. [33]. They investigated the use of PPPs in critical infrastructure development and pointed out the need to set out the legal framework and regulations, including privacy and data protection, by introducing clauses regarding updates as these arise.

Legal risks have been considered in the context of PPPs. Cruz et al. [2] suggested that greater flexibility is necessary, as is a profound change of the existing regulatory and procurement models, to ensure that the private sector will continue to have a pivotal role concerning infrastructure, financing, and management. Chen et al. [4] investigated how and why power relations between government, private investors, users, and citizens in PPP projects distort risk allocation in favor of the stronger party. As a result, risks may not be allocated to the party best able to manage them, as in PPP theory. An analytic framework was set up in which the relative power of parties is founded on the resources they command in the institutional environment. This was then applied to a case study of the Jinlong toll road project in Zhejiang Province, China. Chung et al. [9] proposed that managing commercial risks involving economic decision-making should be allocated for the private sector while unquantifiable social and public risks and those in the domain of public governance are best left with government alone appears to replete with refutable implications. Public perception is a malleable concept and should be managed by both sectors.

Demirag et al. [10] assessed that PPPs' financiers perceive an extensive range of risks as important, and that it is essential to them that many of these risks are either insured or allocated to sub-contractors. When considering participating in PPPs, financiers agree that working with familiar partners on familiar projects and sectors is essential, which may raise barriers to entry and undermine competitive processes. A study by Bing et al. [12] suggested that some risks should still be retained within the public sector or shared with the private sector. Particularly, the main risks of PPP/PFI projects within the micro-level group should be allocated in the private sector. However, there are a few risks where unilateral allocation is not always apparent.

Zhang et al. [34] evolved a legal risk analysis framework for international PPP projects and identified key legal risk factors in each stage of international construction projects. They point out that previous studies have focused on overall risks and have not considered the correlation of legal risk factors. This could lead to inefficient risk management, they say. They developed a multimethod analysis framework by considering articles from various countries. These authors identified several risk categories, including laws and regulations risk, including changes or lack of a suitable framework; legal system risk, including unfair law enforcement or unsatisfactory litigation system; and contractual risk, including flawed and unfair contracts, modification of contracts or inaccurate interpretations. The authors comment that variations in political, market, and legal environments in different countries may produce different results.

A survey of Indian road projects by Iyer and Sagheer [35] referred to a lack of legal awareness as a factor behind significant delays in project approvals and often leads to legal proceedings. Similarly, Likhitruangsilp et al. [36] studied the risk perceptions of public and private sectors involved in PPP transportation projects in Vietnam. They cited frequent changes in laws and regulations as the reason behind delays in project approvals and permits within such projects, along with inadequate transparent legal frameworks. Changes in tax regulation were also a principal risk factor uncovered in the study. Such risks are beyond the control of project participants.

Social Aspect

Palmentieri [5] studied the effectiveness of participatory governance through a survey of citizens in Naples regarding their mobility habits to understand why the public transport sector was inefficient and the consequences of that inefficiency. In seeking an intelligent mobility model, these authors reported upon the strategy of upgrading the railway, road, and IT infrastructures and improving parking areas to make urgently needed reductions in congestion. This author points out that mobility plays a crucial role in improving the quality of life in metropolitan areas. Smart mobility is about traffic management, freight flows, road safety, private vehicles integrated with the public transportation network, security of IT, and a meaningful fares policy. It is no longer sustainable to rely primarily on private vehicles for transportation. Therefore, public transport must be fit for purpose. There also needs to be a holistic approach which systemically considers all types of transportation.

Many articles also concern social aspects, including changing people's behavior and making public transportation more attractive. Thus, Kolobova [37] investigated the necessary renovation of Moscow, including improvements in the public transport infrastructure system. While the main focus of this article is on other aspects of creating a creative city, they did mention the need for additional transport interchanges and infrastructure within walking distance of newly built houses. Halvorsen et al. [6] investigated public transport demand management which can help improve decision-making and reduce crowding. They investigate the main steps for developing such programs regarding problem identification and goals, design, evaluation, and monitoring. The case study used is that of Hong Kong's MTR network smartcard data before and after their intervention. They studied behavioral changes and concluded that customer groups who shifted their behavior were able to be identified. The main factors contributing to their behavioral changes include the length of time a customer had to shift their departure time, savings on fares, and price sensitivity. Targeting specific groups with incentives may be the way to go, given the heterogeneity of passengers unearthed in this study.

A study by Aziz et al. [38] focused on the active transportation modes of walking and cycling. Often research tends to focus on motorized trips. However, non-motorized trips should be included in the analysis. Suitable infrastructure needs to be available for these modes to be supported. The study used agent-based modeling to examine the effect of infrastructure investments. A case study in New York indicated that infrastructure investments such as increasing the network of bike lanes and widening footpaths would significantly affect active transportation choices. Safety is an issue that needs to be addressed and is a risk in rolling out preferences for this kind of transportation. These authors indicate that major factors influencing the choice of active modes include income, age, gender (males are more likely to bike), and education (higher educated people are more likely to choose active modes). Household size and car ownership are also factors. The built environment will influence the availability of these choices. Density is generally positively related to transport use of the more active type. Network layouts will influence transportation. A higher density regarding intersections will increase accessibility to various destinations. Distance to daily needs destinations is also important. Perceptions of safety can come into play as well. The authors noted significant differences between various boroughs in their New York case study. ABM was clearly very helpful in this study.

Natural Environmental Aspect

Lele and Goswami [39] report that the Indian government has earmarked US$ 1054 million to fund the development of 100 smart cities to address the various problems currently being faced in that country. Many of these activities are supported by PPP structures. Transportation is one of the foci of this government initiative. One theme is creating walkable places that reduce congestion and pollution and boost the local economy. Refurbishing or creating new road networks, not only for vehicles and public transport but also for walking and biking, is also highlighted in this initiative. A new transit-oriented development option is being developed within public transport. Pedestrian-focused development is being undertaken near transit terminals for reducing crowding during peak periods. So far, 60 smart cities have been created since the approval for the initiative was granted in 2016. The authors discuss the need for a systemic approach to developing smart cities.

Baruah et al. [40] studied ways of reducing carbon electricity, including in Great Britain's transport services. Using low-carbon electricity can reduce fossil fuel consumption significantly. Given that the UK has signed up for an 80% reduction of greenhouse gases by 2050 through the Kyoto protocol, electrifying the energy system using low-carbon electricity is a crucial way forward. Electrification of both heat and transport systems is investigated in this paper. The study shows that the total annual energy consumption can be reduced, and dependency upon imported natural gas and oil will also be lower. There are many advantages of electrification, including increasing stranded asset pressures on existing infrastructure in the UK. Transport decarbonization does not have many options beyond electrification: biofuel would imply large imports with concerns about the security of supply. There are also some concerns about carbon neutrality and the impact on land use of current biofuels. Also, air transport will probably be given priority for using biofuel.

Hixson et al. [41] investigated future air pollution difficulties by 2030 in the San Joaquin Valley in central California, considering various factors, including land-use policies and investment in transportation infrastructure. Population exposure to particulate matter chemical components was recorded and is focused mainly on major transportation corridors and urban centers. Exposure to these chemicals is related to adverse health effects of people.

3.2 *The Potential Risk of the PPPs' Approach to Smart Infrastructure Development*

Undoubtedly, the 'smart' factor significantly contributes to transport infrastructure development, enhancing local government's ability to solve critical issues in urban cities. Investment in PPPs is generally recognized as a good investment for economic growth [42] and an innovative procurement model whose usage has grown to pursue efficiency and overcome challenges for governments needing to find the money to build this necessary infrastructure. Due to various public debt crises and the need to restrict government spending, investment in infrastructure, including transport infrastructure, has dropped in recent decades in both developed and developing countries [43]. Attracting private capital to a public project may enable projects which would otherwise not be possible. PPPs have been widely used in developed and developing countries [44].

Within PPP arrangements, the public sector and private firms share both risks and profits. Such PPPs are becoming more widespread, particularly where governments have a constrained budget but a great need to expand their public infrastructure. One benefit to the public sector is that resources can now be spent on other public projects. A benefit to the private agency is that they can invest in a profitable enterprise that yields social benefits, including greater access to work and services and reduced pollution and accident rates. PPPs come in a wide variety of shapes and sizes. Such arrangements are complex and involve many agents from many fields. A careful analyses of risks will be necessary, whenever the degree of private participation is higher. Whether large or small, however, there is a clear potential for problems when government/public service agencies interact with private organizations. This interaction, whether at the individual or the organizational level, is prime for communication difficulties, misunderstandings, and opportunism. PPPs are risky operations and should be carefully assessed before being undertaken and monitored throughout their life cycle [45].

This table presents the literature supporting five main risk groups for smart infrastructure projects, and we can see that according to literature, risk factors have been identified within the domains of technological-technical, financial-economical, legal, social, and environmental factors. The risk factors within each domain vary widely and we can see that lack of clarity; lack of transparency; lack of dispute resolution; lack of economic stability; and lack of expertise, standards, and monitoring have been identified by authors. Main risk groups and associated risk factors provide a holistic picture of the potential risks of PPP smart transport infrastructure projects. These factors should be carefully examined in the specific context to rank critical risks and have suffocation risk mitigation plans. The information provided in this table is of critical importance to decision-makers entering into such complex arrangements.

4 Conclusion

This paper analyzes potential risks in PPP smart infrastructure project development. Based on the literature review of selected articles published from 2010 until the present, 19 risk factors were identified that might significantly influence the success of PPP infrastructure project implementation, as detailed in Table 1. These critical risks were then categorized into five main groups (political-legal, technological-technical, economic-financial, social, and environmental); such groups provide a holistic picture of the potential risks associated with the delivery of smart transport infrastructure projects. In line with these results, some suggestions on how the potential risks affect smart infrastructure projects in practice were also presented to clarify the main causes during the project stages.

As there are potential risks during PPP project implementation and operation stages in practice, the identified potential risk factors which prevent the development of such smart transportation infrastructure projects should be monitored. Future work will involve the validation of these risk factors along with the perspective of PPP stakeholders. Challenges and potential solutions to improve PPP projects in smart infrastructure will be analyzed.

Table 1 Potential risk identification for PPP smart infrastructure projects development

List	Groups	Risk factors	Descriptions	References
1	Technological-technical	Lack of PPP project data	Lack of meaningful data creates risks	Fischer et al. [46]
2		Lack of integrated technological platform	Technological integration provides many advantages; however, this is still at an early stage	Hoeft et al. [47]
3		Insufficient standards for PPP projects	Governance and regulatory standards need to be increased in many areas	Torchia and Calabrò [48]
4		Lack of expertise in complex project implementation	Lack of expertise can cause PPP problems and risk	Van Gestel et al. [49]
5	Financial-economical	Unclear tendering processes	It is important for the public client and the private bidders to assess all the potential risks throughout the project	Ose-Kyei and Chan [50]
6		Uncertain macroeconomic conditions	It enables public sector clients to establish more efficient risk allocation frameworks in the early stages of project development	Osei-Kyei and Chan [50]
7		Insufficient fund allocation between private and public partners	Disagreements or inadequacies regarding fund allocation will slow down the project	Rashed and Alam [51]
8		Lack of dispute resolution embedded within PPP contracts	Disputes are inevitable and result from differing expectations, which, if left unresolved, will be very destructive	Sinha and Jha [12]
9	Legal	Lack of government commitment and regulatory framework for PPP projects	Some risks should still be retained within the public sector or shared with the private sector	Osei-Kyei and Chan [50]

(continued)

Table 1 (continued)

List	Groups	Risk factors	Descriptions	References
10		Political corruption	The issue of corruption can arise when there is insufficient shared authority and responsibility	Fleta-Asín and Muñoz [52] Percoco [43]
11		Lack of government policy measures for PPP project outcome	Working with familiar partners on familiar projects and in familiar sectors is important	Prabhudesai and Sarode [53]
12		Lack of collaboration mechanism between government and private sectors involved	Closer collaboration between public and private agencies is important for efficient technology transfer	Fuglie and Toole [54]
13	Social	Value conflicts between parties (e.g., investors, contractors, consultants) involved in PPP	Conflicts of interest are frequent within the complex relationships between partners in PPPs. Good practices, including accountability and transparency, can avoid these	Omobowale et al. [55]
14		Local citizen's resistance to PPP selection	Resistance from the local people can undermine PPP reforms	Tsekos and Hlepas [56]
15		Insufficient stakeholder engagement	Stakeholder management is essential to PPP's success	Jayasuriya et al. [57]
16		Negotiated collaboration and trust in PPP	There is still distrust of government in developing countries, and there is a generally held belief: 'public bad, private good'. Public agencies and private organizations have a mutual advantage when a PPP is characterized by trust, openness, fairness, and mutual respect	Benítez-Ávila et al. [58]

(continued)

Table 1 (continued)

List	Groups	Risk factors	Descriptions	References
1 table 7	Natural environmental	Insufficient natural environmental protection standards for PPP infrastructure projects	PPP firms need to pay attention to environmental issues and protection, and this needs regulation	Ke et al. [59]
18		Lack of environmental monitoring, particularly during the construction stage	It is important to establish an environmental monitoring system, especially during construction	Weizhen et al. [60]
19		Insufficient awareness of the natural environment	Awareness about protecting the natural environment is a critical aspect of PPP's success	Kwak et al. [9]

Acknowledgements This research is funded by Vietnam National University, Ho Chi Minh City (VNU-HCM) under grant number C2023-28-06.

References

1. Musa S (2016) Smart cities-a roadmap for development. J Telecommun Syst Manage 5(3):1–3
2. Cruz CO, Sarmento JM (2017) Reforming traditional PPP models to cope with the challenges of smart cities. Compet Regul Netw Ind 18(1–2):94–114
3. Lee OL, Tay RI, Too ST, Gorod A (2019) A smart city transportation system of systems governance framework: a case study of Singapore. In: 2019 14th annual conference system of systems engineering (SoSE), pp 37–42
4. Yang F, Wen X, Aziz A, Luhach AK (2021) The need for local adaptation of smart infrastructure for sustainable economic management. Environ Impact Assess Rev 88:106565
5. Palmentieri S (2021) Smart cities and the sustainability of urban transport: Strategic directions for the metropolitan city of Naples. J Urban Regener Renewal 15(1):83–94
6. Halvorsen A, Koutsopoulos HN, Ma Z, Zhao J (2020) Demand management of congested public transport systems: a conceptual framework and application using smart card data. Transportation 47(5):2337–2365
7. Chen Z, Daito N, Gifford JL (2016) Data review of transportation infrastructure public-private partnership: a meta-analysis. Transp Rev 36(2):228–250
8. Roehrich JK, Lewis MA, George G (2014) Are public–private partnerships a healthy option? A systematic literature review. Soc Sci Med 113:110–119
9. Kwak YH, Chih Y, Ibbs CW (2009) Towards a comprehensive understanding of public private partnerships for infrastructure development. Calif Manage Rev 51(2):51–78
10. Khanom NA (2010) Conceptual issues in defining public private partnerships (PPPs). Int Rev Bus Res Papers 6(2):150–163

11. Calderini M, De Marco A, Mangano G, Michelucci F (2013) A framework to use public-private partnership for smart city projects. PoliFinance Publishing Ltd. https://www.researchgate.net/publication
12. Sinha AK, Jha KN (2020) Dispute resolution and litigation in PPP road projects: evidence from select cases. J Leg Aff Disput Resolut Eng Constr 12(1):05019007
13. Grimsey D, Lewis MK (2002) Evaluating the risks of public private partnerships for infrastructure projects. Int J Project Manage 20(2):107–118
14. Chung D, Hensher DA, Rose JM (2010) Toward the betterment of risk allocation: investigating risk perceptions of Australian stakeholder groups to public–private-partnership toll road projects. Res Transp Econ 30(1):43–58
15. Demirag I, Khadaroo I, Stapleton P, Stevenson C (2011) Risks and the financing of PPP: perspectives from the financiers. Br Account Rev 43(4):294–310
16. Chen C, Hubbard M (2012) Power relations and risk allocation in the governance of public private partnerships: a case study from China. Policy Society 31(1):39–49
17. Bing L, Akintoye A, Edwards PJ, Hardcastle C (2005) The allocation of risk in PPP/PFI construction projects in the UK. Int J Project Manage 23(1):25–35
18. Kim J, Kim J, Wang Y (2021) Uncertainty risks and strategic reaction of restaurant firms amid COVID-19: evidence from China. Int J Hospital Manage 92:102752
19. Donnici G, Frizziero L, Liverani A, Cacaci L, Costantini N, Pedrielli G (2020) Ides (Industrial design structure) and stylistic design engineering (sde) applied to the mobility of the future. In: Proceedings of the international conference on industrial engineering and operations management, Detroit, MI, USA, pp 10–14
20. Shichiyakh RA, Klyuchnikov DA, Balashova SP, Novoselov SN, Novosyolova NN (2016) Smart city as the basic construct of the socio-economic development of territories. Int J Econ Financ Issues 6(1):157–162
21. Tirachini A, Hensher DA (2011) Bus congestion, optimal infrastructure investment and the choice of a fare collection system in dedicated bus corridors. Transp Res Part B Methodolo 45(5):828–844
22. Moreu F, Kim RE, Spencer B Jr (2017) Railroad bridge monitoring using wireless smart sensors. Struct Control Health Monit 24(2):e1863
23. Stipanovic I et al (2020) Reliability based life cycle management of bridge subjected to fatigue damage. Front Built Environ 6:100
24. Ess J, Luppin J, Antov D (2021) Estimating the potential of a warning system preventing road accidents at pedestrian crossings. LogForum 17(3)
25. Borozan S, Giannelos S, Strbac G (2022) Strategic network expansion planning with electric vehicle smart charging concepts as investment options. Adv Appl Energy 5:100077
26. Katrenko A et al (2020) Development of traffic flows and smart parking system for smart city. In: COLINS, pp 730–745
27. Siripanich A, Rashidi TH, Moylan E (2019) Interaction of public transport accessibility and residential property values using smart card data. Sustainability 11(9):2709
28. Losavio MM, Chow K, Koltay A, James J (2018) The internet of things and the smart city: legal challenges with digital forensics, privacy, and security. Secur Privacy 1(3):e23
29. Abosaq NH (2019) Impact of privacy issues on smart city services in a model smart city. Int J Adv Comput Sci Appl 10(2)
30. Ge Y, Liu X, Tang L, West DM (2017) Smart transportation in China and the United States. Center for Technology Innovation, Brookings Institution, Washington
31. Glancy DJ (2013) Sharing the road: smart transportation infrastructure. Fordham Urb. LJ 41:1617
32. Dewi NK, Putra AS (2021) Law enforcement in smart transportation systems on highway. In: International Conference on Education of Suryakancana (IConnects Proceedings)
33. Selim AM, Yousef PH, Hagag MR (2018) Smart infrastructure by (PPPs) within the concept of smart cities to achieve sustainable development. Int J Crit Infrastruct 14(2):182–198
34. Zhang J, Wang T, Zhang L (2021) Legal risk assessment framework for international PPP projects based on metanetwork. J Constr Eng Manage 147(8):04021090

35. Iyer K, Sagheer M (2011) A real options based traffic risk mitigation model for build-operate-transfer highway projects in India. Constr Manage Econ 29(8):771–779
36. Likhitruangsilp V, Do ST, Onishi M (2017) A comparative study on the risk perceptions of the public and private sectors in public-private partnership (PPP) transportation projects in Vietnam. Eng J 21(7):213–231
37. Kolobova S (2020) Renovation of Moscow as a subprogram of creation of a creative city. IOP Conf Ser Mater Sci Eng 753(6):062003. IOP Publishing
38. Aziz HA, Park BH, Morton A, Stewart RN, Hilliard M, Maness M (2018) A high resolution agent-based model to support walk-bicycle infrastructure investment decisions: a case study with New York City. Transp Res Part C: Emerg Technol 86:280–299
39. Lele U, Goswami S (2017) The fourth industrial revolution, agricultural and rural innovation, and implications for public policy and investments: a case of India. Agric Econ 48(S1):87–100
40. Baruah PJ et al (2014) Energy system impacts from heat and transport electrification. Proc Inst Civ Eng Energy 167(3):139–151
41. Hixson M et al (2010) Influence of regional development policies and clean technology adoption on future air pollution exposure. Atmos Environ 44(4):552–562
42. Melo PC, Graham DJ, Brage-Ardao R (2013) The productivity of transport infrastructure investment: a meta-analysis of empirical evidence. Reg Sci Urban Econ 43(5):695–706
43. Percoco M (2014) Quality of institutions and private participation in transport infrastructure investment: Evidence from developing countries. Transp Res Part A Policy Pract 70:50–58
44. Khasnabis S, Dhingra S, Mishra S, Safi C (2010) Mechanisms for transportation infrastructure investment in developing countries. J Urban Plan Dev 136(1):94–103
45. Kavishe N, Chileshe N (2019) Critical success factors in public-private partnerships (PPPs) on affordable housing schemes delivery in Tanzania. J Facil Manage
46. Fischer K, Leidel K, Riemann A, Alfen HW (2010) An integrated risk management system (IRMS) for PPP projects. J Finan Manage Prop Construct
47. Hoeft M, Pieper M, Eriksson K, Bargstädt H-J (2021) Toward life cycle sustainability in infrastructure: the role of automation and robotics in PPP projects. Sustainability 13(7):3779
48. Torchia M, Calabrò A (2018) Increasing the governance standards of public-private partnerships in healthcare. Evidence from Italy. Publ Org Rev 18(1):93–110
49. Van Gestel K, Willems T, Verhoest K, Voets J, Van Garsse S (2014) Public–private partnerships in Flemish schools: a complex governance structure in a complex context. Publ Money Manage 34(5):363–370
50. Osei-Kyei R, Chan APC (2016) Developing transport infrastructure in Sub-Saharan Africa through public–private partnerships: policy practice and implications. Transp Rev 36(2):170–186
51. Rashed MA, Alam MM (2009) Central PPP development facility: enhancing government obligation in promoting public-private partnership (PPP). In: Proceedings of the international conference of administrative development: towards excellence in public sector performance, Institute of Public Administration (IPA), Intercontinental Hotel, Riyadh, Saudi Arabia, pp 1–4
52. Fleta-Asín J, Muñoz F (2020) How does risk transference to private partner impact on public-private partnerships' success? Empirical evidence from developing economies. Socioecon Plann Sci 72:100869
53. Prabhudesai V, Sarode NG (2018) Critical success factors of public-private partnership for road sector development in India: a private sector perspective. IUP J Bus Strat 15(4)
54. Fuglie KO, Toole AA (2014) The evolving institutional structure of public and private agricultural research. Am J Agr Econ 96(3):862–883
55. Omobowale EB, Kuziw M, Naylor MT, Daar AS, Singer PA (2010) Addressing conflicts of interest in Public Private Partnerships. BMC Int Health Hum Rights 10(1):1–8
56. Tsekos TN, Hlepas NK (2019) Greek municipalities before and during the austerity era: imposed policies, local resistance and unsuccessful reforms. In: Local public services in times of austerity across Mediterranean Europe: Springer, pp 49–71
57. Jayasuriya S, Zhang G, Yang RJ (2020) Exploring the impact of stakeholder management strategies on managing issues in PPP projects. Int J Constr Manage 20(6):666–678

58. Benítez-Ávila C, Hartmann A, Dewulf G, Henseler J (2018) Interplay of relational and contractual governance in public-private partnerships: the mediating role of relational norms, trust and partners' contribution. Int J Project Manage 36(3):429–443
59. Ke Y, Wang S, Chan AP, Lam PT (2010) Preferred risk allocation in China's public–private partnership (PPP) projects. Int J Project Manage 28(5):482–492
60. Weizhen M, Ruwen D, Yu H, Nannan W (2018) The management efficiency of PPP projects in rural China. In: 2018 7th international conference on industrial technology and management (ICITM), pp 179–183

Public–Private Partnership Models for Smart Transport Infrastructure Development

Hoang Long Nguyen, Tiep Nguyen, and Anh Tuan Nguyen

Abstract The advancement of technology in recent years has caused a significant change in the local government approach to the investment in the critical transport infrastructure network. The integration of high technology regarding Internet of Things (IoT), Big Data, and Artificial Intelligence (AI) within infrastructure systems of 'smart cities' has changed the way to deal with significant problems in urban cities including traffic jams, environmental pollution, and rapid population increase. This requires both huge financial investment and expertise in developing 'new' smart transport infrastructure projects. Thus, this study aims to investigate barriers and review Public–Private Partnership (PPP) models for smart infrastructure development. Scopus and Web of Science indexed articles from 2000 until the present were selected for review, and the initial results indicate that the selection of PPP model depends on the characteristics of both 'smart cities' and 'smart infrastructure' defined within the specific context of cities. Furthermore, the outcome of the literature review also shows that service contracts, operation and management contracts, leases, concessions, and build–operate–transfer (BOT) contracts are identified as the most common PPP models applied for smart infrastructure development. In line with that, some important criteria for PPP model selection are proposed for smart transport infrastructure development.

Keywords Public–private partnership · Smart transport infrastructure · Transport infrastructure development

H. L. Nguyen · A. T. Nguyen
Ho Chi Minh University of Transport, Ho Chi Minh City, Vietnam

T. Nguyen (✉)
International University, Ho Chi Minh City, Vietnam
e-mail: nvtiep@hcmiu.edu.vn

Vietnam National University, Ho Chi Minh City, Vietnam

V. T. Ha et al. (eds.), *Proceedings of the 8th International Conference on Sustainable Urban Development*, Advances in 21st Century Human Settlements,
https://doi.org/10.1007/978-981-99-8003-1_3

1 Background

1.1 Smart Transport Infrastructure Fundamentals

Infrastructure is a system network that includes basic facilities such as road, water, wastewater, gas, and telecommunication, all of which are provided to the people living there. A city infrastructure operates efficiently when facilities and services are synchronously connected and instantiated. However, with significant increase in existing networks and population, the existing network is difficult to manage and hence smart infrastructures have been developed quickly. Smart infrastructure is a network that links basic facilities and services effectively with transportation systems coordinated by information technology tools. The smart concept is an urban design principle and practice that promotes basic facilities and services which effectively interact with real-time information communication technology (ICT) to achieve sustainable development. The concept of smart infrastructure is recognizable when upgrading the existing city or developing smart city, and smart city is described by many systems with different elements such as people, government, environmental, economic, mobility, and life circumstances in a place which is effective by information communication technology (ICT) to promote sustainable city development. In the present day, the concept of smart infrastructure is applied in urban planning, and design for sustainable infrastructure development such as integrating ICT, AI, new technologies into smart infrastructure that are cost efficient meet people's requirements for goods transportation, services, improving the environment, and sustainability over a long time. It also includes the integration of all real-time information and activities with effective technology to achieve sustainable living.

According to a UN report (2020), most people will move to cities for working and living by 2050, and this will result in approximately 2.5 billion people living in urban areas. This move to urban areas as people look for jobs will cause many challenges and many countries are struggling with the issue and trying to develop smart cities to achieve the services that people are requesting and to improve the quality of people life. It is like a future picture which describes from urban area to rural area and from developed countries to developing countries in which focus on development becomes 'smart'. In developed countries, governments have attempted to develop their rural areas after completing their attempts to develop smart cities with the purpose of keeping connections between urban areas and rural areas and maintaining information exchange regarding weather conditions or pandemic isolation. For example, in smart projects in Hongkong, smart cities and especially smart infrastructure are of great interest to governments and are a high priority for planners. However, in today's situation, many governments have budgets which have been broken by the pandemic as well as various other challenges and crises, and in consequence, smart infrastructure development projects face very large challenges regarding funding. According to the UN report, the annual global smart city revenue is expected to reach US $88,7 billion by 2025 [1]. In Asian countries, some smart infrastructure project funding is supported through government programs. In addition, funds to invest in smart infrastructure need a huge capital input. PPP is the most suitable way to achieve the large funding requirements for smart infrastructure projects and at the same time contribute to sustainable development goals. In general, infrastructure projects and, in particular, smart infrastructure projects are considered an effective partnership between the government and private sector Fire PPP models with various applications.

To transform a city into a smart city, the first requirement is to develop intelligent transportation infrastructure. Intelligent transportation infrastructure will be an essential component of people's lives in the future. Smart infrastructure will change people's daily habits and behaviors. Residents of a city with an intelligent transportation system will benefit greatly from it. For example, with a smartphone, they can check bus schedules and real-time bus stops, making it convenient for people to use public transportation. They can easily find empty parking spots instead of running around looking for one and paying through a mobile phone app. With an intelligent transportation system, the government can control and automate traffic signals, record security images on the streets through smart cameras, collect data, and analyze it to issue commands in real time. In practice, smartness in infrastructure development has been identified as a central element in transport systems. Infrastructure smartness has described smart technologies which can support economic, environment achieve develop sustainable [2].

A smart technology in transport can be self-operated automatically, computing, corrective system by itself or require a little people involvement. Smart infrastructure working through three layer is as follows [3]:

- Sensors, commands, and control unit.
- Sensing, processing and make decision, and acting.
- Communicating.

The operation of intelligent transportation systems relies on software, artificial intelligence, and machine learning to analyze and capture the behavior of road users in order to make decisions. According to McDonald et al. (2006), the effectiveness and speed of these systems also depend on financial resources. Intelligent transportation systems can integrate multiple modules and perform multiple tasks simultaneously in real time, such as detecting traffic violations, controlling traffic signals, providing information about road conditions for road users, and determining road fees as in Singapore. With the emergence of intelligent transportation systems, cities become more convenient and attractive, not only for residents but also for visitors from other places.

From the literature, there are many definitions about smart transport; and according to Ahmed [4] smart transport can be divided into two categories, they could be standard of smart city developing, they are as follows:

- Physical (smart) transport is described as smart moveable, reducing traffic jams and fostering faster, greener, and cheaper transport; saving energy; and reducing

greenhouse effects. The goal is to optimize journeys that takes place within a city, resulting in saving energy and reducing air pollution.
- Digital (smart) infrastructure [5]: Urban layer, sensor layer, connectivity layer, data analytics layer, automation layer, the Internet of Things, and big data. They are interconnected and operated through information and resource data available.

Drawing from Debnath et al. [6], the standards for a smart transportation system have been defined and established. These standards serve as criteria for assessing the level of intelligence exhibited by transportation systems, and they were utilized to evaluate and rank 26 cities. The study results produced a ranking in order of smart systems with London, Seattle, Sydney, New York, and Melbourne emerging as the top-ranked cities. To achieve such a top ranking, the UK's intelligent transportation system has received special support from both the public and private investment sectors. The UK government's 2012 budget demonstrated strong investment by allocating £200 billion per year until 2030 toward two transport systems that apply intelligent technologies to develop connectivity across all transportation systems, including roads, railways, shipping, and aviation, to alleviate pressure from population growth and travel demands. As a result, both smart cities and intelligent transportation infrastructure are expected to be high-growth areas in the next 10–15 years in the future.

Furthermore, Ahmed [4] aimed to identify and quantify the obstacles encountered in developing intelligent transportation systems. From a different perspective, the mentioned criteria could represent the government's requirements incorporated into the project to estimate the value of the smart transportation system investment package, which is structured as a public–private partnership.

1.2 Public–Private Partnership Approach for Smart Transport Infrastructure Development

Public–Private Partnership (PPP) is a very important part of the development strategy which is applied in almost all national governments around the world to develop infrastructure with an expectation of matching the requirements of fast development of society. PPP is an investment method, wherein governments take advantage of financial skills, people, and transfer the challenges and risks to the private sector. However, there is also an opportunity to enter a huge investment channel. PPP lies between traditional investment and privatization. The term PPP is used to define the relationship between the public sector and private sector, such as private sector participation in government infrastructure projects (private sector participation). A successful PPP requires effective assessment, result transparency, and clarity on the roles and responsibilities of the contracting parties. However, currently there is no standard or exact definition of PPP. The European Commission (EC), Organization for Economic Co-operation and Development (OECD), International Monetary Fund (IMF), World Bank, and others have used various definitions to define PPP. In many countries, depending on the legal or social characteristics, the definition of PPP is different and is made into law. Typically, there are some popular PPP definitions, including

- European Commission [7]: PPP is defined as a relationship between a government and private sector with the purpose of providing a service or developing a project launched by the government.
- The International Monetary Fund (IMF) [8] defines PPPs as arrangements under which private investors provide infrastructure assets and services, based on state-provided infrastructure.
- According to the OECD [9], PPPs are long-term agreements between governments and private partners whereby the private partner provides funds to public services using capital assets, sharing risks with key stakeholders.
- The Australian Government [10, p. 7] has defined a PPP as a service contract between the public and private sectors in which the government pays the private sector (usually 13 consortiums) to provide facilities such as infrastructure and related services in the long term. The private party is the developer that builds the infrastructure and operates or maintains it to specific standards over a long period of time
- In Hong Kong, a PPP is defined as a public–private partnership that leverages private human and financial resources to complement a project, with varying levels of participation and responsibilities to provide more efficient public services [11].
- In Vietnam [12], the PPP law was promulgated in 2020; PPP (public–private partnership) is defined as follows: investment in the form of public–private partnership (hereinafter referred to as PPP for short) is a form of investment made on the basis of a contract between a competent state agency. Rights and investors, project enterprises to implement, manage, and operate infrastructure projects, and provide public services.

According to Erokhina and Brega [13], smart transport is an important part of a smart city which cannot be ignored as an important factor in people's daily lives. Therefore, citizen demand requires a significant change within transport infrastructure as soon as possible. The people living in cities where smart transport can be managed, operated, and supported through intelligent devices enjoy a significant boost to their quality of life. For example, people can use mobile phone apps to access bus, train, or tram timetable or information, including bus stop or distance, and can therefore choose the shortest distance or more convenient location which means saving in travel time. People can be advised about available parking, and they can pay their parking fees, or they can move to business centers free of charge. In terms of security, intelligent devices can record every activity on roads. Smart transport is based on devices such as sensors and digital cameras to provide information and data, and these can be handled by AI, IoT, or machine learning technologies.

Currently, there are many debates about intelligent transportation infrastructure. Debnath et al. [6] believe that the intelligence level of transportation infrastructure can be determined through a scoring system based on criteria such as perception, processing, control, communication, repair, and prevention. On the other hand,

Ahmed [4] defines intelligent transportation infrastructure as systems that rely on digital technologies, including sensor layer systems connected to data analysis centers and automation through high-bandwidth data such as IoT or big data, to make decisions about system operations and exchange information in order to help people save on transportation costs, receive complete traffic information, save fuel, reduce greenhouse gas emissions, and support traffic safety. In general, transportation infrastructure is considered intelligent when the system operates based on digital sensor systems that provide data to processing centers through intelligent technologies combined with available big data and less dependence on human intervention. Such systems are composed of multiple components that are processed by computers using artificial intelligence, IoT, and big data through high-bandwidth data transmission. These systems can be developed independently as modules and constantly updated. The modules will be integrated when they meet technical and financial requirements. Depending on the government's budget situation, there may be a strategy and investment scale for developing intelligent transportation systems or using PPP as a tool to support the development of intelligent transportation systems to keep up with current trends in the world.

Adequate funding is a prerequisite for smart transport development. The public private sector collaboration (PPP) will be the template for developing smart infrastructure in most countries around the world, because of its significant benefits. In Europe, the Croatian economy was enhanced successfully when they applied the PPP model to develop their smart infrastructure [14]. As a result of this initiative, there economy showed improvement through three channels:

- Increase in private capital inflow.
- Increase in social welfare and social benefits.
- Reduction in government's fiscal burdens.

Even though smart infrastructure development is an important future trend, local governments, particularly in developing countries, still have difficulties in selecting the appropriate PPP model. Thus, this research aims to clarify main PPP models and classify them into groups, and highlight their advantages and disadvantages in practice.

1. Research Method

In this study, we used mini scoping technique for conducting the literature review [15]. The literature review was undertaken in three phases. Firstly, a comprehensive literature review identified potential risks during metro railway project implementation. The Scopus, Web of Science, PubMed, and ScienceDirect database were selected, and the following strings of TITLE-ABS-KEY were used, including: 'Smart Transport Infrastructure'; AND 'Project Development' to provide a holistic picture on smart infrastructure project development over the world. The initial search retrieved 129 articles for subsequent refinement. The scope was then narrowed with the following string searched: 'Smart Transport Infrastructure'; AND 'Project Development'; AND 'Investment Models' within the specific time from 2000 to present with popular and well-known publications. This period was selected because of the significant contributions of scholars to smart infrastructure literature during this time. The second phase of the literature review refined the search. Abstracts and introduction sections were analyzed, and articles of a specific technical nature were not considered because it does not reflect the balance between aspects of PPP models within large-scale projects (e.g., technical, technological, financial, environmental, social). This second phase produced a sample of 66 research outputs. In the third phase of the literature review, papers selected in the second phase were read and those that were relevant to the investment models, particularly focusing on public–private partnership (PPP) models were selected. Next a qualitative content analysis was undertaken to identify advantages and potentials risk of PPP model selection for smart transport infrastructure projects. The final database consisted of 20 papers published in international journals, books, conference proceedings, or other research outputs.

2 Initial Findings and Discussions

To attract private investment in transportation infrastructure, various PPP models such as Operation and Maintenance (O&M), Build–Operate–Transfer (BOT), and Design-Build (DB) have been introduced. According to The Asian Development Bank (2008), each model is applied differently depending on the sector, type, and characteristics of the project. The implementation of PPP projects is based on the conditions of the contract between the government representative and the private investor. According to EC (2003), the most appropriate PPP model should be chosen based on market demand, public benefits, contribution levels, and time. Zhang et al. (2016) suggest that the most important decision when considering using PPP is to choose the most appropriate model for the proposed project. Currently, there are various PPP models available that continue to develop to suit different characteristic projects. However, Delmon (2010) found that many projects with different characteristics use the same terminology. According to Kwak et al. (2009), Teker and Teker (2012), the available PPP options for the delivery of transportation infrastructure are based on the degree of private sector involvement and risk allocation. The degree of private involvement needs to be carefully matched to the project's objectives and needs through a cost-benefit analysis of all involved parties to increase the effectiveness of PPP implementation (European Commission 2003). Following the World Bank guidelines, the participation of private investors in PPP projects is crucial. The PPP model is defined based on project duration, objectives, legal status, and risk allocation, as follows:

- Long-term service contracts, O&M, where the public sector bears fewer risks.
- DBB, DB, or EPC contracts. These types of contracts may have short-term collaboration, but risk allocation is shared among participating parties.
- DBF or BT contracts address short-term budget pressures, and the government assumes responsibility and risk for assets in the long term.

- Full or partial privatization: the government transfers 100% or a portion of equity in state-owned companies to private entities.

Through the analysis of the historical usage of the term by international organizations and previous research, the classification of PPP models for transportation infrastructure projects has been subject to varying perspectives. The grouping of PPP models depends on the characteristics of the project. PPP contracts can be based on their funding sources, services, responsibilities, or private sector obligations. The World Bank focuses on project duration, risk allocation, and deliverables. The International Monetary Fund (IMF) categorizes PPP into three groups based on design, construction, and operation. In essence, PPP models are a combination of contracts, which are synthesized in Table 1.

PPP is a scheme of investment that became very popular in early years. After the end of World War II, economies were heavily affected by the war, and many governments used PPP approaches such as appliance strategies to renovation and build up city and infrastructure system. Whether considering either organizations or governments, various types of PPPs can be divided according to different models, with this division creating attractiveness and competition to attract private investors (Table 2).

From The Asian Development Bank (ADB) PPP handbook [17], PPPs are grouped based on the asset ownership according to time period and investment responsibility. There are two main groups: (1) asset ownership and capital investment are public; (2) asset ownership is public or private and investment belongs to a private party. Depending on the purpose of investors/owners, some common contract types have been introduced as depicted in Fig. 1.

Most PPP projects have a lifetime of 20–30 years, and a few projects take longer than 30 years to complete. However, in practice, the PPP type of project is considered and calculated carefully to fit service delivery cost, which is then integrated into the design phase of the project. Service delivery costs include investment costs, maintenance costs, operation costs, and for the benefit of the project these must be efficiently handled. The core of the implementation of PPPs for delivering service to the public sector with private sector participation free PPPs can bring satisfaction to both sectors. Depending on the nature of the project, advantages and disadvantages will vary as shown in the Table 3 that looks at two typical PPP model groups.

Currently, there are few research studies on PPP models applied to smart transportation infrastructure projects. However, according to Zhang, Xueqing [27], a solution is to propose the establishment of criteria, standards, and quality for PPP programs applied to transportation infrastructure development projects. The selection of a suitable PPP model is crucial for investors and project participants, as pointed out by Love, PE, Skitmore and Earl [28]. The specific criteria and requirements for each project must be clearly defined. Additionally, owners should consider other factors to attract more investors to participate and compete in the project. Currently, the selection of PPP models depends greatly on both the project context and the complexity level of smart transport infrastructure. Typically, such projects often contain high risks regarding cost overrun and project delays, so investors or

Table 1 PPP models for smart infrastructure project developments. Adopted from Nassar [16]

List	PPP type	Descriptions	Application boundaries	Referred by
1	BOT	Build, operate, and transfer	Widely application for new project	The International Monetary Fund, The Asian Development Bank
2	BOT	Build, own, and transfer	Widely application for new project	The International Monetary Fund, The Asian Development Bank
3	BOO	Build, own, and operate	Not widely application	The Asian Development Bank
4	BOOT	Build, own, operate, and transfer	Widely application for new project	The International Monetary Fund
5	BLT	Build, lease, and transfer	Project relate to high security	The International Monetary Fund
6	BRT	Build, rent, and transfer	Project relate to high security	The International Monetary Fund
7	BT	Build and transfer	Widely application base on BOT	The International Monetary Fund
8	BTO	Build, transfer, and operate	Widely application for new project	The International Monetary Fund
9	BOR	Build, operate, and renewal of concession	Application base on BOT	The International Monetary Fund
10	DBO	Design, build, and operate	Fixed fee, rate, Lump sum	The International Monetary Fund, The Asian Development Bank
11	DBOM	Design, build, operate, and maintain	Fixed fee, rate, Lump sum	The International Monetary Fund, The Asian Development Bank
12	DBMF	Design, build, manage, and finance	Fixed fee, rate, Lump sum	The International Monetary Fund, The Asian Development Bank
13	DBMF	Design, build, finance, and operate/maintain	Fixed fee, rate, Lump sum	The International Monetary Fund, The Asian Development Bank
14	MOT	Modernize, own, or operate and transfer	Limited application, applied existing project	The Asian Development Bank
15	ROO	Rehabilitate, own, and operate	Limited application, applied existing project	The International Monetary Fund, The Asian Development Bank
16	ROT	Rehabilitate, own, and transfer	Limited application, applied existing project	The International Monetary Fund, The Asian Development Bank
17	O&M	Operate and maintain	Limited application, applied existing project	The International Monetary Fund, The Asian Development Bank

Table 2 PPP group classification. Adopted from Anderson et al. [8]

PPP groups	Modalities
BOO, BDO, DCMF	The private sector designs, builds, owns, develops, operates, and manages an asset with no obligation to transfer ownership to the government
BBO, LDO, WAA	The private sector buys or leases an existing asset from the government, renovates, modernizes, and/or expands it, and then operates the asset, again with no obligation to transfer ownership back to the government
BOT, BOOT, BROT, BLOT, BTO	In this model, the private sector designs and builds a project, after finishing they operate it, and then transfers it to the government when the term of time in contract ends, or because of some reason time in the contract is respecified. The private partner may subsequently rent or lease the asset from the government

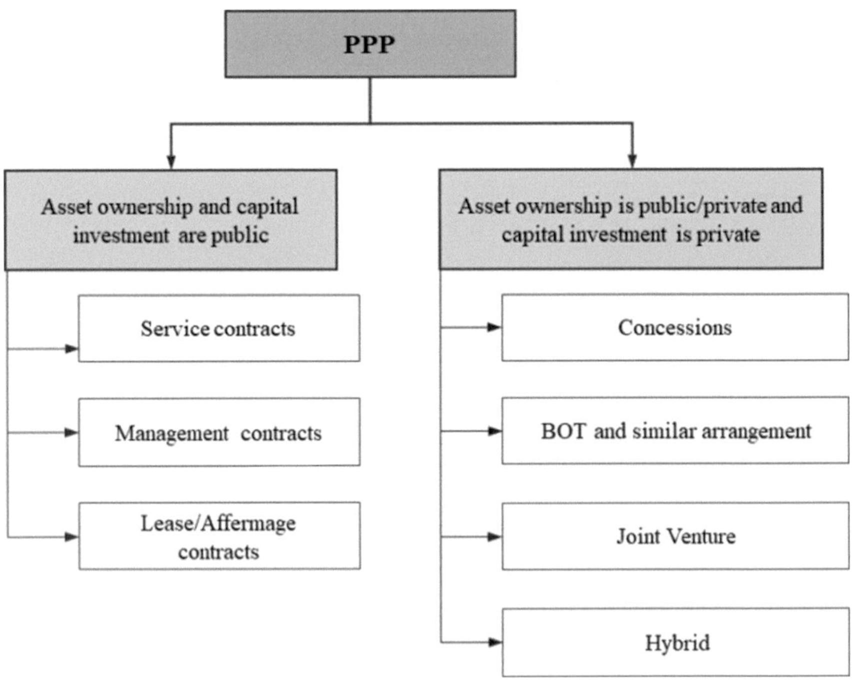

Fig. 1 PPP ownership types. Adapted from ADB (2008)

owners often intend to mitigate risks via carefully analyzing PPP groups and associated contract types before investment model selection and implementation. This work requires the involvement of experts to improve the likelihood of project success.

From the discussion mentioned above, the promising PPP models are

Table 3 Typical PPP models applied for smart transport infrastructure development

PPP models	Advantages	Disadvantages
(BOT), (BOOT), (BTO)	– Allocate risk to the party who better manages them [18] – Fast development – Promote to the private sector and improve value for money – Operation and maintenance efficiency – Solves the problem of public sector budget restraints [19] – Facilitates creativity and encourages innovation in infrastructure development [20]	– Possible conflict between planning and environmental agents – Contracts are complicated and the tendering process takes a long time – Because of long-term finance, which is the reason of lack of attractiveness to the private sector – Lengthy delays caused by political debate – High risk in relying on the private sector [21] – Lack of experience and skills in human resources
Design-Build-Finance-Operate (DBFO), Design-Build-Finance-Operate-Maintain (DBFOM), Design-Construct-Manage-Finance (DCFM)	– As for BOT plus – Interesting to private sector finance – Improve maintainability and buildability [22] – Facilitate creativity and encourage innovation in infrastructure development [23] [24] – Transfer new technologies [25] – Promote organization from overseas for joint completion through joint venture and partnering (Bastin 2003) – Promote local economic growth and employment opportunities [26]	– Possible conflict between planning and environmental issues – The contract could be lengthy because of the complicated tender process – Extra management time is spent in contract transactions – Schedule delays can arise in negotiation [21] – High risk in relying on private sector [21] – Lack of experience and skilled human resources

- The first model: The model includes types of contracts such as Operations & Maintenance (O&M); Operations, Maintenance and Management (OMM), or Management contracts, where the private sector is responsible for the operation and maintenance process to maintain the performance and profitability of the facility for a certain time period. Under this model, the public agency retains ownership of the facility, but the private sector is responsible for managing and operating the facility under a long-term contract. The private operator may invest some of its own capital and will recover the investment and earn reasonable profits according to the contract.

- The second model: The characteristic of this model is that the government invests in assets and owns the assets from the outset. The common forms of this model are DBO, DBM, and DBOM. Design, Build, Operate (DBO) is a type of public–private partnership (PPP) in which the government provides funding, but private contractors design, build, and operate some or all of the operational elements. Design, Build, Maintain (DBM) involves private sector responsibility for designing, building, and maintaining infrastructure under a long-term maintenance agreement. The public sector retains ownership and operation of the infrastructure. Design, Build, Operate, and Maintain (DBOM) is a contract in which the private sector is responsible for designing, building, as well as operating and maintaining an infrastructure for a certain period of time after construction. The project is sponsored by the public sector.
- The third model: The main characteristic of this model is that the project is mostly or entirely funded by the private sector and there is no obligation to transfer ownership rights to the government. This PPP group includes types of contracts such as BOO, ROO, DBFOM, DBFO, and DBFM.
- The fourth model: The hybrid model consists of programs in which the private sector designs, builds, and operates infrastructure for a period and transfers the facility to the government upon the expiration of the concession period or at another predetermined time. The private partner can then lease the asset to the government. The main characteristic of this model is that the project can be partially or fully funded by the private sector and the assets are transferred back to the government when the operating contract ends or at another designated time.

To choose a suitable model, it is essential to identify a set of criteria to determine which one is appropriate for the project proposed by the government. The criteria can be determined through a synthesis of previous research results and practical requirements in practice. The author proposes the following:

Financial source:

- Accessing funds to implement the project, the sources of funding may come from privatization, credit organizations, and the budget. This is also considered a solution for investment when the budget is limited. The impact of the time of providing financial resources can be fast or slow.
- Risk allocation: Under the PPP model, risks in construction investment are shared between the government and the private investor, and the risk allocation ratio depends on the chosen model or contract.
- Impact from government: The level of government intervention in the project from the initial phase to the completion of the project. This intervention leads to the occurrence of delays and setbacks.
- Attracting investors: The financial profitability ratio, low risk, fairness, transparency, and efficiency of the project aim to attract investors.

- Benefits for citizens: Directly, citizens benefit from the project, indirectly, citizens benefit from other infrastructure projects that use funds from the budget.
- Suitability to state policy mechanisms: State policies and mechanisms need to be suitable for the PPP model. This creates a sense of security for investors and minimizes investment risks.
- Scale of the project: The project scale should be appropriate for the PPP model, whether it is large or small. This allows for multiple investors to compete and participate.
- The complexity of the project: The project requires complex technologies, high-level management skills, and high security and safety measures.
- Stable policy and regulatory framework: A consistent policy and regulatory framework throughout the project lifecycle.
- Economic, political, and social stability: The upward or downward trend of the economy can affect the financial return and payback period of the project.
- Collaboration among involved parties: Depending on the PPP model or contract, the number of involved parties in the project is determined.
- Efficiency of the project: PPP model is suitable to bring the project's financial, economic, and social security efficiency.

3 Conclusion

This paper focuses on common PPP models applied for smart transport infrastructure development. Based on the literature review of selected articles published from 2000 until the present, some PPP models were highlighted to provide a holistic picture regarding the use of PPP models around the world. There is no universal acceptance of PPP classifications, but the most common approach to PPP projects can be followed guided by the ASEAN Development Bank (ADB) and the International Monetary Fund (IMF). In line with initial results, the authors attempted to analyze some of the advantages and disadvantages of PPP models. In addition to promising PPP models proposed, the authors also highlight some criteria for model selection. This work is crucial for both decision-makers and practitioners in seeking the right PPP model for smart infrastructure projects in the future.

Acknowledgements This research is funded by Vietnam National University, Ho Chi Minh City (VNU-HCM) under grant number C2023-28-06.

References

1. Park YJ (2019) Strategy for building smart city as a platform of the 4th industrial revolution. J Digital Converg 17(1):169–177
2. Chin H, Debnath A, Yuen B, Haque M (2011) Benchmarking smart and safe cities. Grant No R-264-000-251-112, National University of Singapore, Singapore

3. Akhras G (2000) Smart materials and smart systems for the future. Can Milit J 1(3):25–31
4. Selim AM, Yousef PH, Hagag MR (2018) Smart infrastructure by (PPPs) within the concept of smart cities to achieve sustainable development. Int J Crit Infrastruct 14(2):182–198
5. U. Nations (2016) Smart cities and infrastructure report of the secretary-general economic and social council. United Nations-Economic and Social Council, Commission on Science
6. Debnath AK, Chin HC, Haque MM, Yuen B (2014) A methodological framework for benchmarking smart transport cities. Cities 37:47–56
7. E. Commission (2003) Guidelines for successful public-private partnerships. Erişim Adresi europa. http://www.eu/regional_policy/sources/docgener/guides/ppp_en.pdf
8. Anderson MB, Petrie M, Alier MM, Cangiano MM, Hemming MR (2006) Public-private partnerships, government guarantees, and fiscal risk. International Monetary Fund
9. O. 2012. OECD Principles for Public Governance of Public-Private Partnerships. OECD Publishing
10. A. G. (2008) National public private partnership guidelines overview. Infrastructure and Regional Development, no. Australian Government
11. Lam T-iP, Javed AA (2015) Public private partnership in Hong Kong. In: Public Private Partnerships. Routledge, pp 138–151
12. Law on public private partnership investment (2020)
13. Erokhina O, Brega A (2020) Intelligent transport technologies in "smart" cities. In: 2020 systems of signals generating and processing in the field of on board communications. IEEE, pp 1–5
14. Milenković M, Rašić M, Vojković G (2017) Using public private partnership models in smart cities-proposal for Croatia. In: 2017 40th international convention on information and communication technology, electronics and microelectronics (MIPRO). IEEE, pp 1412–1417
15. Peters MD, Godfrey CM, Khalil H, McInerney P, Parker D, Soares CB (2015) Guidance for conducting systematic scoping reviews. JBI Evidence Implement 13(3):141–146
16. Nassar G (2004) BOT contracts and modern development of concession contract. Dar el nahda elmasryia, Cairo
17. A. D. B. 2008. Public-private partnership handbook. Asian Development Bank, Manila, Philippines
18. Hood J, Mcgarvey N (2002) Managing the risks of public-private partnerships in Scottish local government. Policy Stud 23(1):21–35
19. Ke Y, Wang S, Chan AP, Cheung E (2009) Research trend of public-private partnership in construction journals. J Constr Eng Manage 135(10):1076–1086
20. Cruz CO, Marques RC (2013) Flexible contracts to cope with uncertainty in public–private partnerships, vol 31, no 3, pp 473–483
21. Li B, Akintoye A, Edwards PJ, Hardcastle C (2005) Critical success factors for PPP/PFI projects in the UK construction industry. Constr Manage Econ 23(5):459–471
22. Li B, Akintoye A, Edwards PJ, Hardcastle C (2005) Perceptions of positive and negative factors influencing the attractiveness of PPP/PFI procurement for construction projects in the UK: findings from a questionnaire survey. Eng Construct Archit Manage
23. Kwak YH, Chih Y, Ibbs CW (2009) Towards a comprehensive understanding of public private partnerships for infrastructure development. Calif Manage Rev 51(2):51–78
24. Cruz CO, Marques RC (2013) Flexible contracts to cope with uncertainty in public–private partnerships. Int J Project Manage 31(3):473–483
25. Brzozowska K (2006) Advantages and threats of Public-Private partnerships in larger infrastructure projects. Warszawa: CeDeWu. PL
26. Kwak YH, Chih Y, Ibbs CW (2009) Towards a comprehensive understanding of public private partnerships for infrastructure development, vol 51, no 2, pp 51–78
27. Zhang X (2005) Critical success factors for public-private partnerships in infrastructure development. J Constr Eng Manage 131(1):3–14
28. Love PE, Skitmore M, Earl G (1998) Selecting a suitable procurement method for a building project. Constr Manage Econ 16(2):221–233

Investigating Opportunities of Adopting Sharing Economy in City Logistics in Hanoi, Vietnam

Thanh Thi My Truong

Abstract The sharing economy is becoming increasingly important to the operation of modern cities. By reducing the number of delivery trucks and utilizing greener vehicles, some sharing economy-based solutions can help to promote sustainable city logistics. The potential of the sharing economy for goods deliveries in Hanoi, Vietnam is assessed in this research with a focus on eight truck-sharing and crowd-shipping companies as Ahamove, Giaohangtietkiem, GHN, Kien Vang, EcoTrans, GogoX, sShip, and VTGo. The authors look at two sharing platform-based strategies for light trucks, namely, crowd-shipping and truck-sharing, as well as how Vietnamese customers generally see sharing platforms and their motivation to use them. The findings indicate that Vietnam's truck-sharing sector is still in its early stages. The growth of sharing services predestined for city logistics has been slower than that of passenger transport services. The respondents are intense to adopt business concepts based on the sharing economy. Thanks to the technology that makes the access-based economy efficient, truck-sharing is becoming more and more popular. The potential for financial savings is the main justification for using sharing services, according to 62% of the respondents. Positive reviews and referrals from friends and family members are the biggest motivators for potential users (41%). Transport authorities and logistics managers may learn from the research how potential consumers view sharing economy-based services, and it can also offer advice on how to improve brand recognition and tailor services to match customer needs.

Keywords Sharing economy · City logistics · Truck-sharing · Crowd-shipping · Urban goods delivery

T. T. M. Truong (✉)
University of Transport Technology, 54 Trieu Khuc Street, Thanh Xuan District, Hanoi, Vietnam
e-mail: thanhttm@utt.edu.vn

V. T. Ha et al. (eds.), *Proceedings of the 8th International Conference on Sustainable Urban Development*, Advances in 21st Century Human Settlements,
https://doi.org/10.1007/978-981-99-8003-1_4

1 Introduction

A city that adheres to socio-economic equitable standards, as well as environmental limits, and maximizes economic and social benefits is referred to as being sustainable [1]. The spatial distribution of activity and accessibility to a range of services are crucial components of a sustainable city [2]. Transport, housing, land use, the environment, jobs, and public services are among the aspects of a sustainable city that are characterized by a sectoral approach [3]. Monitoring and assessing a city's sustainability should take into account logistics and related transport activities [4].

City logistics, though necessary for cities to function, has a negative impact on urban life and the environment, contributing to greenhouse gas emissions, air pollution, noise, traffic congestion, and other problems [5–8]. Compared to other vehicle categories, freight vehicles emit more emissions [9]. As a result, there is a need for solutions that will promote the environmentally, socially, and economically sustainable growth of city logistics.

Various sharing economy-based solutions can be introduced to improve the sustainability of urban cargo handling. They can increase logistics efficiency while minimizing pollution and traffic [10, 11].

Numerous problems raised by scholars and administrators regarding the sharing economy in urban freight transport remain unaddressed. A niche is recognized, among others, in the assessment of operators' offerings and bottom-up activities in conjunction with the evaluation of prospective consumers' motivation and interest.

With a large population of 8.5 million habitants (GSO Vietnam, 2021), there is a huge demand for good delivery in Hanoi. According to the statistics of the Ministry of Industry and Trade, there were 123 supermarkets, 455 traditional markets, and more than 7,500 local groceries in Hanoi in 2021. Furthermore, Hanoi citizens have preferences on on-street shopping because of the convenience of the service, for instance, no parking requirements, quick access, and asking for goods rather than finding them. Such unique characteristics facilitated the huge demand for goods delivery in Hanoi for many years. The city logistics, therefore, is complicated in Hanoi and often requires a large volume of truck delivery.

This study aims to assess the sharing economy's potential for city logistics in Hanoi, Vietnam. The emphasis is on conventional trucks, sometimes referred to as delivery vans here, with a gross weight of up to 3.5 tons. Truck-sharing and crowd-shipping services are the two types of options that the researchers have looked at. They both run through a mobile app, a sharing platform, and have two things in common: they are both accessible to users at any time without the need to sign a long-term commitment.

The arrangement of this paper is as follows. In Sect. 2, a review of the literature is provided. The methodology is described in Sect. 3. Section 4 examines two sharing economy-based approaches to city logistics: crowd-shipping and truck-sharing, as well as the findings of a study of residents' attitudes toward the sharing economy. Section 5 concludes with a discussion and recommendations.

2 Literature Review

The ownership based to access-based business model transformation is being seen as a global phenomenon. Many people favor switching from purchasing to renting out items, which explains the rising popularity of collaborative platforms that connect and assist interested parties.

Despite the large number of people involved, the concept's theoretical foundations, which are based on the access-based approach and sharing economy, are still in their early development. Numerous phrases that have been incorporated into everyday speech and commercial practice are used in the scientific literature. It has been classified and defined by several authors [12]. The term "sharing economy" is frequently used to describe a wide range of organizational structures that are changing markets and urban environments and in which commodities and services, talents, and spaces are shared, traded, rented, or leased [13]. A number of phrases that are related to the sharing economy are also in use, though they are frequently used inconsistently because the terminology is continuously changing as new platform variants hit the market. According to Rojanakit et al., while having quite different meanings, the terms "sharing economy," "peer economy," "collaborative economy," "on-demand economy," and "collaborative consumption" are frequently used synonymously [14]. The terminology of services and platforms in the transport sector is based on the term sharing (car-sharing, ride-sharing, bike-sharing, etc.).

The conventional network of urban logistics providers is supported by the sharing economy. The two systems, the traditional network and the sharing economy, that operate simultaneously and complement one another are the foundation of the two-track model illustrated in Fig. 1 [15]. The traditional network consists of businesses, particularly shops or supermarkets, who run their logistic networks to deliver goods to clients (shipper-operated), as well as expert logistics providers like DHL or UPC (carrier-operated). Because of the top-down organizational structure of this type of network, carriers must follow the instructions of their superiors when delivering goods.

Deliveries made using the sharing economy concept supplement the operation of the traditional system. Sharing vehicles and routes expedites deliveries, lowers

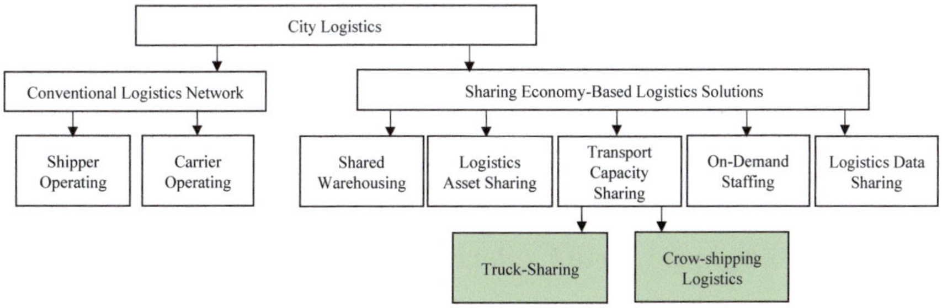

Fig. 1 Conceptual framework of city logistics with examples of sharing economy-based areas

their costs, and improves their environmental impact [16]. Truck-sharing and crowd-shipping are the two main sharing strategies used in city logistics. In both cases, a smartphone app is necessary to use the service. Customers can use shared warehouses when their volume is inadequate to support a dedicated facility, when they need flexibility due to seasonal demand peaks or unpredictable demand, or when sharing is more cost-effective than doing it alone [17]. Consolidation points [18] and urban discreet warehouses [19] are also sometimes used in this way. Shared warehouses can also be combined with shared use of trucks for end-customer delivery [20]. In labor-based businesses, for example, logistics, on-demand staffing portals connect job searchers and recruiters that need temporary personnel as soon as possible [21, 22]. By integrating information resources, sharing information, and connecting infrastructures and services, logistics data sharing on an open platform enhances visibility, traceability, interoperability, and collective decision [23].

The logistics approaches highlighted in green in Fig. 1 are studied further in this research. Vehicle-sharing with the use of delivery vans can be called truck-sharing, as in this model, customers are provided trucks (delivery vans) with a gross weight of up to 3.5 tons. Larger loads that cannot be transported in passenger cars can be moved using this method. Crowd-shipping, also known as crowd logistics or crowdsourced delivery, is another sharing economy-based solution that focuses on freight transport for the requirements of urban residents [4]. Crowd-shipping is a part of the wider sharing economy concept that uses excess capacity in goods and services to redistribute, share, and reuse which allows individuals to better utilize resources [11]. The idea of sharing services, such as transport, intends to increase the effectiveness and sustainability of how people and goods are transported, stored, supplied, and used around the globe by integrating ideas from online data transfer with actual shipping procedures. Because technology enables people to make greater use of the space in their vehicles, including by carrying packages for others, crowd-shipping also entails a network [24]. Collaborative consumption is made possible by the sharing and trading of resources without needing to purchase them [11, 20, 23]. The shared use of extra vehicle capacity for delivery is essential to crowd-shipping [25]. In crowd-shipping, shippers use technological platforms to outsource logistical work to a large number of drivers who use their own vehicles to deliver logistics services as independent contractors [26]. The concept is to encourage drivers to use their extra cargo space on their vehicles, such as cars, trucks, and motorbikes, to transport packages for other people [27]. It is implemented using the on-demand "instant deliveries" formula [28] and can be carried out using either the B2C (retailers are engaged, such as Amazon Prime Now) or Peer2Peer (individuals are the only parties involved) business models.

3 Methodology and Data Collection

In order to assess the possible application of sharing economy in city logistics in Vietnamese cities, this article will examine the following three research questions:

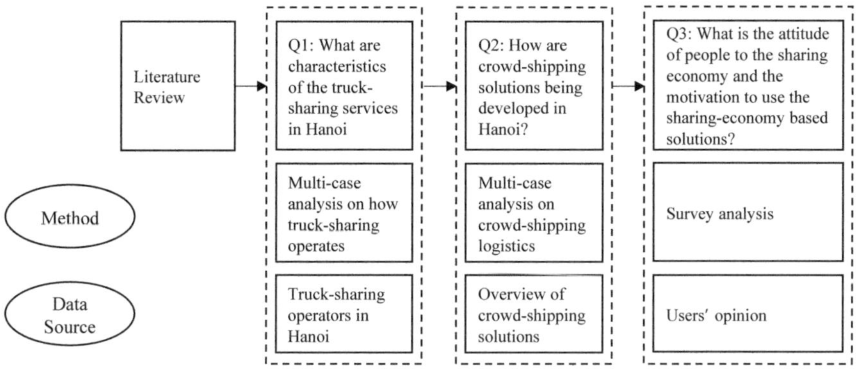

Fig. 2 Research framework

- Question 1. What are the characteristics of truck-sharing services in Vietnamese cities?
- Question 2. How are crowd-shipping solutions being developed in Vietnamese cities?
- Question 3. How do people perceive the sharing economy and why do they choose to employ sharing economy-based solutions?

The research methodology was divided into four steps (see Fig. 2). The choice of efficient research techniques, data collection methodologies, and research instruments was necessary for each step.

Multi-case study and survey analysis are the two approaches used in the research. Based on a comparative analysis of the truck-sharing market's offerings, the multi-case study was developed. Only the operators who provide delivery trucks are considered in the analysis of truck-sharing. Apart from one company, all the businesses under investigation provide delivery trucks and shipping motorcycles. To compare the services, the following criteria were used: the offered truck's qualities, the pricing strategy, and trip costs. These parameters were chosen specifically for this study.

Based on information from the market, a comparison of the potential advantages of truck-sharing versus long-term leasing has been conducted. The calculations were made using GogoX's price list, a truck-sharing company that operates in major Vietnamese cities. Since leasing vehicles is one of the most popular ways to buy assets on the market, it has been used as a benchmark. Although the availability of leasable vehicles can increase the number of newer, greener cars on the road, vehicle leasing is less environmentally friendly than truck-sharing because it doesn't eliminate all the delivery vehicles currently on the road.

A description of the services functioning under the crowd-shipping model has been added to the examination of the sharing model in city logistics. The study focuses on the crowd-shipping services offered in Vietnam, many of which are still just start-up ventures making their imprint on the industry.

The necessity to move goods or passengers, the operator's competitive price, the service's accessibility, and the user's confidence in the sharing economy as a resource-efficient model are some of the elements that have been considered to affect their decision to adopt a sharing economy solution. A survey was used to investigate the final factor. Before deciding to employ a specific operator's offering, potential clients must have confidence in the sharing economy solutions. The survey results allow us to determine whether or not the sharing economy solutions are able to attract the attention of potential users, and if so, why. They are an important component of determining the sharing economy's potential for cargo transports in Vietnamese cities like Hanoi, particularly because of the unique characteristics of the local inhabitants [29].

In order to conduct the research, a computer-assisted web interviewing technique was used with the diagnostic survey method. E-mail and social media were used to distribute the survey form. The authors supposed that respondents might be interested in using sharing platforms. The survey was limited to Hanoi residents and was conducted in July and August 2022.

Its objective was to determine how participants think about sharing economy solutions. They were asked to describe their interest in sharing platforms as well as the considerations that influenced their decision when selecting a service. In total, the questionnaire was completed by 485 individuals. Table 1 displays the characteristics of the samples.

Table 1 Characteristics of respondents (N = 485)

Respondent attribute	Description	Percentage of total respondents (%)
Gender	Male	49.1
	Female	50.9
Age	18–29	28.3
	30–39	32.6
	40–49	18.5
	50–59	19.4
	60 and over	1.2
Education	College/associates	26.2
	Undergraduate	59.2
	Graduate	14.6
Monthly household income	<3 M VND/mo	7.2
	3–6 M VND/mo	16.5
	6–10 M VND/mo	29.6
	10–20 M VND/mo	29.8
	20–40 M VND/mo	12.6
	>40 M VND/mo	4.3

4 Results

4.1 Characteristics of Truck-Sharing in Hanoi

The sharing economy-based municipal logistics solutions must be less expensive than conventional services, practical to use, and widely accessible to generate demand. User's awareness of their environmental friendliness may not be a convincing argument. Therefore, the level of costs assessed for truck-sharing services, the methodology used to determine prices, and the accessibility of the service are all examined in this section. One can evaluate opportunities for the evolution of the complete model by being aware of the terms of usage.

Only present in three main cities in Vietnam—Hanoi, Da Nang, and Ho Chi Minh—truck-sharing services are still in the early stages of establishing their market existence.

There are various potential customer types, as well as various scenarios in which they might employ truck-sharing services. One set of users consists of people who periodically move large or awkward goods that are challenging or even impossible to transfer in a passenger automobile, such as when relocating or remodeling. Some truck-sharing companies (such GogoX, 2020) keep parking places at supermarkets for their convenience.

The services are also intended for business owners whose sporadic transport requirements do not warrant maintaining a fleet of vans of their own. The services are a choice for companies who do not find it economically feasible to support the costs of fleet ownership, not a replacement for the company's own fleet.

The truck-sharing companies in Hanoi are listed in Table 2. They all meet the two primary requirements for the truck-sharing service: they run a smartphone app that makes it easier to book, start, and pay for the service, and they let customers pay by the minute. The operating models of many traditional van rental companies that do not adhere to the established standards have not been included in Table 2. Trucks can be rented from the majority of the firms analyzing vehicle rentals as an add-on service to passenger cars. Only GogoX offers passenger automobiles; all other companies only specialize in renting delivery vans on a minute-by-minute basis. GogoX first entered the Vietnamese market in 2017 and has since restricted its business to delivery vehicles.

Even though the business models of all the truck-sharing companies under investigation are similar, there are some variances between them, particularly in the pricing strategy. There are various methods for calculating rental costs. The most typical estimate takes into account two factors: the length of the rental and the distance traveled. We analyzed the costs charged by the operators for hiring a van for a covered distance of 30 km while taking into account the various billing methods.

Comparisons show that different providers charge different fees. The most expensive rates are levied by EcoTrans when taking into account the specified trip characteristics (about 30 km). However, sShip has the most affordable renting rates.

Table 2 Features of truck-sharing companies in big cities in Vietnam

Name of operator	Name of service	Fleet type	Pricing strategy (truck 1.25–1.5 tons)	Fee per trip* [VND]
EcoTrans Shipping Joint Stock Company	EcoTrans	Combined trucks from 500 kg to 10 tons, conventional fuel	First 4 km: 500.000 VND; From 5th km: 29.000 VND/km Waiting fee: 50.000 VND/h; Loading and unloading fee: 230.000 VND/ person	1,254.00
Tuc Thoi Joint Stock Company	Ahamove	Combined trucks from 500 kg to 2 tons, conventional fuel	First 4 km: 320.000 VND; From 5th km: 29.000 VND/km Waiting fee: 60.000 VND/h; Loading and unloading fee: 110.000 VND/ person	1,074.000
SGLOBAL	sShip	Combined trucks from 500 kg to Container 40 ft, Conventional fuel	First 4 km: 165.000 VND; From 5th km: 19.000 VND/km Waiting fee: 30.000 VND/h; Loading and unloading fee: 100.000 VND/ person	659.00
Gogox Vietnam Corporation	GogoX	Combined trucks from 500 kg to 2 tons, conventional fuel	First 4 km: 300.000 VND; From 5th km: 23.000 VND/km	898.000
Kien Vang Group Joint Stock Company	Kien Vang	Combined trucks from 500 kg to 2.3 tons, conventional fuel	First 4 km: 250.000 VND; From 5th km: 21.500 VND/km Waiting fee: 60.000 VND/h; Loading and unloading fee: 350.000 VND/ person	809.000

*The truck weight is about 1.25–1.5 tons. A trip that covers 30 km, no waiting time, and no loading–unloading service

Between the lowest and highest fee, there is a noticeable gap that ranges from 659.000 VND to 1.254.000 VND. In Vietnam, hiring a truck costs 809.000 VND on average. GogoX, which ranks in the middle and only focuses on cargo trucks, deserves special consideration. The fact that the price is the same for both smaller and larger trucks gives it a competitive edge. GogoX is the most affordable option when merely looking at prices in the segment of the largest trucks that are readily available. Additionally, it offers services in most cities around Vietnam. It is important to note that not all of the places where the company first established its presence still offer truck-sharing services through GogoX; for instance, they no longer operate in Hanoi. The business strategy of GogoX, which is only devoted to urban freight transport, is riskier than that of its rivals. Financial issues arise for the entire business as a result of insufficient demand for the truck-sharing service. Other operators can use the more developed passenger car-sharing industry's activity to make up for any drop in demand in the freight market.

4.2 Crowd-Shipping Solutions for City Logistics in Hanoi

Services for crowd-shipping are offered both internationally and locally. Local services are being offered on a growing number of online crowd-shipping platforms in recent years. In the Vietnamese market, several new projects are currently offering solutions for crowd-shipping.

Platforms for "crowd-shipping" connect those who want to send packages with people who want to transport them. In Hanoi, you can use the following crowd-shipping services:

1. Ahamove.com—This website connects shippers with drivers who are heading to specific locations. As a result, the shipper receives a timely delivery while the driver lowers travel expenses. A price can be given by both the shipper and the driver. The website gives users the choice of utilizing their ID card or bank account to verify their identification (Ahamove, 2022).
2. Giaohangtietkiem.vn—Another website that connects people looking to deliver cargo (of any size, from an envelope to a pallet). There is no alternative for identification verification, thus those who want to register must give their phone number. The delivery is the carrier's responsibility. There are no fees associated with using the website, and online insurance for the package is not an option. A database contains the carriers, which can be accessed (Giaohangtietkiem, 2020).
3. Grab.com—The first online marketplace for transport services in Vietnam. With the help of recognized carriers and consumer evaluations and comments, it connects customers with reliable carriers and free to customers. A description of the package (such as furniture, a box, a bike, or motorcycle) and the desired location must be posted in order to get transport. The customer can choose the most appropriate carrier from among the prospective carriers' offerings. Up to 60% of transport expenditures might be saved thanks to the platform. There is a chance

to locate a firm that is already driving to the required location and can deliver the cargo there because the post is seen by thousands of transport businesses. The customer is also given the best rate possible because the prices are set up in a declining price auction. Customers may easily search for cheap and secure transport on Grab.com. It gives carriers quick access to fresh orders, which helps them find more load and cut down on empty runs (Grab, 2020).

4. Ninjavan.co—The website was initially intended to act as a platform for transport service intermediaries. NinjaVan intended to deliver unusual, foreign packages (like animals to Vietnam, for instance). But ultimately, NinjaVan adopted the Uber-like business model. It only trades on the Hanoi market at the moment. Couriers are auto-driven. Although the business focuses on the B2B sector, it also takes orders from private customers. Numerous buddies and acquaintances make up the group of couriers (NinjaVan, 2020).
5. Spx.vn (Shopee Express)—A shopping website run by the Shopee e-commerce. Customers can create a shopping list and share it publicly with their location using the Shopee mobile app. They can also ask someone else to shop for them and have the items delivered to their home. Currently, Shoppe Express exclusively serves the Shopee e-commerce platform in Vietnam; however, there are plans to expand this service to include all major cities in Vietnam as well as complete good delivery in the Hanoi and Ho Chi Minh area (ShopeeExpress, 2020).

By providing affordable delivery options, these services enable retailers to gain a substantial competitive edge and speed up the shipping process. The continual creation of new solutions based on the concept shows that crowd-shipping is becoming more and more important. Another concern is whether or not the packages arrive on time and undamaged, as well as whether or not the carriers can be trusted, as well as the protection of the contracting parties' personal data and occurrences like theft, fraud, damaged products, and late deliveries.

The administrators and owners of the aforementioned platforms must make every effort to guarantee that the services are of adequate quality and safety to draw in potential clients. Like any novelty on the market, this kind of service is likely to go through dynamic changes and gradually fall into place, balancing customer expectations with supplier capabilities, particularly in Western countries where societies are known for a higher level of trust in people and innovative ideas.

4.3 Major Success Element for the Model: An Analysis of Survey Data

An examination of potential sharing economy applications for urban freight transport must include an examination of user attitudes. People will only support the model and value the benefits it provides above conventional transport networks if they believe in it. The findings of a survey looking at how Hanoi locals see the sharing economy are presented in this section. By responding to Question 3, we can determine whether and why the sharing economy solutions are able to draw in potential users. The sharing hospitality market is leading the way and has the greatest average annual growth rate in Vietnam as the sharing economy gains traction in Hanoi [30]. Platforms for renting trucks and cars are a novelty in the city. The age distribution of those who responded to the question about their desire to use sharing platforms is shown in Fig. 3. "Yes," "No," and "Perhaps" were the possible responses. Only 3.8% of the respondents said they would not use sharing platforms again. There were more "perhaps" responses than "yes" in each age group, albeit the differences in the numbers are not very noticeable.

The respondents who use or may use sharing platforms in the future have responded to a further question about the reasons for which they are willing to participate in the business model (see Fig. 4).

Because the services given through sharing platforms are less expensive than those that can be bought through traditional channels, 62% of the respondents either

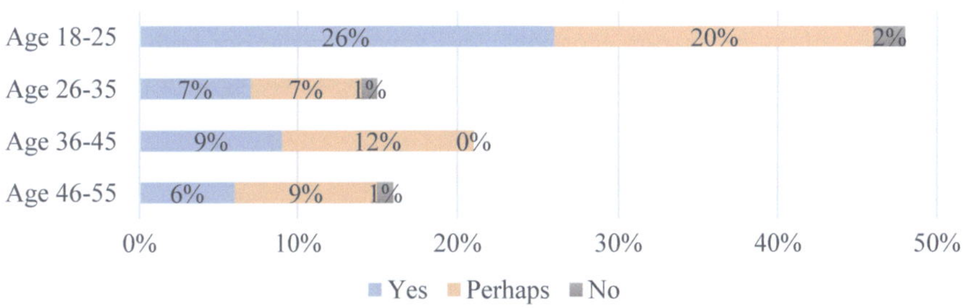

Fig. 3 Current and future users of sharing platforms, by age [%]

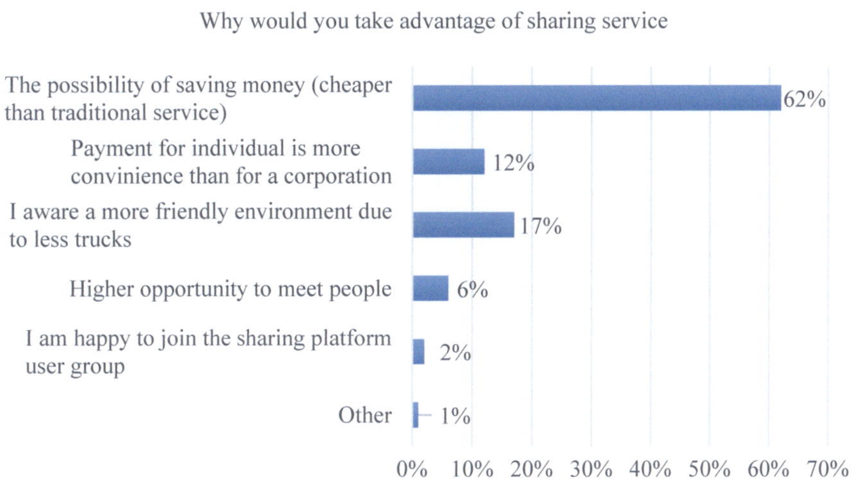

Fig. 4 Reasons for using sharing platforms [%]

use them or plan to in the future. 17% of respondents use sharing platforms to safeguard the environment, while another 6% see them to meet people and discover other lifestyles. Another 12% of respondents prefer to pay individuals rather than businesses for the services.

A discussion of factors that may encourage potential consumers to adopt sharing platforms was conducted. The responses are displayed in Fig. 5.

Most survey participants (31%) and those who would utilize a service in conjunction with a friend or family member (26%) would do so through a sharing platform if it came highly recommended. Up to 23% of them would be inspired by user reviews that were put online, and 15% would be inspired by favorable press or online publicity. It's important to remember that recommendations found online have considerably less weight in the decision-making process than those gathered in an old-fashioned manner, from friends and family.

The responders are generally excited about utilizing new company concepts. Only 4% of them claim to have no interest in the sharing economy, while the rest are either very enthusiastic or at the very least do not completely rule out utilizing solutions that are based on the sharing economy. This is encouraging for both new market entrants and sharing operators that are thinking about expanding their operations to Hanoi.

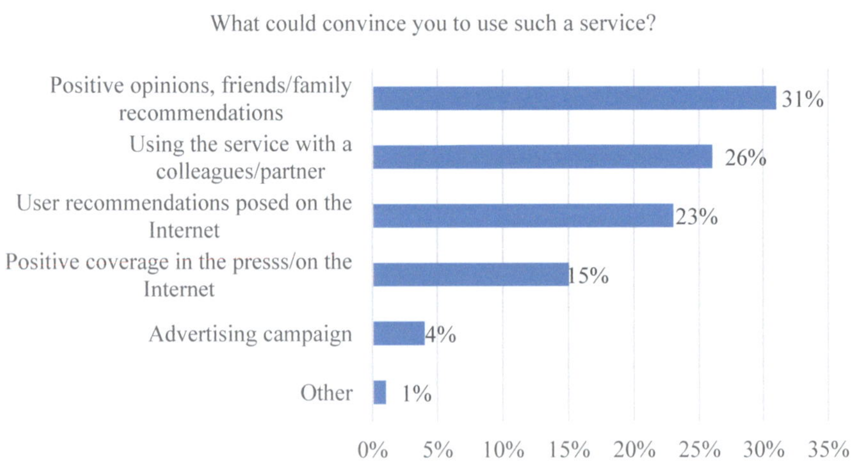

Fig. 5 Factors encourage potential consumers to adopt sharing platforms [%]

5 Discussion and Conclusions

Environmental friendliness, which consists of more intense and communal use of owned items and a refusal to buy new ones, is arguably the benefit most stressed by sharing economy solutions. In this regard, it is worthwhile to draw comparisons between urban freight transport employing truck-sharing and sharing systems specifically designed for people. The services are offered similarly because they are both based on the sharing economy. However, there are glaring discrepancies between them as a result of various consumer segments, vehicle uses, and vehicle kinds. Different factors should be considered when determining how much truck-sharing and passenger car-sharing contribute to urban sustainability. Due to the fact that multiple drivers utilize the same vehicle, passenger car-sharing is occasionally viewed as one of the potential options to enhance urban sustainability. However, there are contemporary alternatives that are considerably more advantageous for addressing the needs of urban transport in a sustainable manner. It is best to view car-sharing as an additional and supplementary solution. An effective and convenient public transport system is an alternative that is more environmentally friendly for travelers. It might potentially lead to a decrease in the number of passenger cars in the city while ensuring sustainable urban growth. Public services are not an alternative for freight transport. The use of transport company services, owning a van, or hiring one through another, a less convenient system, for a longer duration, without utilizing a mobile app, are alternatives to truck-sharing. The fact that operators have a modern, low-emission fleet, including electric cars, is another argument in favor of thinking of truck-sharing as an environmentally good business. In actuality, the cargo bike is the only truly environmentally beneficial mode of transport in the freight transport industry. It is worthwhile to suggest cargo motorcycle-sharing for modest and light cargoes. Although cargo motorbikes are becoming more and more widespread in many Asian cities, they are still hardly ever seen in Vietnam.

Our research concludes with multiple findings incorporating different points of view by combining three research methodologies, including literature research, multi-case study, and survey analysis, as well as various data sources.

First of all, the Vietnamese cities' truck-sharing markets are just getting started. Urban freight transport-specific sharing services have grown more slowly than those geared toward passenger transport. Only available in Vietnam's largest cities, truck-sharing services come in a range of price points. The rates for a trip of 30 km range from 659.00 VND to 1.254.000 VND, with the average being 809.00 VND, demonstrating the stark discrepancies between the minimum and maximum charges. Price balancing will undoubtedly be aided by rising competition, expanding market penetration, and the promotion of truck-sharing as a "must" for sustainable urban development.

Another sharing economy-based approach targeted at participants in the city logistics industry, crowd-shipping, has drawn some noteworthy interest. Both locally and globally, crowd-shipping services are offered; however, platforms that offer local crowd-shipping services are becoming more and more popular. Crowd-shipping solutions are currently being offered by a number of start-up businesses in Vietnam, and the idea is becoming more popular in the cities.

The study's findings indicate that respondents are eager to accept business concepts based on the sharing economy. Only 5% of them indicate no interest at all, but the remainder is either interested in using a service based on the sharing economy or does not completely rule it out. Thanks to the technology that makes the access-based market function efficiently, car-sharing is becoming more and more popular. The potential for financial savings is the main justification for using sharing services, according to 62% of the respondents. Positive reviews and referrals from friends and family members are the biggest motivators for potential users (31%). It's important to remember that recommendations found online have considerably less weight in the decision-making process than those gathered in an old-fashioned manner, from friends and family.

The research findings discussed in this work have consequences for management as well as cognitive characteristics. They give management insight into how potential customers view sharing services and can also offer advice on how to improve brand recognition and tailor services to match client needs.

Once consumers have more faith in the sharing economy-based model, truck-sharing systems will undoubtedly become more popular. Given that sharing platforms are new in Vietnamese cities, including Hanoi, the enthusiasm shown by survey respondents in them encourages us to have optimism for the future. However, a risk analysis should always be conducted before the start of a truck-sharing initiative in a new city. The qualities, expectations, and confidence in the operator's business model of the target consumers should all be investigated in this study.

Crowd-shipping's prospects are heavily reliant on the elements that local communities need to establish the sharing economy concepts, such as participant trust in one another, faith in recommendation systems, and awareness of potential benefits. Naturally, questions come up regarding whether the delivery will arrive on schedule, in what condition, and whether we can trust the courier. The intended consumer profile should be taken into account while creating a sharing platform's brand. To determine whether specific business solutions and new projects have a likelihood of succeeding, more thorough research is needed on particular types of platforms.

Acknowledgements This work was supported by the Ministry of Transport, Vietnam (Grant ref: DT203022). Thanks are due to interview participants from delivery companies, such as Ahamove, Giaohangtietkiem, GHN, Kien Vang, EcoTrans, GogoX, sShip, and VTGo, as well as to my colleagues at the University of Transport Technology who provided comments on the whole article and the methodology section. The authors also like to thank the interviewees for their contribution.

References

1. Mori K, Yamashita T (2015) Methodological framework of sustainability assessment in city sustainability index (CSI): a concept of constraint and maximisation indicators. Habitat Int 45:10–14. https://doi.org/10.1016/j.habitatint.2014.06.013
2. Ahvenniemi H, Huovila A, Pinto-Seppä I, Airaksinen M (2017) What are the differences between sustainable and smart cities? Cities 60:234–245. https://doi.org/10.1016/j.cities.2016.09.009
3. Maclaren VW (1996) Urban sustainability reporting. J Am Plann Assoc 62(2):184–202. https://doi.org/10.1080/01944369608975684
4. Buldeo Rai H, van Lier T, Meers D,. Macharis D (2018) An indicator approach to sustainable urban freight transport. J Urban: Int Res Placemaking Urban Sustain 11(1):81–102. https://doi.org/10.1080/17549175.2017.1363076
5. Thanh Truong TM, Friedrich H, Charoenngam C (2020) Success factors for financial sustainability of toll road projects: empirical evidence from China. Transp Res Procedia 48:1848–1860. https://doi.org/10.1016/j.trpro.2020.08.219
6. Thanh Truong TM, Ngoc AM (2020) Parking behavior and the possible impacts on travel alternatives in motorcycle-dominated cities. Transp Res Procedia 48:3469–3485. https://doi.org/10.1016/j.trpro.2020.08.105
7. Thanh Truong TM, Ngoc AM (2020) The city-airport connection, implications for urban transport planning in motorcycle-dominated cities. Transp Res Procedia 48:3232–3244. https://doi.org/10.1016/j.trpro.2020.08.155
8. Kijewska K, Iwan S (2016) Analysis of the functioning of urban deliveries in the city centre and its environmental impact based on Szczecin example. Transp Res Procedia 12:739–749. https://doi.org/10.1016/j.trpro.2016.02.028
9. Perera L, Thompson RG, Wu W (2020) A multi-class toll-based approach to reduce total emissions on roads for sustainable urban transportation. Sustain Cities Soc 63:102435. https://doi.org/10.1016/j.scs.2020.102435
10. Paloheimo H, Lettenmeier M, Waris H (2016) Transport reduction by crowdsourced deliveries—a library case in Finland. J Clean Prod 132:240–251. https://doi.org/10.1016/j.jclepro.2015.04.103
11. Arslan AM, Agatz N, Kroon L, Zuidwijk R (2019) Crowdsourced delivery—A dynamic pickup and delivery problem with Ad Hoc drivers. Transp Sci 53(1):222–235. https://doi.org/10.1287/trsc.2017.0803
12. Bucher E, Fieseler C, Lutz C (2016) What's mine is yours (for a Nominal Fee)—Exploring the spectrum of utilitarian to altruistic motives for internet-mediated sharing. Comput Hum Behav 62:316–326. https://doi.org/10.1016/j.chb.2016.04.002
13. Mont O, Palgan YV, Bradley K, Zvolska L (2020) A decade of the sharing economy: concepts, users, business and governance perspectives. J Clean Prod 269:122215. https://doi.org/10.1016/j.jclepro.2020.122215
14. Rojanakit P, Torres de Oliveira R, Dulleck U (2022) The sharing economy: a critical review and research agenda. J Bus Res 139:1317–1334. https://doi.org/10.1016/j.jbusres.2021.10.045
15. Jäger-Roschko M, Petersen M (2022) Advancing the circular economy through information sharing: a systematic literature review. J Clean Prod 369:133210. https://doi.org/10.1016/j.jclepro.2022.133210
16. Bachofner M, Lemardelé C, Estrada M, Pagès L (2022) City logistics: challenges and opportunities for technology providers. J Urban Mobil 2:100020. https://doi.org/10.1016/j.urbmob.2022.100020
17. Franklin R, Spinler S (2011) Shared warehouses—Sharing risks and increasing eco-efficiency. Int Commer Rev 10(1):22–31. https://doi.org/10.1007/s12146-011-0070-3
18. Antheaume N, Thiel D, de Corbière F, Rowe F, Takeda H (2018) An analytical model to investigate the economic and environmental benefits of a supply chain resource-sharing scheme based on collaborative consolidation centres. J Oper Res Soc 69(12):1888–1902. https://doi.org/10.1080/01605682.2017.1415638

19. van Duin R, Wiegmans B, Tavasszy L, Hendriks B, He Y (2019) Evaluating new participative city logistics concepts: the case of cargo hitching. Transp Res Procedia 39:565–575. https://doi.org/10.1016/j.trpro.2019.06.058
20. Zissis D, Aktas E, Bourlakis M (2018) Collaboration in urban distribution of online grocery orders. Int J Logist Manag 29(4):1196–1214. https://doi.org/10.1108/IJLM-11-2017-0303
21. He P, Zhang S, He C (2019) Impacts of logistics resource sharing on B2C E-Commerce companies and customers. Electron Commer Res Appl 34:100820. https://doi.org/10.1016/j.elerap.2018.100820
22. Bulatov NK, Balabaev OT, Arpabekov MI, Bobeev AB (2020) Formation of a transport and logistics center within the boundaries of a transitive economy. Res Transp Bus Manag 37:100556. https://doi.org/10.1016/j.rtbm.2020.100556
23. Wang X-X, Liu X-Y, Li Z-Q (2018) A social collaborative urban distribution integration platform. J Interdiscip Math 21(5):1109–1113. https://doi.org/10.1080/09720502.2018.1493038
24. Behrends S, Lindholm M, Woxenius J (2008) The Impact of urban freight transport: a definition of sustainability from an actor's perspective. Transp Plan Technol 31(6):693–713. https://doi.org/10.1080/03081060802493247
25. Cohen B, Muñoz P (2016) Sharing cities and sustainable consumption and production: towards an integrated framework. J Clean Prod 134:87–97. https://doi.org/10.1016/j.jclepro.2015.07.133
26. Castillo VE, Bell JE, Rose WJ, Rodrigues AM (2018) Crowdsourcing last mile delivery: strategic implications and future research directions. J Bus Logist 39(1):7–25. https://doi.org/10.1111/jbl.12173
27. McKinnon SL. Transnational publicity, gender-based violence, and central American women's asylum cases. In: Gendered Asylum, University of Illinois Press, pp 19–38
28. Dablanc L, Morganti E, Arvidsson N, Woxenius J, Browne M, Saidi N (2017) The rise of on-demand 'instant deliveries' in European cities. Supply Chain Forum: Int J 18(4):203–217. https://doi.org/10.1080/16258312.2017.1375375
29. Wagner N, Strulak-Wójcikiewicz R, Landowska A (2019) Trust in sharing economy business models from the perspective of customers in Szczecin Poland. Sustainability 11(23):6838. https://doi.org/10.3390/su11236838
30. Truong VD (2018) Tourism, poverty alleviation, and the informal economy: the street vendors of Hanoi, Vietnam. Tour Recreat Res 43(1):52–67. https://doi.org/10.1080/02508281.2017.1370568

Floating House and Green Farming Model Adapted to Climate Change for Rural Coastal Areas in Can Gio District, Ho Chi Minh City, Vietnam

Minh-Le Ngo and Vu-Hoang-Nguyen Le

Abstract Can Gio is a suburban coastal district located in southeast of Ho Chi Minh City (HCMC). Climate change has affected the Can Gio district in recent years, with low-lying terrain and average elevation from 0 to 1.5 m above sea level. That is the loss of residential land and arable land in some coastal areas, along rivers and canals, causing unwanted migration of people. In addition, the emergence of the Covid-19 epidemic has caused a double negative impact on people's lives and economic activities. This study aims to understand and propose a model of floating houses and green farms that can adapt to climate change and epidemics in rural coastal areas (RCA) in Can Gio district. The content of the article includes assessing the impact of climate change on RCA houses in Can Gio, thereby proposing solutions to adapt to climate change and epidemic diseases by the model of floating houses and green farming in the RCA of Can Gio district. Solutions include raised floors, floating floor designs, and applications that combine VAC and Aquaponics (vertical farming). The applied research methods are analyzing and synthesizing data, observing with recording images—drawings, and modeling-simulation. The research results will ensure the self-service ability of people when epidemics occur and, at the same time, improve the adaptability to climate change for RCA houses in Can Gio district, HCMC.

Keywords Floating house · Green farm · Climate change · Can Gio—Ho Chi Minh City

M.-L. Ngo (✉)
Faculty of Civil Engineering, Ton Duc Thang University, Ho Chi Minh City, Vietnam
e-mail: ngoleminh@tdtu.edu.vn

V.-H.-N. Le
Faculty of Architecture and Art, Ho Chi Minh City University of Technology (HUTECH), Ho Chi Minh City, Vietnam

V. T. Ha et al. (eds.), *Proceedings of the 8th International Conference on Sustainable Urban Development*, Advances in 21st Century Human Settlements,
https://doi.org/10.1007/978-981-99-8003-1_5

1 Introduction

Ho Chi Minh City is on the list of the top 10 cities most affected by climate change and high sea level rise (SLR). According to the United Nations, the sea level will rise by 1.0 m by 2100, and nearly 20% of the area of HCMC will be in flood [1]. It can be said that climate change is and will be a great challenge for HCMC, as it constantly faces inundation due to high tides in the rainy season and deep saline intrusion in the dry season. Climate change is inter-regional. In HCMC, the area is low-lying and interlaced rivers, such as districts in the south of the city (Nha Be, Binh Chanh, and Can Gio District). Can Gio is a suburban coastal district in the Southeast of HCMC with a low-lying topography. The average elevation is from 0 to 1.5 m above sea level (see Fig. 1). This area[1] is directly affected by climate change in human life in two aspects of living environment and daily activities. This area is forecasted to be increasingly flooded when the SLR is high. The main consequences of climate change and SLR for Can Gio district are flooding, saline intrusion, temperature changes, and landslides [2]. Coastal erosion occurs in most coastal areas, with an intensity of several meters per year. Rising sea levels and river flows are the main causes of coastal erosion in Can Gio district (see Figs. 3 and 4). Negative impacts from riverside landslides and coastal erosion directly impact people's lives in this area. In which housing projects, especially rural houses in coastal areas, are most affected. Rural house architecture in coastal areas faces the risk of encroachment and even relocation, which greatly affects the existence and development of residential areas that have existed since ancient. At the same time, the phenomenon of flooding damages the housing works of local people and reduces the source of residential land. It directly affects the lives of people and the economic zones (see Figs. 2 and 5). The inevitable consequence is increased poverty, no more opportunities for farming, and forced migration to the central areas, adding already great pressure to the city center. The district has strengths in developing fisheries and marine economy and has potential eco-tourism. However, natural resources are gradually being depleted and directly affected by climate change and SLR in recent years, so the residential areas of Can Gio district are facing the loss of arable land, even a small amount of residential land in An Thoi Dong, Ly Nhon, and Thanh An island communes [3]. In addition, the appearance of the Covid-19 epidemic in the past time has caused a double negative impact on people's lives and economic activities (agricultural production has been stopped, broken food supply chains).

[1] Can Gio district borders the East Sea and is separated from Nha Be district to the west by the Soai Rap River. This is the only district bordering the sea with a land area of 704 km^2 and a population of about 80,000 people of HCMC. The district is about 50 km from the city center by road, has a dense primary mangrove forest ecosystem as well as rivers and canals along with typical diverse flora and fauna in the coastal region of Vietnam. In 2000, Can Gio was recognized by Unesco as a "World Biosphere Conservation Area". Can Gio district used to hold an important strategic position, the base of Saigon—Gia Dinh in the resistance war against the French, and a springboard base, an important logistical base of the enemy on the outskirts of Saigon during the resistance war against the US, the Ho Chi Minh campaign. Has an important historical role in defense.

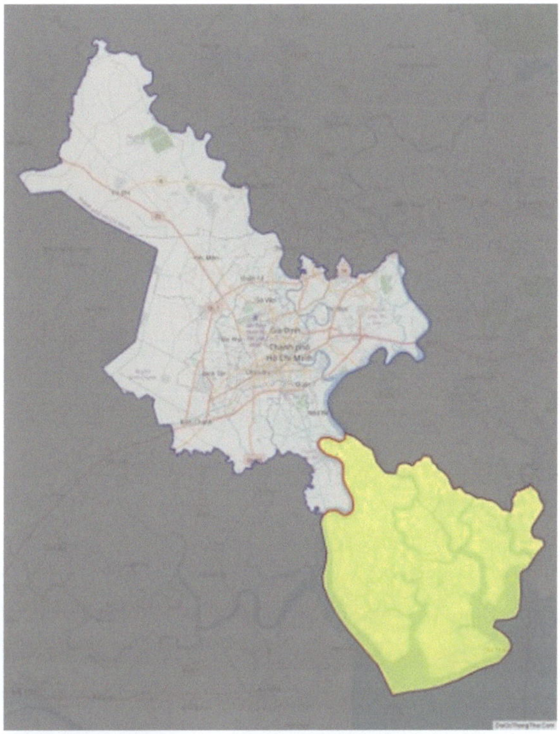

Fig. 1 Can Gio is a suburban district on the southeast coast of Ho Chi Minh City. *Source* [4]

Fig. 2 The high foundation to avoid flooding in Thanh An island, Can Gio district, HCMC. *Source* [5]

Fig. 3 Embankment protection for a residential area in Binh Khanh commune, Can Gio district, HCMC. *Source* [2]

Fig. 4 Saline water intruding inland, building stone embankment along the coast of Can Thanh, Can Gio district, HCMC. *Source* [2]

Fig. 5 Residential area of Can Thanh town in Can Gio district. *Source* [5]

So, how to protect and sustainably develop residential areas in Can Gio district, HCMC, against the impacts of flooding caused by climate change is an important issue in the city's urban development and management program. This question is becoming more and more necessary in the coastal district of Can Gio, when in the future, the existing coastal rural house models may no longer meet the living and production needs of the people. With limited income from livelihood activities such as agriculture, fishing, aquaculture, and seafood trade, along with living and production habits attached to the sea, it is not easy for people in coastal areas to move or emigrate to another place [6]. So far, there has not been an in-depth study in the field of architecture and planning that is suitable for the RCA houses model of Can Gio district. Therefore, it is essential and meaningful to research, learn, and propose a model of floating houses and green farms that can adapt to climate change and epidemics in the RCA of Can Gio district, HCMC. This mission comes from the need to protect the lives of residents, create new career opportunities for local people, and promote the value of the available potential in the locality towards the existence and sustainable development in the future.

2 Domestic and International Experience

In Vietnam, Quang Binh province, there are Tan Hoa floating houses, using light structures (wooden/iron frame, wall, corrugated iron roof) and many plastic or iron drums to make the house float. The advantage of this structure is that when the water is flooded, the house will float up and not be submerged in floodwater. In fact, these floating houses have been used by people in the flood areas of Tan Hoa, and Quang Binh during periods of bad weather and high flood water overflowing into the residential area [7]. In addition to the above-mentioned structural system, it is also combined with anchoring measures to ensure the safety of the house when there is high wind or strong water flow, the floating house may drift or collide, causing danger (see Fig. 6).

House model in a flood area in Thanh My, Dong Thap province, with an area of 48 m^2 having a hallway in front of the house, 2 bedrooms, 1 bathroom, 1 living room, and an integrated dining room [7] . The house is built with an assembled steel frame, and the wall covering the house uses panels with durability and longevity of up to 50 years. It also uses recycled materials, which are dismantled from old buildings such as wooden doors, corrugated iron, and steel boxes. The house is implemented with a green architectural solution combined with wind and water conditions, cooling the building with a natural lighting ventilation system (see Fig. 7).

The application combines the model of Garden-Pond-Cage (VAC- Vietnamese abbreviation for Garden-pond-cage) and Aquaponics, a system of growing plants and raising fish. This integration brings the benefits of practicality and uniqueness to the model. Instead of adding fertilizers and chemicals to grow plants, Aquaponics uses fish waste by converting microorganisms into nutrients necessary and sufficient for plant growth. In contrast, instead of treating and then releasing water from fish

Fig. 6 Model of floating houses in Tan Hoa, Quang Binh province

Fig. 7 Model of houses in flood areas in Thanh My, Dong Thap province. *Source* [7]

farming into the environment, Aquaponics uses plants to clean the water and return it to the aquarium. This water can be reused indefinitely and only needs to be replaced when it is evaporated. This closed recirculation system is being applied in Vietnam (see Fig. 8).

Outside of the country, in the Netherlands, the floating house model is designed as a kind of boat which can float on the water in all weather conditions. The company Dura Vermeer (Netherlands) designs offer a "floating village" plan for about 12,000 people near Schiphol Airport, not far from Amsterdam. This design consists of floating houses forming a housing cluster with the following characteristics: a waterproof hollow concrete structure that can float with the house. The submerged concrete basement is fixed below the water level, where the booster pistons are located. When the water rises, the house can float on high water. Some other types of houses such as houses with high foundations, houses built on high ground with earth and creating canals around, these canals make up a network of transport systems in the whole

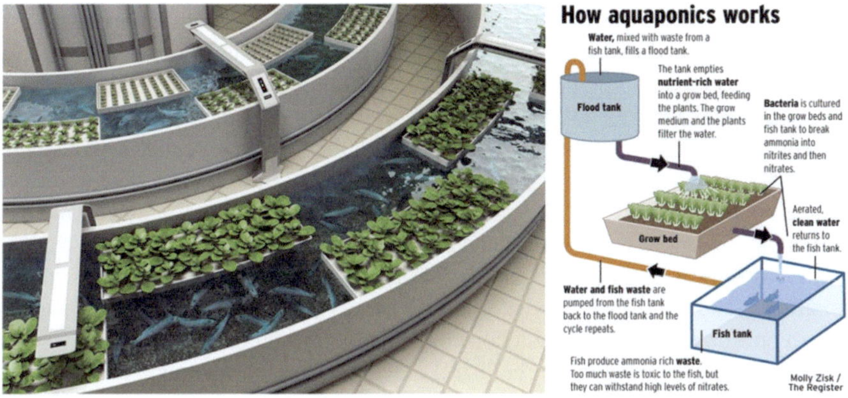

Fig. 8 Combined application of Garden–Pond–Stable (VAC) and aquaponics. *Source* https://www.pinterest.com

village, boats are the main vehicles. Typically, the ancient ecological village in the province of Overijssel, the Netherlands, was formed in the 1230 s and still exists today (see Fig. 9).

Studiomobile Design Studio has proposed a floating greenhouse to improve food security in areas with little arable land in Italy. Jellyfish Barge consists of a floating glass garden house, supported by wooden support, covering an area of 70 m^2. It can

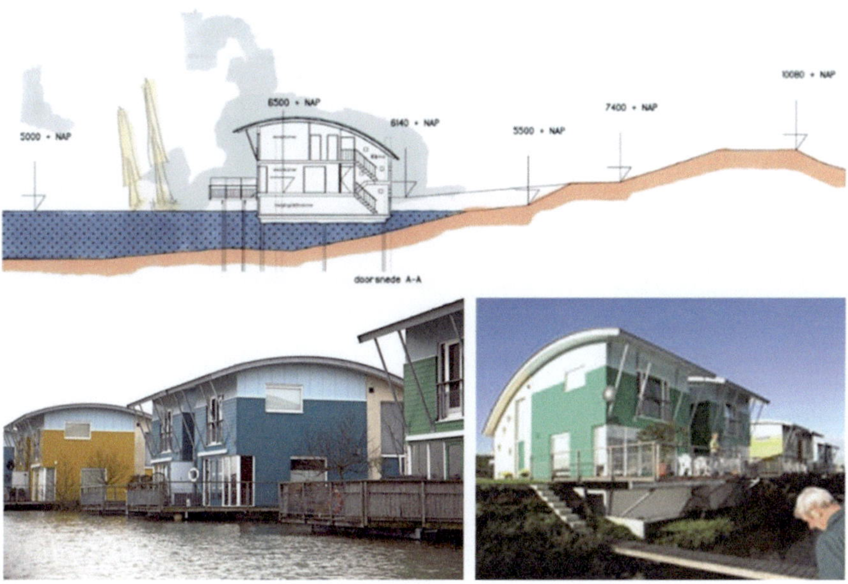

Fig. 9 Model of a floating house in the Netherlands. *Source* https://www.archdaily.com

float thanks to 96 recycled plastic drums taped to the bottom and is modular, so it can also be attached to any number of other Jellyfish Barge units.

The greenhouse can pump the water underneath for irrigation, whether it is river water, salt water, or even polluted water. The water is made suitable for irrigation by treating it with 7 types of solar collectors on the floating greenhouse. There is also a rainwater harvesting system that can produce a total of 150 L of freshwater per day. Jellyfish Barge also has a hydroponic system, which is very efficient and can save up to 70% more water than a standard hydroponic system. This system can use about 15% seawater to increase efficiency. An automated system that provides remote monitoring and control of the hydroponic system. The model currently focuses on system optimization and construction cost reduction (see Fig. 10).

In the US, lift houses were built in Louisiana, adapted to flood water and are very sustainable, resistant to storms, hot and humid weather, high maneuverability, suitable for local climate and lifestyle, efficient use of energy, and reducing costs during long-term use. Fiber cement wall cladding, and wooden plank structure are treated with waterproofing and anti-termite treatment. The roof overhangs a large

Fig. 10 Model of the jellyfish barge floating greenhouse in Italy. *Source* https://www.archdaily.com

space with a rain screen arrangement. Design against wind loads through successive load-lines. The design uses energy efficiently, reducing solar radiation and increasing natural ventilation.

3 High Adaptability Housing Model in Coastal Areas

(a) In terms of the development orientation of HCMC, since 2014, the city has had a strategy to adapt to climate change for the whole city, including Can Gio district [8]. The marine economic benefits of HCMC can be exploited from the 17 km long coastline in Can Gio and hundreds of kilometers of waterways. These are favorable conditions for the development of seaports, along with the rise and fall of high tide, which is favorable for ships. However, developing toward the sea means that HCMC must deal with climate change, especially flooding caused by high tides and SLR [6].

(b) In the process of choosing a place to live, the ancients recognized the importance of rivers and canals or coastal areas in stabilizing their lives later. Choosing a place to live along rivers, canals, or coastal areas is not random, but it is all practical experience from interacting with the natural environment. Living along the river and along the coast has facilitated movement by waterway, irrigation of fields, and crops, a cool place to live, and many conveniences of daily life such as bathing, washing, fishing, catching seafood, exchanging goods, and wholesale, … In the process of building their own living spaces, with the ingenuity of their hands and minds, people have created a living space that adapts to natural conditions. Each housing project reflects an understanding of people about the relationship between shape and function. Therefore, the housing architecture of the Vietnamese is deeply and strongly influenced by the climate and soil conditions and forms the local architecture. Residential space is not simply a house, but, more than that, a whole space around the house [9]. Thus, the human habitation space has stretched from the house (residential space of a family), temple, and village (residential space of the community). Traditional residential spaces are always associated with the natural environment. Humans have coped and taken advantage of nature through solutions to build and organize indoor and outdoor spaces [10]. Nature has not only brought people great incentives but also brought many disasters on the other hand. Because of the duality of nature, the construction of houses and the organization of new residential spaces require appropriate measures to deal with nature (see Figs. 11 and 12).

(c) Requirements in advance of the new situation on the ability to adapt to natural disasters and epidemics.

According to the World Health Organization (WHO), in the global context of facing natural disasters and unpredictable epidemics, people's living space will be changed in the coming time [12]. Housing not only ensures privacy, comfort, and accessibility to services but also must meet new requirements for environmental

Fig. 11 The architecture of the house is spare empty on the first floor, with a flood relief yard. *Source* https://www.archdaily.com

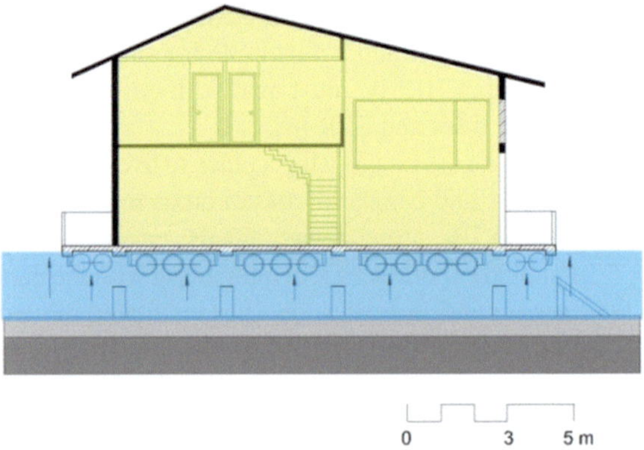

Fig. 12 Solution for floating house model when there is flooding. *Source* [11]

sanitation, the density of residences, and safety against natural disasters. That means housing needs to be equipped with self-protection, self-regulation, or self-adaptation to the surrounding environment, allowing people to live more comfortably and safely. The Covid-19 epidemic has caused people to re-examine themselves and their place of residence in the context of being isolated at home for a long time. The epidemic has most truthfully revealed the reality of the living conditions of workers and poor people with cramped, inadequate, and unsafe living spaces in high-density residential areas. Most workers live in residential areas, suburban neighborhoods, and suburbs—where the cost of living is low and suitable for their income. During the time of social distancing, because of the epidemic, most people lived in full-time accommodation, resulting in high population density, narrow, temporary, inadequate living space, and living conditions that did not guarantee the 5 K rule, so there have been many infections.

In summary, to adapt to a life with many upcoming changes, it is necessary to adjust several factors related to the sustainability of housing architecture and the organization of living spaces that can adapt to natural conditions and diseases.

4 Proposing Model of Floating House and Green Farm for Can Gio District

To prepare for the scenario of high SLR causing negative impacts on rural houses in Can Gio district, this study has investigated and collected information on the level of impacts of SLR causing flooding on houses in the research area. At the same time, determine the scientific basis to propose the application possibilities of the floating house model in Can Gio district, HCMC. The proposed solutions are based on the results of in-depth interviews with local people, officials, and village heads in Can Gio district so that the proposed solutions are suitable to people's needs and the flood situation of each area. Solutions to adapt to climate change and epidemics for the RCA of Can Gio district include:

(a) *House solution on the reinforced concrete frame*

The solution is to raise the floor, which aims to raise the essential living area above the highest flood level due to SLR in Can Gio district (according to the climate change scenario to 2050). This measure is solid and can help people cope with rising water levels for many days, but it brings many inconveniences in daily life when there is no flood. The area built on the empty frame must be enough to serve basic and essential living needs for a household of 2–5 people during the flood (see Fig. 13). Scope of application can be applied to almost every area and different types of terrain and geology. In addition, with climate change and SLR expected to cause flooding in the worst-case scenario, which is 2.0 m deep, the elevation of the raised floor on the frame needs to exceed 2.0 m above the ground so that the house can adapt to SLR within the next 30 years. In the case of use in combination with other solutions, the floor on the reinforced concrete frame is proposed for the minimum functional space required of a house, such as a bedroom, kitchen, and toilet. There have been many awards for the previously proposed method using houses on frames or piles; however, those solutions simply help people move to an empty, elevated floor and avoid contact with floodwater but forget their minimum needs such as eating, cooking, bathing, washing, sanitation, etc.

(b) *Solution using the floating floor*

Floating floor solutions offer greater flexibility for adaptive architectural modeling. The principle is that wherever the floodwater rises, the floor rises. The structural solution is quite simple. It is possible to use lightweight floor tiles placed on a simple structural frame; underneath are buoys that can float when meeting water. Using floating floors should be noted in combination with the solution of anchoring and balancing the floor plates through special structures. Application range: With Can Gio district's geological characteristics, this solution needs to be researched and re-evaluated. Its suitability: when there is water, the floor will rise, and when there is no water, the floating buoy systems will lie on thick alluvial soil. When this alluvial soil layer dries, it will form a hard mass that surrounds and holds the buoy system, so when the water has not risen, the floating floor is stuck in the soil, and when the

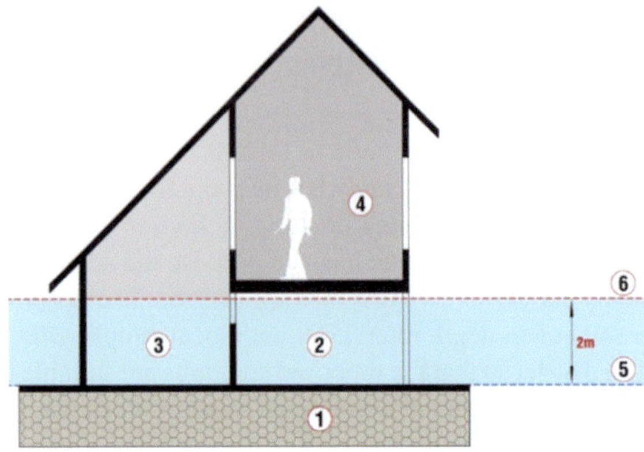

Fig. 13 Solutions for houses on reinforced concrete frames. *Source* [13]. (1) Natural ground; (2) Flooded area; (3) Non-essential living space; (4) Essential living space; (5) Existing ground elevation; (6) The highest sea level due to climate change by 2050

water rises, the floating floor cannot emerge on its own. The technical solution to this problem is to arrange the buoy system separately from the ground. The float system and floating platform will be raised on a support frame, from which it can be raised and lowered easily without being affected by mud (see Fig. 14). At the same time, the floating floor system, although highly flexible, is still only a temporary solution. Floating decks are only effective when properly loaded, and the mooring systems are in good working order. Considering the flexibility and instability of this flooring system, it is recommended to apply only to non-essential functional spaces such as external access to the house, small-scale farming, or shelter to protect livestock from being flooded.

(c) *Small self-service scale vertical farm solution*

The proposed solution is a small-scale, ecologically oriented, and self-catering vertical farm for the coastal rural household. This model is adaptive to climate change, and disease, based on lessons learned and combining Aquaponics and Jellyfish Barge models (presented in Sects. 2 and 3). The authors propose two solutions: A floating

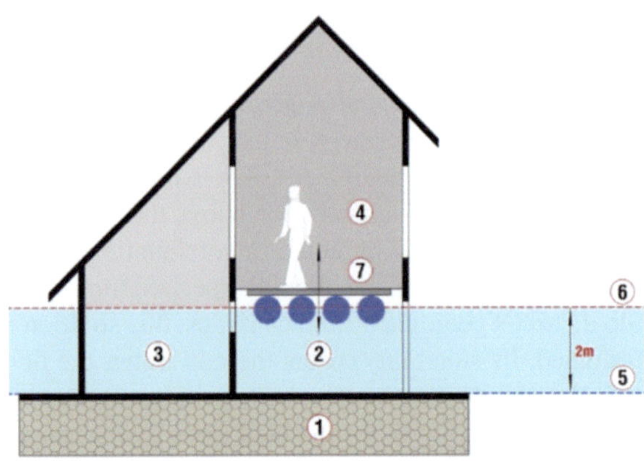

Fig. 14 Using floating floor. *Source* [13]. (1) Natural ground; (2) Flooded area; (3) Non-essential living space; (4) Essential living space; (5) Existing ground elevation; (6) The highest SLR due to climate change by 2050; (7) The floating floor is raised and lowered flexibly when the water level rises

Green-floor and a Floating Greenhouse (see Figs. 15 and 16). In summary, there will be solutions to organize the architectural model of RCA to adapt to climate change and epidemics in Can Gio district (see Fig. 17). In the process of applying these solutions, it is necessary to consider the specific nature, economic and social conditions of Can Gio district (each residential area will have different conditions) to calculate the number of applications to best suit the characteristics of each residential area and bring the highest efficiency.

(d) *Model of rural houses and green farms for the coastal area of Can Gio district*

The rural house model applies to households with 4–5 people, adjacent planning along lines, with the average area of each land plot being 100 m^2 (taking 5 m × 20 m as a typical), construction density is 50–80%, construction density for the main

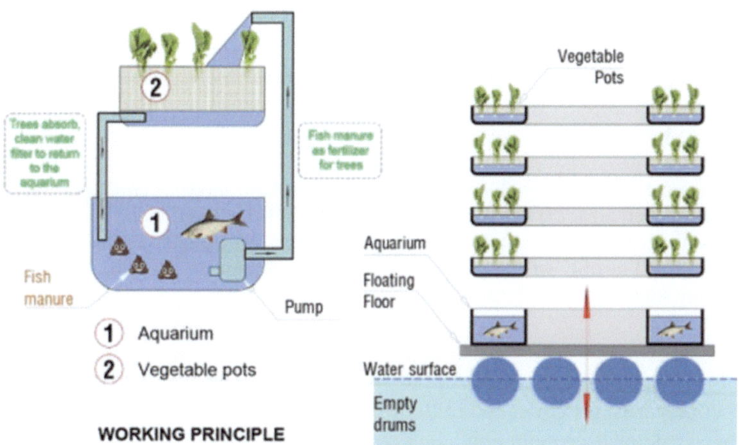

Fig. 15 Floating green-floor solution. *Source* [13]

Fig. 16 Floating greenhouse solution. *Source* [13]

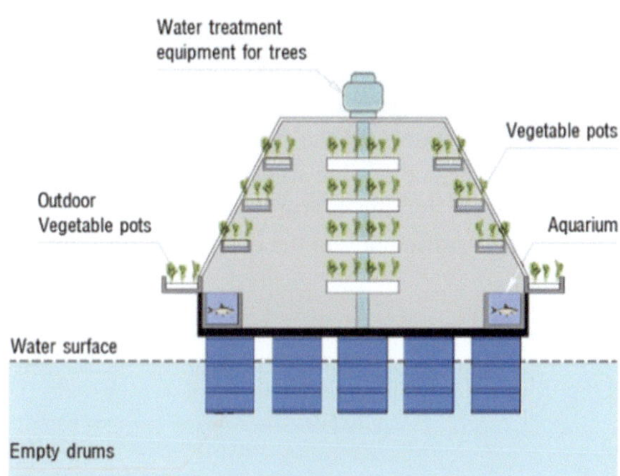

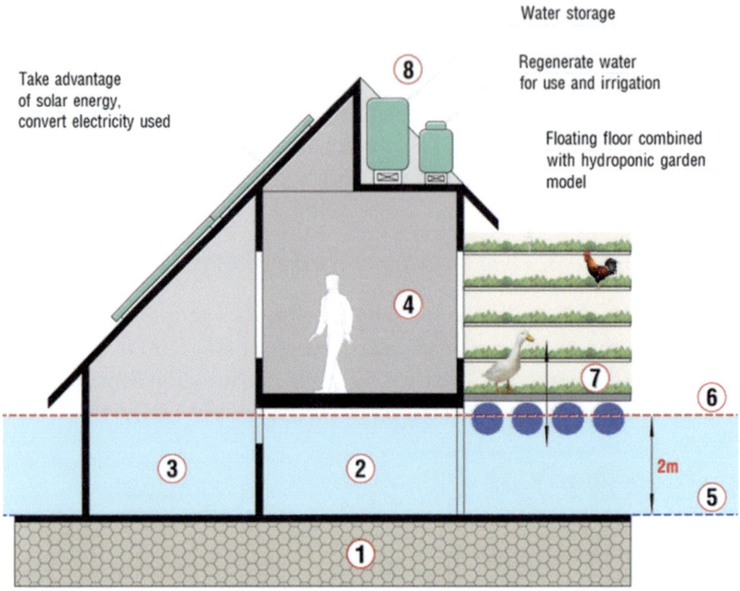

Fig. 17 Integrated solution for floating house and green farm model. *Source* [13]. (1) Natural ground; (2) Flooded area; (3) Non-essential living space; (4) Essential living space; (5) Existing ground elevation; (6) The highest SLR due to climate change by 2050; (7) The floating floor is raised and lowered flexibly when the water level rises; (8) Equipment for optimal exploitation of natural energy sources

house is 50%, equivalent to 50 m^2 (5 m × 10 m). The rest are functional areas such as courtyards, gardens, and outbuildings. Reinforced concrete structure system and other structural components such as the green brick wall, light wall partition like cement board and industrial wood. The first floor arranges non-essential spaces such as a living room, common space, parking, and storage, Toilets are arranged separately, and the area behind the house is at the end of the wind direction; The 2nd floor arranges essential spaces such as bedrooms (when needed, it can be converted into a kitchen-dining room) and toilet. It is necessary to arrange 2 bedrooms for households of 4–5 people.

The empty spaces in the house, as well as the balcony, can apply green solutions, using vertical garden solutions (hydroponics) to bring aesthetic beauty, create a microclimate for the house, and provide a source of food and dairy products, ensuring self-sufficiency in the event of natural disasters or epidemics (see Figs. 18 and 19).

Fig. 18 Model of a semi-detached house with floating green-floor in normal state. *Source* [13]

The "Floating Greenhouse" model is installed separately or in groups of modules. The main components of the model are bamboo, wood, ETFE (ethylene tetrafluoroethylene), or polycarbonate, and the base is made of recycled plastic drums. The model takes advantage of solar radiation to evaporate water, cool and then extract purified water. Solar energy heats water to the point of evaporation. As the water evaporates, the steam rises, condenses back into clean water when it cools, and drops down, accumulating. Many types of solar collectors include large-scale concentrated solar collectors, box and pit collectors, and water bottle or plastic bag collectors. The

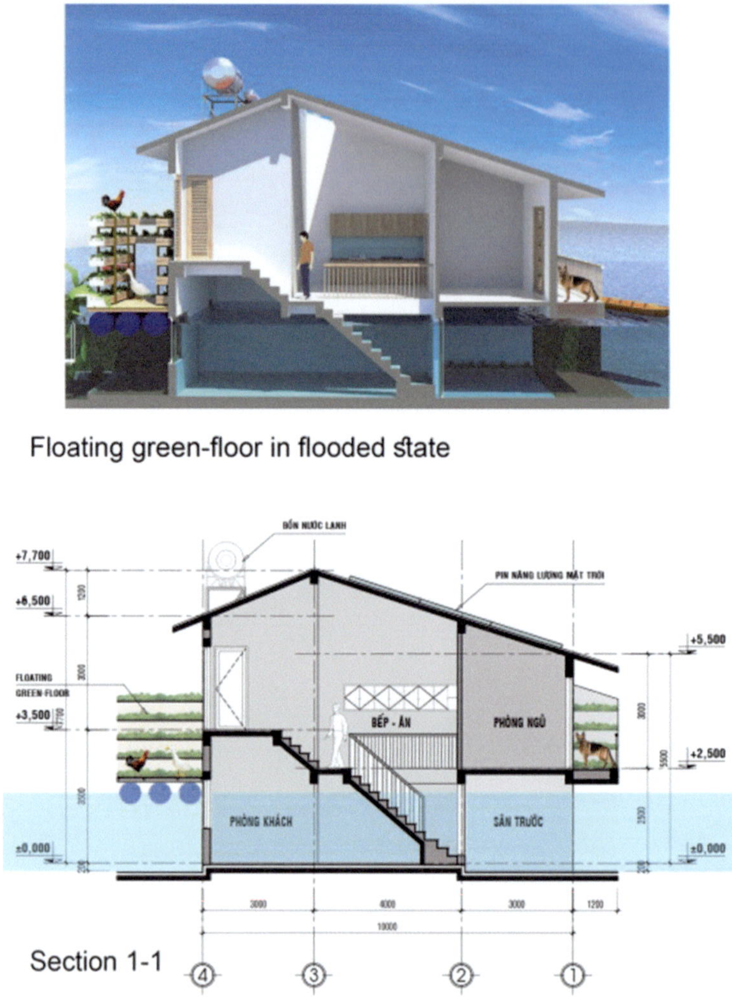

Fig. 19 Model of a semi-detached house with floating green-floor in flooded state. *Source* [13]

Floating Greenhouse model requires that fan and pump power are provided by renewable energy harvesting systems integrated with the structure (low power consumption is provided by solar panels and other renewable energy systems) (see Fig. 20).

Fig. 20 Floating greenhouse farm model. *Source* [13]

5 Conclusion

Under the influence of climate change, HCMC and surrounding areas are experiencing many negative changes. Air humidity decreases while sea levels and high tides increase yearly [14]. This issue needs to be studied and researched early in many fields, such as planning, construction, and housing architecture, including the RCA house of Can Gio. This area is predicted to be most affected by climate change, especially saline intrusion and flooding caused by climate change for people's houses.

According to the study of the current situation of residential areas, RCA houses in Can Gio district are suffering from the negative impacts of climate change and epidemics. This study explores the experiences of floating houses and green farms adapting to climate change and epidemics, thereby suggesting some solutions for Can Gio district, including:

- For a house, optimal adaptive solutions are building houses on reinforced concrete frames and using floating floors. The application of these solutions is suitable for the region's actual situation and economic and technical conditions.
- In terms of disease adaptability, the combination of VAC and Aquaponics (vertical farming) models integrated into the house helps ensure self-sufficiency in food for people. This area is not only where epidemics occur but also where climate change is extreme.

This study's initial and practical contributions are the scientific basis for proposing a rural house model to adapt to climate change and epidemics in the RCA of Can

Gio district. There are flexible and adaptive living space designs to ensure residents' safety, assets, and livelihoods. The proposed solutions for the RCA of Can Gio district in this study aim to adapt to climate change and epidemics, helping reduce the risk of injury to communities in coastal areas. Given the solutions' limitations, research is necessary to find more opportunities to deal with the increasing threats of climate change, SLR, and epidemics. Moreover, it is also necessary to consider and gradually upgrade the local capacity in forecasting and preventing natural disasters and epidemics. Finally, the proposed solutions can be considered and applied to other RCA in Vietnam and for future research reference.

References

1. Asian Development Bank: Ho Chi Minh City_ Adaptation to Climate change. Study Report Volume 2 (2010)
2. Ngo LM (2021) Assessment of the impact of flooding due to climate change on housing in Nha Be, Can Gio (HCMC), and propose planning and architectural solutions to respond. Scientific project of Ho Chi Minh City Department of Science and Technology
3. Mai TA (2019) Report on environmental protection and response to climate change. Department of Meteorology, Hydrology and Climate Change, Department of Natural Resources and Environment, Ho Chi Minh City
4. DiaOcThongThai. The map of HCMC districts 2021. https://diaocthongthai.com/ban-do-cac-quan-tp-hcm/#Huyen_Can_Gio. Ngo L-M, Hoang H-Y (2019) Climate change and sea-level rise response solutions for Can Gio District, Ho Chi Minh City: potential to adapt ideas from selected developed countries. In: Huong LTT, Pomeroy GM (eds) AUC 2019—Proceedings of the 15th international Asian urbanization conference. Advances in 21st century human settlements. Springer, Singapore (2019). https://doi.org/10.1007/978-981-15-5608-1
5. Nguyen HN (2020) Sổ tay nhà an toàn. Chương trình nhà chống lũ. Quỹ Hỗ trợ Phát triển Cộng đồng Sống bền vững (Handbook for safe home. Flood resistant housing program. Sustainable Living Community Development Support Fund)
6. Frank S (2013) Adapt-HCMC. Handbook on climate change adapted urban planning and design for Ho Chi Minh City, Vietnam. Brandenburg University of Technology Cottbus
7. DongThap Newspaper: Độc đáo "Nhà lưỡng cư" của thanh niên 9X (2020). https://www.baodongthap.vn/xa-hoi/doc-dao-nha-luong-cu-cua-thanh-nien-9x-89109.aspx. Last Accessed 28 Jan 2020
8. Ministry of natural resources & environment–MONRE: Vietnam Climate change scenario. Vietnam (2016)
9. Bora MC. The flood situation of Assam—A case study. Presented at the BALWOIS 2010, Ohrid, Republic of Macedonia. https://onlineasdma.assam.gov.in/kmp/pdf/1578984939The_Flood_Situation_of_Assam_-_A_Case_Study.pdf
10. Le TTH, Trần QD (2018) Experience in housing adaptation to flood reduction and application opportunities for townhouses in HCMC. J Construct 3:145–148
11. General Statistics Office: Population and housing census (2019)
12. World Ocean Review: Coping with Rising Sea Levels. World Ocean Review. https://worldoceanreview.com/en/wor-5/improving-coastal-protection/coping-with-rising-sea-levels. Last Accessed 13 Jul 2019

13. Le NH-V (2021) Mô hình kiến trúc nhà ở nông thôn ven biển thích ứng biến đổi khí hậu tại huyện Cần Giờ, TP. Hồ Chí Minh (Research on the architectural model of coastal rural houses adapting to climate change in Can Gio district, Ho Chi Minh City). Architectural master's Thesis. Ho Chi Minh University of Architecture. Ho Chi Minh City
14. Ngo LM, Kieu LT, Hoang H-Y, Nguyen H-B (2020) Experiences of housing adapted to sea level rise and applicability for houses in the Can Gio District, Ho Chi Minh City, Vietnam. Sustainability 12:3743. https://doi.org/10.3390/su12093743

Sustainable Tourism Development on the Basis of Research on Coastal Urban Morphology, Case Study Nha Trang, Viet Nam

Binh Hai Nguyen and Minh Le Ngo

Abstract This article studies the relationship between urban morphology and tourism development in the coastal city of Nha Trang. The studies show how urban morphology attracts tourists in the context of competition between tourist cities nowadays as well as the role of tourism in changing the urban landscape. Can the orientation of tourism urban planning based on the study of urban morphology determine what is in the city's tourism development? This research investigates the natural condition and topography of the coastal urban Nha Trang, and the development process of urban space through tourism service buildings; especially, high-rise resort buildings to identify the characteristics of a coastal tourist city and their influence on the city's tourism development. Specifically, the research method is carried out through the collection of data, maps of city development over time, and general planning of Nha Trang city. Through the analysis of urban morphology associated with the development of coastal high-rise buildings and resort projects, the interrelationships between them are established. The obtained results can help local managers have a better view of tourism development based on urban morphology, to orient sustainable tourism development for the city.

Keywords Sustainable development · Tourism · Urban morphology · Nha Trang

1 Introduction

The development of tourism in many regions is often based on the diversity of the natural environment and the uniqueness of the local culture. At the same time, the level of tourism development also affects the environment, and natural and cultural heritage [1]. Thanks to nature's gift of beautiful beaches, cities and towns have the conditions for rapid tourism development, leading to rapid coastal land changes [2]. People have always been attracted to beautiful coastlines since they (residents and

B. H. Nguyen (✉) · M. L. Ngo
Faculty of Civil Engineering, Ton Duc Thang University, Ho Chi Minh City, Vietnam
e-mail: nguyenhaibinh@tdtu.edu.vn

V. T. Ha et al. (eds.), *Proceedings of the 8th International Conference on Sustainable Urban Development*, Advances in 21st Century Human Settlements,
https://doi.org/10.1007/978-981-99-8003-1_6

tourists) are concentrated in the sea area, so large concentrations of the population are often located in coastal areas. This is reflected in the morphology of coastal urban areas which are increasingly growing in both width and height. The exponential development of resorts and coastal tourism services is accompanied by increasing revenues from tourists [3]. The development of diversified means of transport (railway, air, waterway, and road) has contributed to increasing time for tourism and entertainment activities. The face of coastal tourist cities develops from a natural state to increasingly urbanized through tourism development in a short time. The growth of tourism offers many economic opportunities for people who are looking for work and business. Coastal tourism brings foreign exchange to the government and leads to social, environmental, and political changes [3]. This development has contributed to encouraging the coastal property market, with resorts along the coast. Coastal tourism also affects the urban landscape by restricting people's visual access to the natural (marine) landscape of the accommodation facilities, and at the same time, the rapid development, the architectural form of the buildings didn't attract enough attention, leading to urban areas without identity [4]. From the point of view of urban planning, the study of urban morphology, to orient the development of a tourist city, needs to be associated with an understanding of the characteristics of the urban morphological structure.

Therefore, the objective of this study is the relationship between urban morphology and tourism development in the coastal city of Nha Trang, a famous coastal tourist city in Vietnam as a particular case. It also examined morphological changes in Nha Trang city, with a major focus on accommodation and resort projects, especially high-rise buildings. Through the qualitative method, analyzing the city development map through the stages and synthesizing relevant data helps the research team identify urban morphology, which is associated with tourism development issues in Nha Trang.

2 The Relationship Between Urban Morphology and Tourism

2.1 Urban Morphology

Morphology comes from the Greek word "morph", which means "shape". It studies all the physical elements of the city [5] including streets, squares, buildings, and plots of land; and its theoretical concerns include the relationship between morphology and function, how structure operates, and how features of urban landscapes adapt and change over time [6]. It began to take shape in the late nineteenth century as a field of study related to the landscape [7]. In recent decades, urban morphology has become increasingly linked with interdisciplinary archaeology, architecture, geography, landscape architecture, urban planning, and design. Figure 1 shows the elements that make up the urban morphology, including the constant group: nature, heritage, and

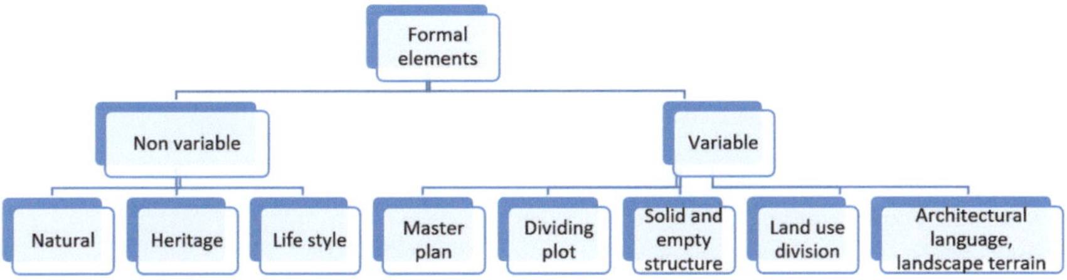

Fig. 1 Elements of urban morphology. *Source* Doan M Khoi, 2017 [9]

lifestyle, these are the main factors that attract tourists when they come to a tourist destination. The inconstant group includes physical structures such as the overall plan, the division of plots, fragments, the solid and hollow structures of the town, the division of land use, the architectural language, and the topography of the landscape. In addition, variable elements in the urban spatial structure shape urban characteristics to exploit the unique characteristics of tourism in each city. In contrast, a range of tourism-induced impacts on urban morphology such as socio-cultural, economic, environmental, and spatial patterns. The rise and importance of morphology are structured around the relationship between tourism effects, forms, and functions [8].

2.2 *The Relationship Between Urban Morphology and Tourism*

Morphology studies the physical form of landscapes, including how landscape structures work and function, the adaptability of forms, and how functions and forms change over time. Every city and town is physically distinct from each other and their spatial significance is uneven including a combination of geography, physics, site morphology, and infrastructure. Tourism has reshaped the landscape through land tenure patterns and attractive landscapes. Urban morphological models and methods were applied to understand the touristic characteristics of a destination and the complex relationships between tourism and morphological changes in destinations. Urban morphological features often form the focus of tourism promotion and several important tourist attractions. Tourism spatial affects land use, regeneration, and the development of new urban spaces. The consequences of travel also lead to urban fragmentation, referring to the morphological division of urban plots and their dispersion in urban space. Summarizing, urban morphology and tourism are interwoven [8]. Morphology can be utilized as a tool to assess coastal resorts [3].

Fig. 2 Sustainable tourism development model, based on 03 elements: Environment, Culture—Society, and Economy. *Source* International Ecotourism Society (2004)

2.3 Sustainable Tourism

The concept of sustainable tourism development has been introduced by researchers over different periods. Butler defines sustainable tourism as tourism-based activities for local people, tourists, and the environment [10]. Muller has defined that sustainable tourism development is achieved through satisfaction, economic development, social welfare, and protection of natural resources and local culture [11]. The United Nations Environment Program (UNEP) defines sustainable tourism development as a condition for tourism development in which activities are carried out in a way that minimizes negative impacts on society, economy, and natural environment [12]. Therefore, according to most experts, sustainable tourism development is defined as taking advantage of the natural environment and cultural values to meet the needs of tourists, taking into account long-term economic benefits and environmental protection, and helping to improve local living standards. Referring to sustainable tourism in cities and coastal areas, the International Ecotourism Society, and Asia Pacific Economic Cooperation said that sustainable tourism development has 3 main components, sometimes referred to as a "tripod" [13, 14] (Fig. 2).

Within the scope of this research, we focused on the current issues of Nha Trang during its development, with prominent environmental issues along with socio-economic issues. Special attention was paid to the visual environmental factor, which is related to the image of the city. Many cities have focused on their image as a potential tourist destination [15, 16]. In most cases, high-rise buildings are the most dramatic element of the city [17, 18]. By recognizing the increasing influence of high-rise buildings on the aesthetic quality of the urban environment, a trend has been overviewed to develop the principle of considering aesthetic effect as an important factor. For the coastal area in Nha Trang, high-rise buildings were the main factors affecting this environment. The development of resorts and coastal high-rise accommodations; especially, when there were not enough plans and strict management regulations, tended to spread out along the coastline, leading to the "overload" of the urban landscape. Meanwhile, urban landscape is a factor that contributes to the branding and competitiveness of cities [19, 20].

3 Method

The research method applied in this paper is qualitative. We analyzed city development maps through different stages and synthesized relevant data to help the research team to identify urban morphology. The research phase was a boom period for tourism development in Nha Trang, with high-rise accommodations and coastal resort projects from 2008 to the present. Documents and maps were drawn from research articles in journals and are accessible through online databases which were published between 1960 and 2022. Criticism and evaluation seminars by reputable domestic and foreign authors were also considered to update and study the process of coastal urban development as well as changes in the natural environment. Urban landscape elements were analyzed based on linking to the argument of urban morphology with data from design documents of high-rise accommodation and resorts projects, upcoming projects in the future, planning maps using city land, planning maps of 1/500 detailed planning projects, 1/2000 zoning plans, and 1/5000 general planning adjustments, combining with satellite images (from Google map, Google Earth). They were combined with local tourism planning data. In addition, research data was collected from interviews and discussions with experts, including architects, planners, and investors of high-rise resort projects.

4 Urban Morphology of Nha Trang City

Urban morphology analysis allows us to understand the variability of urban structure. Nha Trang is a famous tourist coastal city in the South-Central region of Vietnam, with diverse terrain; including bays, mountains, hills, and rivers. At the same time, the city has many peaks, where there are many architectural heritages and beautiful landscapes.

(a) The urban morphology of the research areas: Based on natural topographic factors, administrative boundaries of wards, structure, and urban development process, the whole coastal area was divided into three main segments as follows (See Table 1).

In the coastal area of Nha Trang, in the first stage of development with the Nha Trang Lodge Hotel (14 floors), it was the first high-rise hotel in Nha Trang, which was built on the foundation of large plots of land serving previous religious purposes. After that was the development phase of a series of other high-rise buildings in the central area such as Havana, The Costa, Sheraton, etc. The remaining areas with the first phase of development included the construction of residential buildings or renovation into simple buildings for low-budget tourists, then would be upgraded as visitor numbers increase and then built by developers into hotels to meet the increasing demand [21]. From the spatial development of the above-mentioned high-rise buildings, with the stable development area in the city center, there is still a

Table 1 Basic morphology of the system of streets, street blocks, high-rise accommodation, and resorts in the coastal area of Nha Trang

Segment		
The southern area of the city	The center area of the city	The northern area of the city
Development stage		
from 2006-2020	from 1996-2019	from 2010-present
Spontaneously developed area from the resident's houses, then merged and split the plot to build a high-rise hotel	The development area was controlled according to the general planning and construction regulations	the resident's houses (follows detailed planning, then merged and split the plot to build a high-rise hotel (according to the permission of the management agency to adjust the planning)

	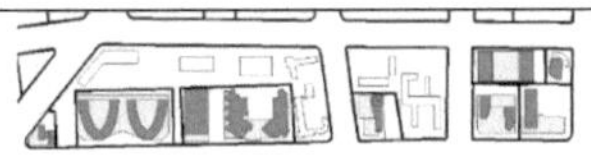	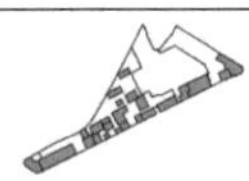
Back - to back plot. Division of lots in the form of townhouses. Some cases of lots were divided according to local adjustments of planning. FAR of high-rise buildings form 13-18	Large streets on the foundation of previous administrative, educational and religious lands, were divided into large-sized lots. FAR of high-rise buildings <13	Back - to back plot. Division of lots in the form of townhouses. Some cases of lots were divided according to local adjustments of planning. Street lots had a thin depth, the depth of which is from 10 to 18 meters. FAR of high-rise buildings form 18-24.
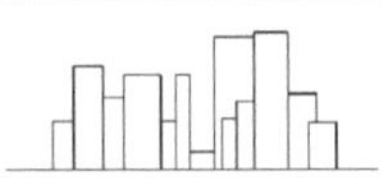	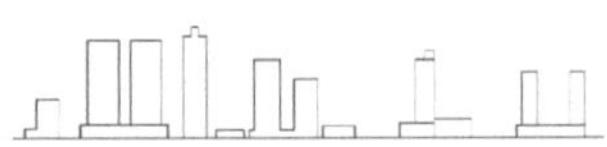	
High-rise hotels developed on the townhouse. There were no gap between the buildings. the skyline had a change in height	High-rise hotels were built on large land plots, there were the distances between them and the skyline had a change in height.	High-rise hotels developed on the townhouse. There were no gap between the buildings. the skyline had very little height variation

Source Based on master general plan of Nha Trang, and author's field survey, July 2022

gap between the buildings; in contrast, spontaneously developed areas or adjusted planning, high-rise buildings like a dense wall blocking the view to the sea from behind (Table 2).

(b) Changing landscape morphology and topography in coastal areas, expanding, and exploiting coastal space for tourist purposes.

5 Discussion

The analysis results on the urban morphology of Nha Trang showed that the transformation of urban morphology in Nha Trang city took place according to the rules of the market economy, like other developing countries in Southeast Asia. At the same time, it illustrated that the morphology of many coastal resorts was gradually changing according to the tourism industry. Tourism development also started

Table 2 The changing landscape morphology and topography in coastal areas through sea encroachment of tourist projects. Source: Figure b and c: Google map and Figure a: drawn by the author from the general planning map of Nha Trang to 2025

Analysis areas		
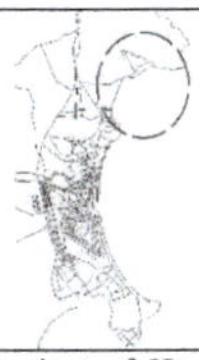	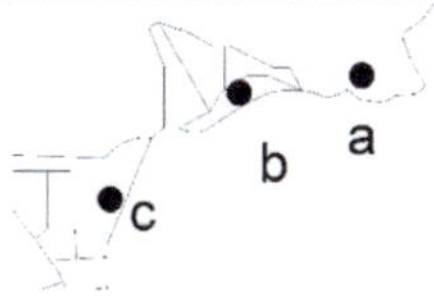	- The area was selected for analysis was in the northern of the city. It was an area developed after 2002 when Tran Phu Bridge connected two banks of the Cai River along the coast. - Tran Phu route was extended to the National Highway 1A. Following it were tourism development projects, high-rise accommodation, and resort projects, in order to exploit the advantages of the sea surface and reduce the load on the downtown area
a. The project of Ho Tien tourist service area consists of four different projects		
Before 2002	After 2002	Consequences
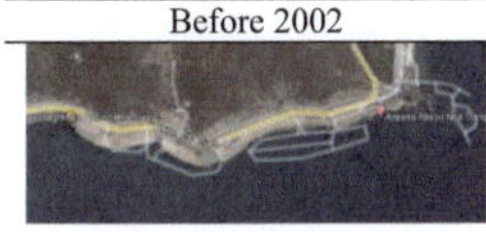		- Resort project creates a new landscape through encroaching on Nha Trang Bay - The coastal promenade has been privatized. The elevation of the sections of the project is different, the coastal promenade is not synchronized
b. Vinh Hoa sea encroachment's residential area		
Before 2002	After 2002	Consequences
		36-storeys La Swiss Marina Hotel (according to the local adjustment plan), located in Vinh Hoa residential area, where there are no beach as well as accompanying services. Besides, this place was formerly a place to anchor boats and trade seafood. The construction of resorts here pushed the anchorage and trading areas aside and created conflict between local fishermen and resort developers. Therefore, it was not reasonable to invest here.
c. Ba Lang sea encroachment's residential area		
Before 2002	After 2002	Consequences
		- Fully connected traffic on Tran Phu - Pham Van Dong route (from Nha Trang port to National Highway 1A) - Due to difficulties in relocation, the sea has been encroached to build embankments and solid roads, causing the loss of the sand and beaches in this area. - The appearance of many new high-rise buildings through local adjustment from residential land to commercial and service land for economic development has formed walls separating space and vision.

around the downtown area and expanded to the surrounding areas [22]. Nha Trang Lodge Hotel appeared first and started a race to develop high-rise buildings. Since then, some positive impacts were brought to the city which created many new job opportunities and developed the local tourism economy. However, the city lost its own identity, when administrative, religious, and research buildings were converted into tourism and services.

Issues that need to be addressed are:

(i) The city has plenty of accommodation, but lacks recreation areas [23][1]

[1] The data of Khanh Hoa Tourism Department (2019) and author's survey results on the number of high-rise buildings in Nha Trang in July 2022: "*in year 2019, Khanh Hoa has about 49.000 rooms*". In Nha Trang there were about 160 high-rise accommodations (July 2022). "*However, there were only three major amusement parks: Vinpearl, Diamond Bay and Hon Tam*"; four shopping malls in

(ii) Create a clear separation of urban space between high-rise building development areas and residential areas (the northern area of the city)
(iii) Traffic structure with the coastal road (Tran Phu, Pham Van Dong) has a large cross-section with high speed, making it difficult for tourists and residents to approach the sea. Also, with the construction of seaside structures, visual access to the sea is limited.
(iv) The local adjustment of the planning to satisfy the needs of investors who always want to maximize the land fund for profit, has changed the urban landscape and the stability in people's lives (especially, fishermen's lives in the northern area of the city)
(v) Buildings and administrative functional areas are still located in the tourist development areas, leading to inefficient exploitation and use of land, reducing the value of land use and the landscape of the surrounding areas.

Therefore, there is a need for improvements in urban management and tourism development based on understanding studies of urban morphology; especially in the coastal area, where there is a lot of potential for tourism development but is also very difficult to control. This is only effective when the city clearly defines the responsibilities of each level of local government, and regularly supervises the implementation process.

6 Conclusions

The development of tourism in Nha Trang City had positive and negative impacts on socio-economic and environment. Through studying the urban morphology of the coastal tourist city of Nha Trang, we have realized the process of developing high-rise accommodation and resort projects in this city. The initial high-rise construction situation occurred in the downtown area. The demand for tourism development increased, which required the adjustment of planning to serve the interests of investors and their spontaneous construction problems. They made the planned construction not well controlled. They affected the urban landscape and the lives of local people, especially in the northern part of the city. This research proposed a master tourism development plan that was suitable for the urban form of the city, to limit the negative impacts of the construction of high-rise accommodations and resorts on the environment and local community. It helped balance the socio-economic development and environmental protection, working towards sustainable development.

the base of high-rise buildings: Nha Trang Center 1, Nha Trang Center 2, Vinpearl Empire, Vinpearl Beachfront and two big shopping malls: Lottemart and Winmart.

References

1. Chen C-F, Chen P-C (2010) Resident attitudes toward heritage tourism development, tour. Geographies 12(4)
2. Andriotis K (2006) Researching the development gap between the hinterland and the coast—Evidence from the island of Crete. Tourism Management 27
3. Gu K, Xie PF (2013) Morphological changes of coastal tourism: a case study of Denarau Island, Fiji. Tour Manag Perspect 5:75–83
4. Luong PT (2021) Promoting Vietnam's coastal tourism development. Tourism Magazine, Ha Noi, pp 74–76
5. Karen and Frank (1994) Ordering space: types in architecture and design, Van Nostrand Reinhold. Washington DC, pp 289–311
6. Liu J, Wall G (2009) Resort morphology research: history and future perspectives. Asia Pacific J Tour Res 4(14)
7. J. W.-6th I. S. Syntax and undefined 2007, Conzenian urban morphology and urban landscapes, spacesyntaxistanbul.itu.edu.tr (2007). http://spacesyntaxistanbul.itu.edu.tr/papers/invitedpapers/Jeremy_whitehand.pdf. Last Accessed 07 Mar 2022
8. Xie PF (2022) Towards a framework for the morphology of tourism. Tour Plann Dev
9. Doan MK (2017) Urban morphology. Construction Publisher, Ha Noi
10. Butler RW (1991) Tourism, environment, and sustainable development. Environ Conserv 18(3):201–209
11. Müller H (1994) The thorny path to sustainable tourism development. J Sustain Tour 2(3):131–136
12. U.-U. E. Program, Making tourism more sustainable: a guide for policy makers | UNEP—UN Environment Program. https://www.unep.org/resources/report/making-tourism-more-sustainable-guide-policy-makers. Accessed 18 Aug 2022
13. B. A. International Ecotourism Society, Sustainable tourism - Module 10, The simple user's guide to certification for sustainable tourism and ecotourism (2004)
14. de Bruyne C (2013) Sustainable development of tourism destinations. Asia Economic Corporation, Singapore
15. Masoud H, Mortazavi M, Torabi Farsani N (2019) A study on tourists' tendency towards intangible cultural heritage as an attraction (case study: Isfahan, Iran), City. Cult Soc 17:54–60
16. Ruiz EC, de la Cruz ERR, Vázquez FJC (2019) Sustainable tourism and residents' perception towards the brand: the case of Malaga (Spain). Sustainability 11(1):292
17. Rød J, van der M-E D, P B, undefined (2009) Visibility and dominance analysis: assessing a high-rise building project in Trondheim, journals.sagepub.com. https://journals.sagepub.com/doi/abs/10.1068/b34118. Accessed 05 Apr 2022
18. Short M (2007) Assessing the impact of proposals for tall buildings on the built heritage: England's regional cities in the 21st century. Prog Plann 68(3):97–199
19. Trịnh HV (2017) The advantages and disadvantages of high-rise architecture in coastal cities in the South-Central Coast, Architecture Magazine, Vietnamese Architecture Associates, no. 4
20. Nguyen TL (2013) Urban landscape, an element that contributes to brand building city, Construction Planning Magazine, VIUP
21. Nordstrom, Beaches and dunes of developed coasts. Cambridge University Press, New York (2009)
22. Wall G, Mathieson A (2006) Tourism: change, impacts, and opportunities. Pearson Prentice Hall, New Jersey, p 412
23. Nguyen T (2019) Too much sleep, no place to play, Khanh Hoa online magazine. https://www.baokhanhhoa.vn/du-lich/201901/du-lich-lo-cho-ngu-thieu-cho-choi-8101784/. Last accessed: 2023/5/26

Suburbanization in Ho Chi Minh City Metropolitan Area in the Context of City's Peripheral Industrialization: A Case Study of Thuan Dao Industrial Park

Huong Thi Thu Le, Nghia Tuan Le, and Kieu Thi Le

Abstract In response to their dynamic economic growth, countries in the Global South and developing economies have undergone a rapid process of peripheral industrialization that causes considerable changes in the suburban areas, especially those of the big cities. This so-called "suburbanization" process is characterized by a higher population mostly due to rural-to-urban migration, a transformation of land use to urban functions, changes in housing, increase and diversity of urban services and activities, etc. On one hand, this suburbanization motivates the speedy development of these areas and creates numerous economic opportunities. However, on the other hand, it causes numerous issues, especially under poor management in different aspects. Intending to reflect on these phenomena in the context of Vietnam's industrialization, this paper selects the residential areas located along the edge of Thuan Dao Industrial Park in the Ho Chi Minh City metropolitan area for a case study. Through literature reviews, mapping, observation, and a quick interview, this paper takes advantage of primary and secondary data to reveal the diversity in urban structure, housing and architecture, socio-economic activities, infrastructure, and the inadequacy in living quality caused by several environmental issues. Even though this paper's findings provide some perspectives into the suburbanization occurring around a single industrial park, they can generally contribute as an example of this phenomenon in developing countries. Moreover, it can also be a source of reference for the local authorities in perceiving the characteristics of suburbanization in the city.

Keywords Suburbanization · Industrial park · Urban periphery · Changes · Ho Chi Minh City

H. T. T. Le (✉) · N. T. Le
Faculty of Engineering, Vietnamese-German University, Binh Duong, Vietnam
e-mail: huong.ltt@vgu.edu.vn

K. T. Le
Institute of Constructive Design and Building Construction, Technical University of Darmstadt, Darmstadt, Germany

V. T. Ha et al. (eds.), *Proceedings of the 8th International Conference on Sustainable Urban Development*, Advances in 21st Century Human Settlements,
https://doi.org/10.1007/978-981-99-8003-1_7

1 Introduction

Since the launch of Doi Moi (economic reform) in 1986, Vietnam has been a magnet for large-scale Foreign Direct Investment (FDI) inflows. As a result, the industrial growth led by export-oriented manufacturing and the FDI sector has grown rapidly over time. The industrialization has been accompanied by the spatial economic transformation that is especially concentrated in and around the metropolitan areas of Hanoi and Ho Chi Minh City (HCMC) [1]. With this process, a major labour force has been reallocated from agriculture to industry and from rural to urban and suburban areas, making the rural areas outside the two metropolitan regions grow significantly.

In HCMC metropolitan area, two-thirds of the land development occurred beyond the urban communes within 1990–2012 [2]. This phenomenon reflects the first significant suburbanization process in Vietnam, which happened due to the development of industrial estates and the movement of small industries from the city to the suburban areas. An increasing population characterizes the so-called suburbanization process in the suburban areas mostly due to rural-to-urban migration, a transformation of land use to urban functions, changes in housing, increase and diversity of urban services and activities, and so on. This suburbanization, recently, has also happened due to the rise of many large-scale housing projects around the metropolitan.

From the observed phenomena and challenges, it has been necessary to review the industrialized suburban areas of HCMC metropolitan area, focusing on its current situations and issues. Accordingly, this paper used a mixed approach to investigate a case study of the residential areas surrounding Thuan Dao Industrial Park, about 30 kms from the city centre. The urban structure, housing and architecture, socio-economic activities, infrastructure, environmental conditions and living quality in the areas are explored. The findings reflect the characteristics of suburbanization in HCMC metropolitan in the context of industrialization. This case study, therefore, can be seen as an example of suburbanization phenomenon in developing countries.

2 Materials and Methods

The authors used a mixed approach for data collection and analysis to take advantage of secondary and primary data. Literature was mostly collected using common databases including Google Scholar and Elsevier to develop the research background. Reputed journals were prioritized among the found sources to gather the data with high reliability. The materials focused on various disciplines in suburbanization and industrialization as well as these phenomena in the Global South and at the urban scales related to the study area. Besides, some other secondary data were referred from the recent reports on relevant statistics from United Nations, World Bank, and other organizations.

Mapping was used to demonstrate the study area and visualize the characteristics of the suburbanized areas considering land cover and land use. The maps were developed based on the Open Street Map, chronological satellite images collected from Google Earth and illustrated with the support of QGIS software. Outputs from the maps were processed within Microsoft Excel for chart creations. More importantly, site visits were carried out in December 2019, including observation and quick interviews as the main activities. The observation was implemented with a focus on urban structure, architectural landscape, tangible infrastructure and socio-economic activities, and environmental conditions. Simultaneously, photos were taken when relevant for a later investigation, many of which were reported later in this article. In addition, quick interviews were conducted with ten locals to learn comprehensively about their social-economic activities and their perceptions of the potential and challenges of the area.

3 Suburbanization in the Global South

3.1 Terminology

Urbanization is the term used to describe the process of inhabitant increases, multidimensional structural changes in urban areas such as those associated with resource uses (i.e., land and water), socio-economic conditions (i.e., demography, employment, production, and service delivery), and political decisions in relation to urban governance [3]. According to United Nations, 55% of the world population lived in urban areas in 2018, and this number keeps increasing to 68% in 2050 [4]. Therefore, studies on urbanization and its relevant aspects are vital to the collective knowledge of the current global situations and development visions in the future.

Suburbanization, also peri-urbanization [2, 5], refers to a similar process to urbanization to some extent, in which the suburbs of cities are increasingly populated [6] and characterized by the extension of residential areas [7]. While the Anglo-American suburbanization in the Global North is associated with clear, stable, and conventional models, the patterns of suburbanization in the Global South seem challenging to comprehensively theorize due to the new developments and changes [7].

Industrialization is a process associated with multidimensional exploration to achieve higher efficiency in value creation [8]. Industrialization is the central part of a larger and more complex process termed "modernization" that both together firmly connect to the urbanization rate [9] and majorly shapes the economic landscape of the world today [8].

3.2 Terminology Causes and Effects of Suburbanization in Relation to Industrialization

While urbanization in some developing countries closely connects to their natural material export activities and therefore has fewer impacts from manufacture and services, this process in many other countries is believed to have a visible relationship with industrial presences and is termed the "production cities" [10]. This intimate relation is reflected in historical reviews and is commonly seen in rapidly developing countries [9].

The more dynamic the non-tradable sectors are, the more labours are witnessed to work the manufacturing areas and finance, insurance, real estate and business services [10]. The attractiveness of job opportunities drives the massive influx of migrants from the countryside to the suburban [11] and urban areas where factories and their relevant services are typically located. Together with the locals and citizens from the central districts associated with the peripheries in light of consumption possibilities, they establish the patterns of suburbanization processes in these areas [11].

Industrialization challenges the suburbs in a wide range of aspects. For example, the development of industrial sectors causes the loss in organic carbon soil along the suburbanization process [12], the increase in carbon dioxide emission due to the dynamics of transport services [13], and the high concentration of atmospheric metals due to production activities [14]. The mentioned effects due to industrialization result in a degradation in ecological services and severe health-related risks for the peripheries.

Due to their location advantages, the surroundings of the industrial zones become the catchment areas for emerging residential areas that possess similar characteristics and challenges to the suburbs. Accordingly, besides the mentioned challenges, some of the additional major issues of these areas are the misuse of land [15], insufficient urban infrastructure and services [16, 17], lack of affordable housing [18], unauthorized settlement formation [15] and informal economic development [16, 18].

3.3 Industrialization and Suburbanization in the Global South

Globally, 600 cities accounted for 60% of the overall GDP, 70% of which are located in the South [19]. Therefore, the shift of economic hubs into these areas and their consequences are of great interest. Resulting from the dynamic economy, noticeable urbanization processes have been recognized in the South [20]. Observing those occurring in Asia, Africa and Latin America, Myers found five common themes within these processes: land transformation, rapid urban sprawl, formation of informal housing and trading markets, remaining traces from imperialism and colonialism, and shifts in governance and service diversity [20]. From the evidence found in several countries in the South, many of these themes have certain connections with industrialization, such as the relation of industrial sector development to the informal economy formation [5] and the expansions of residential areas, especially the informal settlements, into the natural lands at urban fringes [21]. The latter also reflects an apparent linkage between industrialization and suburbanization, in which the concentration of labour surrounding the industrial zones contributes to the emerging housing need for low-income migrants in the suburban areas.

With the driving forces from industrial activities, the suburbanized areas have been dealing with several challenges. In India, Bengaluru, formerly Bangalore, is known for its variety of heavy, light, and especially IT industries [22]. On one side, the development of industrial infrastructure as a foundation for its production activities and service delivery significantly contributed to its urban competitiveness [23]. On the other side, the investments were not equally distributed, which causes certain social discriminations in basic urban areas services [24]. After 30 years of undergoing a suburbanization process, the peripheries were observed to have fewer possibilities to access the healthcare and education system as well as water supply services than the central districts [24]. In Vietnam, the urban fringes in the major cities such as Hanoi and Ho Chi Minh City (HCMC) have been no longer led by their central areas in the last decade but driven by urban sprawl and rapid industrialization [25]. Besides certain achievements acknowledged to urban gentrification, regeneration and restoration of the central areas, there is a wide range of issues that these cities have been facing in their peripherals. Among these, high population density in the suburbanized areas with insufficient technical and social infrastructure as well as lack of affordable housing supply is considerable [25]. As a result, quality of life of the workers living around the Export Processing Zone or Industrial Parks in HCMC were reported as low concerning the level of satisfaction of life (2.63/5) and happiness (2.8/5) [26]. Water and air pollution is another concern at the industrial zones, as documented in [27, 28], due the poor management in wastewater treatment and high concentration of heavy trucks.

4 Case Study of Thuan Dao Industrial Park: Suburbanization Characteristic and Issues of Concern

4.1 Thuan Dao Industrial Park

Thuan Dao Industrial Park is located at the centre of Ben Luc Town, Ben Luc District, Long An Province. Long An is locally considered as a "backyard" of HCMC and is the key and only trading gateway between the Southern key economic region (HCMC, Binh Duong, Dong Nai, Ba Ria - Vung Tau, Binh Phuoc, Tay Ninh, Long An and Tien Giang Province) and key economic region of the Mekong Delta (Can Tho City, An Giang, Kien Giang and Ca Mau Provinces). It connects the national transportation

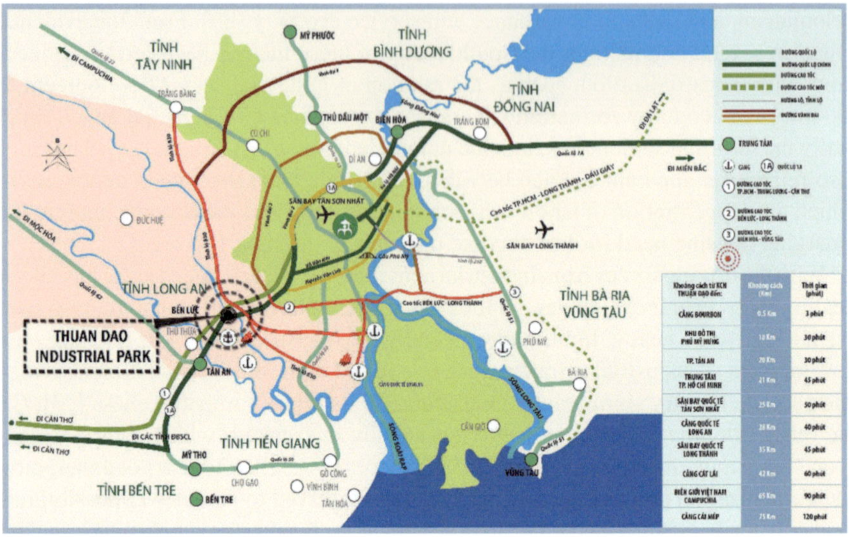

Fig. 1 Thuan Dao Industrial Park and transportation network. *Source* E-Brochure Thuan Dao IP-VN, 2018

system (road, air, waterway, and railway) easily and conveniently through different highways, ring roads, interprovincial roads, which link to the international airports and many domestic ports (Fig. 1).

This industrial area has an expected total area of 765 hectares that includes:

- The existing Thuan Dao Industrial Park (2003–2007) in Ben Luc Town, Ben Luc District, Long An Province, with an area of 113.8 hectares (100% of the land has been sublease).
- The expansion of the Thuan Dao industrial area in Long Dinh Commune, Can Duoc District, Long An Province, is divided into two phases: Phase 1 (2012–2017) and Phase 2 (2017–2022), with an area of 189.8 ha and 461 ha respectively.

The main study area of this paper focuses on four residential areas adjacent to the existing industrial park and its extension in Phase 1 until 2017 (see Fig. 2) to explore the peri-urbanization phenomenon. Several changes in the areas are discovered, including the urban structure, housing and architecture, and infrastructure and associated environmental problems. Besides, this paper also describes the general socio-economic characteristics of these residential areas to show the study context more comprehensively. In the next session, the general features of the suburbanization process at the study area will be presented, and key findings on the studied aspects will be discussed with relevant data and information.

Fig. 2 Research areas around Thuan Dao Industrial Park. *Source* Google earth, edited by the authors, 2021

4.2 Suburbanization

In the study area, suburbanization has gradually happened since the establishment of Thuan Dao Industrial Park in 2003, which attracted a lot of migrant workers to come to live around the factories with the associated urban services to serve them. This migration has caused a major impact on the population growth in urban areas. From 2009 to 2019, the mechanical population growth rate of Ben Luc District was 1.49% (2,543 people/year) and the population density of Ben Luc District in 2019 was 631 people/km2 (increased by 115 people/km2 compared to 2009). Particularly, compared to the whole district, Ben Luc Town, where Thuan Dao IP is located, has the highest population density at 3,101 people/km^2 (increased by 517 people/km2 compared to 2009).[1]

In addition, the development of the transportation network and infrastructure system in and around the area has created numerous economic opportunities for the local people and newcomers. However, on the other hand, it caused several issues, especially under poor management of the area in different aspects. More specifically, the changes in urban structure, housing and architecture, infrastructure and environmental issues, and socio-economic activities will be presented in the following sub-sessions.

[1] Preliminary results of the 2019 Population and Housing Census: Population of Ben Luc District over 10 years of census (April 1, 2009-April 1, 2019). Link: http://cctkbenluc80.blogspot.com/2019/09/ket-qua-so-bo-tong-ieu-tra-dan-so-va.html.

4.2.1 Urban Structure

Thanks to its location near 1A Highway (connected by two main roads, as shown in Fig. 3), Thuan Dao area is easily connected to the surrounding provinces. It is also near the ports to facilitate the functional needs of many factories, while at the same time, workers can also easily commute from different areas to their workplaces. However, the transportation of several trucks, even container trucks, to and from the factories across the residential areas creates an extreme danger to the locals.

In terms of land-use, although the industrial park was built in 2003, the land has been substantially transformed to serve urban functions since 2012 due to the development of the transportation system. This paper, therefore, presents the land-use changes in this area from 2012 to 2020, as shown in Fig. 4 below. In 2012, when Thuan Dao Industrial Park began to be built with 113 hectares, a small residential area appeared in the North close to National Highway 1A. By 2015, when the phase 1 expansion of Thuan Dao began to take shape and a series of factories were built close to Vam Co Dong River, the mass of construction increased quickly. It clearly shows

Fig. 3 Main roads around Thuan Dao Industrial Park. *Source* Google Earth, edited Research Team, 2019

(a) 2012: a small residential area appeared in the North.

(b) 2015: The mass of construction increased quickly.

(c) 2020: land-use has changed along the main roads at both sides of the industrial area.

Fig. 4 The land-use transformation from 2012 to 2020 (black and white). *Source* Google Earth, edited by the authors, 2021

that the land-use has changed along the main roads at both sides of the industrial area, especially on the riverside where people can easily access river-based transport. Finally, after phase 1 was leased nearly 98% in 2020, where most agricultural lands have been transformed into residential land.

Figure 5 shows a zoomed-in area where the current land use is mostly for commercial and residential functions. The agricultural land remains only about 12% while before the appearance of the industrial estate, it was a totally agriculture-based area. In the context of city development, where most of the industrial parks are outside the urban areas (with about 30 km on average), the formation of residential and service areas right next to the industries is inevitable. This phenomenon of industry-led development is explainable as the workers, mostly migrated from the rural areas to the city, prefer to live around their factories for easy commuting. However, although the density of industrial and residential lands increases, there is inadequate provision of social infrastructure associated with public buildings to serve people's basic living activities, especially in the health care sector. Some facilities such as schools and hospitals are far from the site (about 4 km minimum).

In short, the transformation of land use due to infrastructure development and increasing population has created favourable conditions for residential development synchronously with the industrial area and services. Several changes in the housing sector have been observed by the research team in terms of housing types, characteristics, and functions.

4.2.2 Housing and Architecture

The type of buildings that mostly occupy this area are residential. Due to suburbanization, the housing types and characteristics have become diverse. They can be either

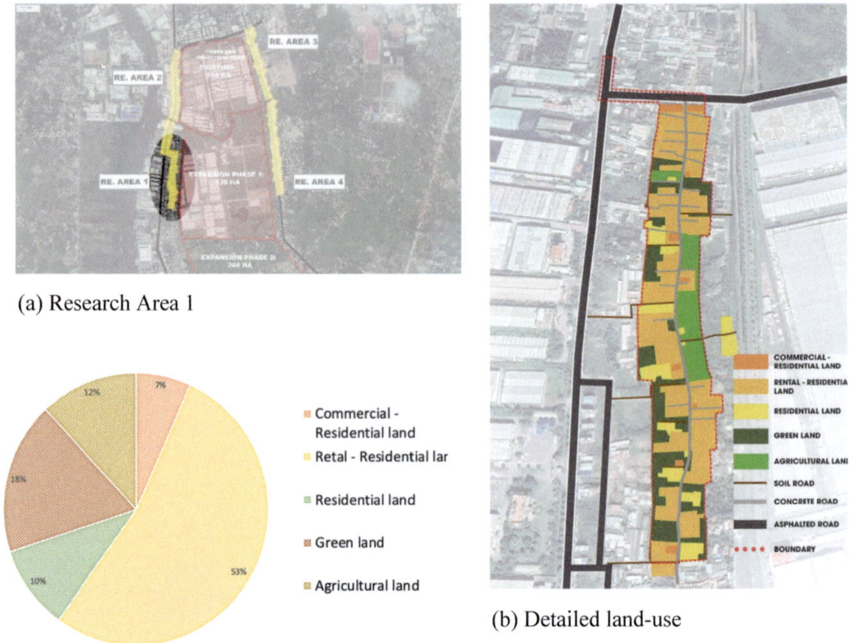

(a) Research Area 1

(b) Detailed land-use

Fig. 5 Existing land-use of research area 1. *Source* Research Team, 2019 measured and estimated based on google map

pure housing of local people, mixed-use with other commercial and service functions, or rental building blocks built to meet the housing needs of industrial workers. In general, the architectural form of housing is relatively simple, without any unique features, and most of the buildings are low-rise with a single storey. However, as more functions have been added to the local housing to serve the new and temporary residents i.e., industrial workers, the housing composition, arrangement, and facades are varied depending on the owners.

Specifically, the local residential housing is diverse in quality (such as semi-permanent houses, grade 4 houses,[2] newly built houses) and the construction density is low. Most houses are in the alleys, far from main traffic axes and roads connecting to industrial zones. The design and quality of houses are also different depending on the financial condition of the owners, in which many newly built houses have good architectural features, and they are owned by the new residents who come from the city. Figure 6 shows the co-existence of the old rural houses and newly built houses with diverse designs due to suburbanization.

[2] According to the Government's Decree No. 209/2004 / ND-CP of December 16, 2004, grade-4 houses are the houses with floor space of less than 1,000 m^2 or less than 3 storeys in height. According to this Decree, most houses in Vietnam today are grade 4.

(a) Old rural house

(b) New rural house (Grade-4 house)

(c) New modern house

(d) New two/three-storey house

Fig. 6 Residential buildings. *Source* Research Team, 2019

Besides new housing construction on existing land, a series of housing projects are being developed near the study area (Fig. 7). After Decree 82/2018/ND-CP stipulating the management of industrial zones and economic zones was issued, the synchronous planning between industrial parks, service areas and residential areas is a new model attracting the attention of the investors. It has promoted the trend of developing residential areas with industrial parks, combining industrial development with urbanization, and ensuring industrial parks' synchronous and sustainable development. In the study area, Thuan Dao Industrial Park creates a favourable condition for the development of residential projects near the site. These will provide a variety of housing solutions and utilities, attracting residents to set up a sustainable community that resonates with values for the industrial park. In addition, it also ensures the feasibility of an expansion plan for the industrial park in the future.

In terms of functional changes, many local households run small businesses combined with stores, restaurants, and cafes to meet the necessities of citizens and workers living nearby. The business space is quite simple, mostly renovated or utilized from the space in front of the house. Some façade elements are described in Figs. 8 and 9.

(a) Map shows new residential project near Thuan Dao IP

(b) Example of Project information board near Thuan Dao IP

Fig. 7 Residential projects are being developed near the study area. *Source* Research Team, 2019 and Google earth, edited by the author, 2021

(a) Grocery store

(b) Cafe or restaurants

(c) Wholesale shop of grass

Fig. 8 Mix-used residential buildings. *Source* Research Team, 2019

Fig. 9 Mock-up shop house. *Source* Research Team, 2019

More importantly, rental building blocks are the most common form of accommodation in the residential areas near the industrial park, built on residential land converted from agricultural land. The architectural form is relatively simple, using light materials and easy to construct but low-quality finishing materials and living facilities (Fig. 10). They are usually divided into three main spaces:

- The access space is a long corridor with one entrance for the whole area.
- The living space is divided into separated small rooms of 12 to 15 m2 each, of which a small toilet occupies about 2 m2.
- The common space is outside for drying clothes or for parking.

4.2.3 Infrastructure and Environmental Issues

Although the industrial park itself does not face industrial waste pollution due to good infrastructure investment and management as required by the authorities for all industrial estates in the city, the pollution problem in the study comes from other causes. According to the research team's observation and the complaints from residents, pollution does not come from the industrial park itself but the transportation and construction activities in and surrounding the residential areas. In Thuan Dao industrial zone, two major sources of pollution are discovered as follows:

Firstly, a lot of dust, smog, and air pollution come from transportation activities supporting industrial zones, concentrated along the main traffic routes and access roads to the factories. It also causes noise pollution due to the high traffic of heavily loaded vehicles. Safety of the small vehicles and pedestrians is also a critical concern here. (Fig. 11).

Secondly, the pollution is from the living and trading activities of residents around the industrial zone. Garbage is stored and accumulated in the workers' accommodation in the alley, affecting the aesthetic, streetscape, and environment of the residential areas. Garbage is spontaneously burned when overloaded due to poor garbage collection service, causing heavy pollution in the air (Figs. 12 and 13).

4.2.4 Socio-Economic Characteristics

As mentioned above, the main cause of suburbanization in Thuan Dao area is industrialization which has pulled many residents from the neighbouring provinces to migrate to this area for job opportunities. In addition to the changes in urban structure, housing, and infrastructure, the rural-to-urban migrants and new residents have created several changes in terms of socio-economic characteristics.

Due to population increase, a variety of service activities have been developed. Many households run different commercial activities to increase income. Moreover, more and more households have converted agricultural land into residential land to build rental rooms for workers, as presented in 4.2.2. According to the interviewees, the number of rooms and room quality (i.e., size, facilities, etc.) very much depend on the available land and investment ability of each owner. In general, the quantity of rental rooms is high, but the quality of rooms is very poor, especially since the support services, security, and environmental issues are not guaranteed. Figure 14 shows an example of one household that earns benefits from rental room service.

(a) Rental rooms along alley with one entrance

(b) Shared space

(c) Example of room

Fig. 10 Rental housings. *Source* Research Team, 2019

(a) Risk for small vehicles and pedestrians due to no proper lanes and pathways

(b) Air and noise pollution from heavy traffic

(c) Dust and smog come from the traffic

Fig. 11 Air pollution from transportation activities near Thuan Dao Industrial Park. *Source* Research Team, 2019

(a) Garbage is spontaneously burned (b) Scrap purchasing shop (c) Unhygienic drying area

Fig. 12 The pollution from residential and trading activities. *Source* Research Team, 2019

Fig. 13 Overloaded trash bin in the dense rental-room area affects the streetscape. *Source* Research Team, 2019

Fig. 14 The owner of ten rental rooms. *Source* Research Team, 2019

(a) Vacant land for rent.

(b) House for rent.

(c) Land for sale (not house).

(d) Real estate service.

Fig. 15 Real estate activities around Thuan Dao Industrial Park. *Source* Research Team, 2019

The 50-year-old man, who is a current resident, said that: "I have ten rooms for rent, the cost is 800.000 VND/month/room. The land price here ranges from 2 to 3 million/1 m^2".

After deducting expenses for rental housing construction, according to some of the owners, the income earned is higher than that gained from saving money in the bank.

Besides, the increase in real estate values in the industrial zones has inadvertently created great impacts on the development of the whole area. When urban periphery districts develop industrial zones, the real estate market in that area begins to develop. This phenomenon has inevitably happened around Thuan Dao industrial area. Since its development, many land-related business types have appeared near the industrial park, including real estate business and other services, as shown in Fig. 15.

The number of labourers and jobs creates a wave of housing and service demands. As the government and the industrial park cannot meet these demands, many local households showed their interest in investing in this business channel to make profits, according to the interview results. However, such spontaneous activities lead to chaos in the market and disorder of real estate agents or brokers. In other words, although the business activities have brought additional income for the locals, they create another concern about the planning and development of the area.

In addition, like in many other industrial areas in Vietnam, social problems are one of the concerns in Thuan Dao area, such as fights between workers, conflicts between residents and migrants, gamblers, thieves, etc. Therefore, since May 2020, the local government has established a security team in the residential area under the guidance of the local police to manage and minimize social problems. This was narrated by a local in an interview.

5 Discussion

In the process of periphery-oriented industrialization, many industrial parks have been formed in the suburbs of Ho Chi Minh City. Besides their economic benefits and environmental concerns, this peripheral industrialization has led to suburbanization with the increasing population and land-use transformation from agricultural to residential and commercial functions. However, the common features of many industrial areas around the city are lack of medical, educational, cultural, and other service facilities. Industrial zones, especially invested by domestic enterprises, have almost no mechanism for investment in such social infrastructure. There are very few industrial zones that set up synchronous construction projects for housing, schools, hospitals, and other facilities for their workers.

In addition, in the process of land-use transformation, real estate near the industrial zone has become a "hot spot" for investors. Landowners mostly want to take advantage of the attraction of industrial park projects and related infrastructure to benefit from the increasing land price. Unfortunately, this leads to chaos in the real estate market as the land price has been pushed up spontaneously without government guidance or control.

The case of Thuan Dao industrial area in this paper clearly shows the suburbanization phenomenon due to industrial development. Besides the changes in urban structure and development of road systems, many private investors have spontaneously built rental houses around the industrial park, causing construction chaos, and disrupting urban planning. In exchange for low rent, workers must live in a poor-quality environment near the industrial zone, suffering pollution and limited social services for their lives. Regarding the environment, thanks to the investment in infrastructure and the proper planning of industrial parks, the pollution of industrial substances has been significantly reduced. However, due to the lack of management in the areas around the industrial parks, pollution comes from industry-supporting activities and residential activities, as shown in the case study of Thuan Dao. If this problem is not handled in time, it can negatively cause hazardous impacts on human health.

The residents are also concerned about the increasing social problems arising from the complexity of mixed, even temporary, migrant residents who come from various places in the country. In this case study, the biggest beneficiaries are the local households who benefit not only from the increasing land value but also from the commercial activities and room-rental services they provide for the new population, mainly the industrial workers. Nevertheless, without proper guidance and management by local authorities, the general feature of the area is challenging, and it raises a concern on the ways to move towards more sustainable suburbanization around the city.

6 Conclusion

Industrial parks have made many important contributions to the socio-economic development of HCMC region, especially creating jobs and incomes for workers and locals. On one hand, this peripheral industrialization comes together with infrastructure development and modernization of the suburbs through land-use transformation; on the other hand, it also faces several challenges in different aspects of urban structure, housing, environment, socio-economy, and so on. Although this paper's findings provide only some perspectives into the suburbanization realized around a single industrial park, they can generally contribute as an example of this phenomenon in developing countries. Moreover, it can also be a source of reference for the local authorities in perceiving the characteristics of suburbanization in the city to take necessary actions. Especially concerning the living quality in such suburbanization zones, the management of security, social order, environmental sanitation, and quality of rental rooms for workers is important to avoid turning those areas into new kinds of slums in the periphery of HCMC in the long term.

Acknowledgements A part of the primary data of this paper was collected by the SUD2019 students of a Master's course entitled Urban Development and Architecture of Cities taught by the corresponding author at the Vietnamese-German University in December 2019 (referred to as Research Team in this paper).

References

1. World Bank (2020) Vietnam's Urbanization at a Crossroads : Embarking on an Efficient, Inclusive, and Resilient Pathway. Washington, DC
2. Kontgis C, Schneider A, Fox J, Saksena S, Spencer JH, Castrence M (2014) Monitoring peri-urbanization in the greater Ho Chi Minh City metropolitan area. Appl Geogr 53:377–388. https://doi.org/10.1016/j.apgeog.2014.06.029
3. Batishcheva S (2013) World Urbanization Prospects and the Problem of Its Infrastructural Provision. Econ Anal 46:72–81
4. United Nations (2019) Department of Economic and Social Affairs, Population Division: World Urbanization Prospects: The 2018 Revision. United Nations, New York
5. Bolay, J.-C.: Urban Facts. In: Urban Planning Against Poverty. pp. 7–55. Springer International Publishing, Cham (2020). https://doi.org/10.1007/978-3-030-28419-0_2.
6. Tzaninis Y, Boterman W (2018) Beyond the urban–suburban dichotomy: Shifting mobilities and the transformation of suburbia. City 22:43–62. https://doi.org/10.1080/13604813.2018.1432143
7. Mace A (2020) Suburbanization. In: International Encyclopedia of Human Geography. pp. 121–125. Elsevier. https://doi.org/10.1016/B978-0-08-102295-5.10343-9
8. Simandan D (2020) Industrialization. Presented at the February 3 (2020). https://doi.org/10.1016/B978-0-08-102295-5.10086-1.
9. Spence M, Annez PC, Buckley RM (eds) Urbanization and Growth. Commission on Growth and Development. Commission on Growth and Development : World Bank, Washington, DC
10. Gollin D, Jedwab R, Vollrath D (2016) Urbanization with and without industrialization. J Econ Growth 21:35–70. https://doi.org/10.1007/s10887-015-9121-4

11. Shen J, Wu F (2013) Moving to the Suburbs: Demand-Side Driving Forces of Suburban Growth in China. Environ Plan A 45:1823–1844. https://doi.org/10.1068/a45565
12. Luo Y, Shen J, Chen A, Tao Q, Li Q, White PJ, Li T, Li B, Chen L, Li H, Gao X, Xu Q, Wang C (2021) Loss of organic carbon in suburban soil upon urbanization of Chengdu megacity. China. Science of The Total Environment. 785:147209. https://doi.org/10.1016/j.scitotenv.2021.147209
13. Rossolov A, Lobashov O, Kopytkov D, Naumov V (2020) Sustainable suburban supply chain. Transportation Research Procedia. 45:795–802. https://doi.org/10.1016/j.trpro.2020.02.092
14. Hsu C-Y, Chi K-H, Wu C-D, Lin S-L, Hsu W-C, Tseng, C.-C., Chen, M.-J., Chen, Y.-C (2021) Integrated analysis of source-specific risks for PM2.5-bound metals in urban, suburban, rural, and industrial areas. Environmental Pollution. 275, 116652. https://doi.org/10.1016/j.envpol.2021.116652.
15. Baye F, Wegayehu F, Mulugeta S (2020) Drivers of informal settlements at the peri-urban areas of Woldia: Assessment on the demographic and socio-economic trigger factors. Land Use Policy 95:104573. https://doi.org/10.1016/j.landusepol.2020.104573
16. Mapunda DW, Chen SS, Yu C (2018) The role of informal small-scale water supply system in resolving drinking water shortages in peri-urban Dar Es Salaam. Tanzania. Applied Geography. 92:112–122. https://doi.org/10.1016/j.apgeog.2018.02.001
17. Teschner N, Sinea A, Vornicu A, Abu-Hamed T, Negev M (2020) Extreme energy poverty in the urban peripheries of Romania and Israel: Policy, planning and infrastructure. Energy Res Soc Sci 66:101502. https://doi.org/10.1016/j.erss.2020.101502
18. Adam AG (2014) Informal settlements in the peri-urban areas of Bahir Dar, Ethiopia: An institutional analysis. Habitat Int 43:90–97. https://doi.org/10.1016/j.habitatint.2014.01.014
19. Dobbs R., Smit, S., Remes, J., Manyika, J., Roxburgh, C., Restrepo, A.: Urban world: mapping the economic power of cities. McKinsey & Company (2011)
20. Myers, G.: Urbanisation in the Global South. In: Shackleton, C.M., Cilliers, S.S., Davoren, E., and du Toit, M.J. (eds.) Urban Ecology in the Global South. pp. 27–49. Springer International Publishing, Cham (2021). https://doi.org/10.1007/978-3-030-67650-6_2
21. Bolay, J.-C.: An Intermediate City in Brazil: Between Inequalities and Growth: The Case of Montes Claros. In: Urban Planning Against Poverty. pp. 121–165. Springer International Publishing, Cham (2020). https://doi.org/10.1007/978-3-030-28419-0_5
22. M. N., S., Gowda, K., M. V., S.: CITIES IN TRANSISTION: Impact of Information Technology Industries in Bangalore. IJSRP. 3, (2013)
23. Anfinsen M, Jacobus S, Johnston C, Jones U, Böbel I (2014) Bangalore information technology cluster. IJEPEE. 7:191. https://doi.org/10.1504/IJEPEE.2014.065248
24. Sridhar, K.S., Manasi, S., Latha, N.: Suburbanization and spatial inequality in the distribution of urban services in Indian cities. In: Urbanization in the Global South. pp. 99–123. Routledge India, London (2021). https://doi.org/10.4324/9781003093282-6
25. Quertamp, F., De Miras, C.: Periurbanization and governance of large metropolises in Vietnam. In: Ton, N.Q.T., Quertamp, F., De Miras, C., Nguyen, Q.V., Le, V.N., and Truong, H.T. (eds.) Trends of urbanization and suburbanization in Southeast Asia. Ho Chi Minh City General Publishing House, Ho Chi Minh City (2012)
26. Pham QT, Pham, BPMH The life quality of workers in industrial zones of HCMC, Vietnam. Current Politics and Economics of South, Southeastern, and Central Asia. 25
27. Duc HN, Truong TP (2003) Water resources and environment in and around Ho Chi Minh City, Vietnam. EGJ. 1, (2003). https://doi.org/10.5070/G311910533
28. Lo Dinh D (2021) The impacts of transportation vehicles on environment: The case of Ho Chi Minh City. E3S Web Conf. 234, 00072. https://doi.org/10.1051/e3sconf/202123400072

Climate Change Resilience Based Transforming Growth Models

Cooling Effects of Blue and Green Infrastructure on Urban Microclimate: A Case Study at Thu Duc District, Ho Chi Minh City

Kieu Thi Le and Stefan Schäfer

Abstract Blue and green infrastructure are potential nature-based solutions to the critical climatic impacts in dense city areas. However, the presence of practical assessment, compared to theoretical research, is still a limit on this topic. This study aimed to numerically and spatially examine the potential effects of blue and green infrastructure on urban microclimate using a case study with scenario analysis. An existing residential area in Thu Duc District, Ho Chi Minh City, was taken for thermodynamic simulations with data from the open sources. Findings were generated by the calculation and demonstration of the air temperature, relative humidity, and heat index values presented at different locations and elevation levels within the given district. The findings confirmed certain cooling effects of blue and green infrastructure in mitigating severe microclimatic conditions and therefore reducing health-related risks. Taking these effects up to the pedestrian level, blue infrastructure can be significantly efficient in heat stress reduction with almost 40% less heat-related danger time and approximately 10% more safe time. The effects were presented at and close to the water body, as well as following the prevailing wind direction. Green infrastructure cooled down the areas along the greenery surfaces due to the reduction of radiant heat flux up to 56.79 W/m^2, in combination with shading effects. The outputs of the cooling effects imply the significance of blue and green infrastructure in mitigating severe urban microclimatic conditions. Besides, those measures can also improve land-use efficiency, diversify urban ecosystems, and provide further health-related benefits.

Keywords Blue infrastructure · Green infrastructure · Microclimate · Cooling effect · Thermodynamic simulation

K. T. Le (✉) · S. Schäfer
Institute of Constructive Design and Building Construction, Technische Universität Darmstadt, Franziska-Braun-Straße 3, 64287 Darmstadt, Germany
e-mail: kieu@kgbauko.tu-darmstadt.de

S. Schäfer
e-mail: sts@kgbauko.tu-darmstadt.de

V. T. Ha et al. (eds.), *Proceedings of the 8th International Conference on Sustainable Urban Development*, Advances in 21st Century Human Settlements,
https://doi.org/10.1007/978-981-99-8003-1_8

1 Introduction

The rapid urbanization of Ho Chi Minh City (HCMC) is reflected in the population growth rate, as well as the significant transformation in spatial development. However, the process is believed to be shaped by market forces rather than comprehensive planning [1]. Therefore, the lack of urban planning on urban development level causes many of its facets, such as urban climate, to be neglected. Consequently, the rapid urban transformation causes considerable changes in its urban surfaces, exposing the city to severe urban climatic issues, including heat island effects [2, 3]. For instance, an increase in air temperature of 0.033 °C was observed from 1991 to 2005 in HCMC [4]. From 1989 to 2009, the 2 m temperature increased by 0.64 °C, resulting from monitoring at Tan Son Hoa weather station, and 0.31 °C from urbanization calculation models [5]. From this evidence, urbanization likely contributed to half of the overall temperature rise in the city [5].

Besides addressing the situation, a growing body of literature, e.g. [4, 6–8], has been carried out recently calling for measures to reduce heat island mitigation. Accordingly, nature-based solutions such as blue and green infrastructure were proposed thanks to their ecological characteristics. The practical study on this topic is, however, still limited.

For this reason, this study aimed to numerically and spatially examine the potential impacts of blue and green infrastructure on urban microclimate using thermodynamic simulations at a small urban scale. An existing residential area in Thu Duc District was taken as a case study for scenario development and analysis. The housing block is composed of the most typical settlement patterns in the city, characterized by townhouses facing narrow streets. The air temperatures of such areas were recorded among the highest in HCMC [9]. Data from several open sources were employed and processed using the methods presented below.

2 Methodology

To investigate the effects of surface conversions, this study focused on urban blocks in a developed residential area in Ho Chi Minh City with an open-accessed space such as a pocket park or similar. A site of 100 m × 100 m in Thu Duc District was selected based on these criteria (Fig. 1).

Based on the data from Google Earth and Google Street Maps, the 3D model of the site was created with its current surface characteristics. It was called the baseline model in scenario analysis. In the second scenario, a 4 m-high water fountain with a diameter of 10 m was added close to the center of the pocket park to feature the blue infrastructure. All brick outer walls and flat roofs were converted into green walls and green roofs for the last scenario.

To proceed further with the models, the meteorological characteristics of the site were defined. Based on the historical weather data from the cited source in Fig. 3,

Fig. 1 Location of the study site (Authors, 2022, based on data from https://www.igismap.com/vietnam-shapefile-download-country-boundaryline-polygon/ and Google Earth)

the date 15.05.2022 was selected for the investigation as it was the hottest day in the dry season of the year. the significant indicators in microclimate study, which include temperature, light, wind speed, and moisture [10] were taken into account, besides the available weather data, humidity, air temperature, and wind direction. On the selected day, hourly data recorded the peak temperature (36.11 °C) and the almost lowest humidity (53%) at 02:00 PM. Therefore, this time instant was selected in the study concerning the scenario analysis.

Lastly, thermal effects were examined at different elevations with factor k, a given indicator in the thermodynamic simulations in ENVI-met (Fig. 4). The lowest elevation level (k = 0) was close to the ground, followed by the average pedestrian height (k = 2), which was the most meaningful level concerning outdoor comfort assessment. Two common living heights and the maximum height (k = 5–7) were also studied, based on an observation of the median and maximum building heights on the site.

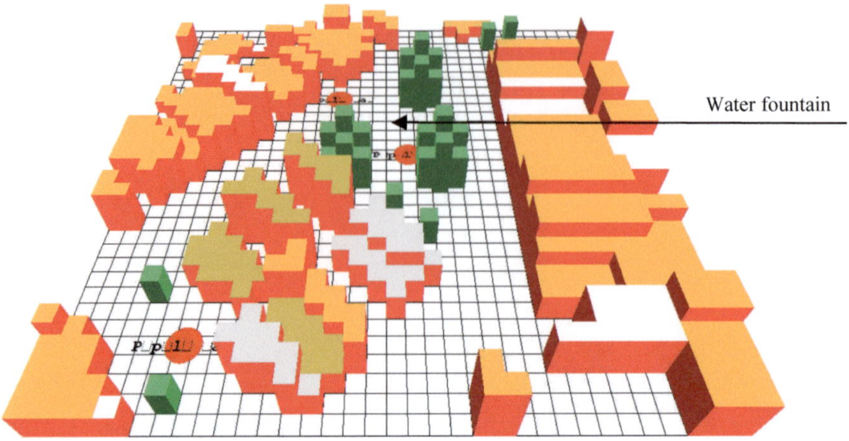

Fig. 2 A 3D model of the study site (Authors, 2022, based on data from https://www.igismap.com/vietnam-shapefile-download-country-boundaryline-polygon/ and Google Earth)

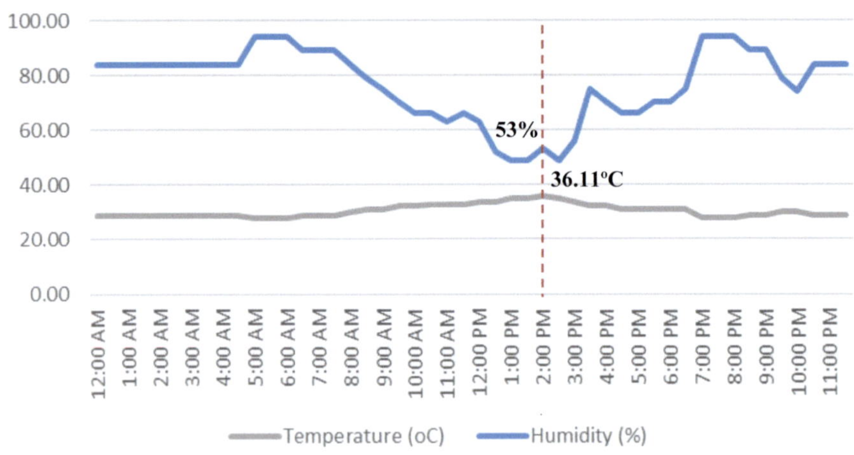

Fig. 3 Meteorological characteristics observed on 15.05.2022. (Authors, 2022, based on data from https://www.wunderground.com/history/)

The heat index chart (Fig. 5) was used to assess the thermal comfort level. The chart is a measure developed by Robert G. Steadman and is commonly applied in environmental health studies [11]. The values from the matrix, so-called apparent temperatures, help to predict the potential heat stress that people might experience due to their exposure to certain air temperatures and relative humidity levels (Table 1).

Results from the thermodynamic simulations indicated the potential air temperature (°C) which was used to measure the thermal effects of blue and green infrastructure at the study site. Within this study, the potential air temperature can be considered as the absolute air temperature [12]. Its values were classified using a

Fig. 4 An illustration of the selected elevations for the study (Authors, 2022)

NOAA national weather service: heat index

Relative humidity \ Temperature	80 °F (27 °C)	82 °F (28 °C)	84 °F (29 °C)	86 °F (30 °C)	88 °F (31 °C)	90 °F (32 °C)	92 °F (33 °C)	94 °F (34 °C)	96 °F (36 °C)	98 °F (37 °C)	100 °F (38 °C)	102 °F (39 °C)	104 °F (40 °C)	106 °F (41 °C)	108 °F (42 °C)	110 °F (43 °C)
40%	80 °F (27 °C)	81 °F (27 °C)	83 °F (28 °C)	85 °F (29 °C)	88 °F (31 °C)	91 °F (33 °C)	94 °F (34 °C)	97 °F (36 °C)	101 °F (38 °C)	105 °F (41 °C)	109 °F (43 °C)	114 °F (46 °C)	119 °F (48 °C)	124 °F (51 °C)	130 °F (54 °C)	136 °F (58 °C)
45%	80 °F (27 °C)	82 °F (28 °C)	84 °F (29 °C)	87 °F (31 °C)	89 °F (32 °C)	93 °F (34 °C)	96 °F (36 °C)	100 °F (38 °C)	104 °F (40 °C)	109 °F (43 °C)	114 °F (46 °C)	119 °F (48 °C)	124 °F (51 °C)	130 °F (54 °C)	137 °F (58 °C)	
50%	81 °F (27 °C)	83 °F (28 °C)	85 °F (29 °C)	88 °F (31 °C)	91 °F (33 °C)	95 °F (35 °C)	99 °F (37 °C)	103 °F (39 °C)	108 °F (42 °C)	113 °F (45 °C)	118 °F (48 °C)	124 °F (51 °C)	131 °F (55 °C)	137 °F (58 °C)		
55%	81 °F (27 °C)	84 °F (29 °C)	86 °F (30 °C)	89 °F (32 °C)	93 °F (34 °C)	97 °F (36 °C)	101 °F (38 °C)	106 °F (41 °C)	112 °F (44 °C)	117 °F (47 °C)	124 °F (51 °C)	130 °F (54 °C)	137 °F (58 °C)			
60%	82 °F (28 °C)	84 °F (29 °C)	88 °F (31 °C)	91 °F (33 °C)	95 °F (35 °C)	100 °F (38 °C)	105 °F (41 °C)	110 °F (43 °C)	116 °F (47 °C)	123 °F (51 °C)	129 °F (54 °C)	137 °F (58 °C)				
65%	82 °F (28 °C)	85 °F (29 °C)	89 °F (32 °C)	93 °F (34 °C)	98 °F (37 °C)	103 °F (39 °C)	108 °F (42 °C)	114 °F (46 °C)	121 °F (49 °C)	128 °F (53 °C)	136 °F (58 °C)					
70%	83 °F (28 °C)	86 °F (30 °C)	90 °F (32 °C)	95 °F (35 °C)	100 °F (38 °C)	105 °F (41 °C)	112 °F (44 °C)	119 °F (48 °C)	126 °F (52 °C)	134 °F (57 °C)						
75%	84 °F (29 °C)	88 °F (31 °C)	92 °F (33 °C)	97 °F (36 °C)	103 °F (39 °C)	109 °F (43 °C)	116 °F (47 °C)	124 °F (51 °C)	132 °F (56 °C)							
80%	84 °F (29 °C)	89 °F (32 °C)	94 °F (34 °C)	100 °F (38 °C)	106 °F (41 °C)	113 °F (45 °C)	121 °F (49 °C)	129 °F (54 °C)								
85%	85 °F (29 °C)	90 °F (32 °C)	96 °F (36 °C)	102 °F (39 °C)	110 °F (43 °C)	117 °F (47 °C)	126 °F (52 °C)	135 °F (57 °C)								
90%	86 °F (30 °C)	91 °F (33 °C)	98 °F (37 °C)	105 °F (41 °C)	113 °F (45 °C)	122 °F (50 °C)	131 °F (55 °C)									
95%	86 °F (30 °C)	93 °F (34 °C)	100 °F (38 °C)	108 °F (42 °C)	117 °F (47 °C)	127 °F (53 °C)										
100%	87 °F (31 °C)	95 °F (35 °C)	103 °F (39 °C)	112 °F (44 °C)	121 °F (49 °C)	132 °F (56 °C)										

Key to colors: Caution   Extreme caution   Danger   Extreme danger

Fig. 5 Heat index chart. *Source* https://bigladdersoftware.com/epx/docs/9-6/engineering-reference/resilience-metrics.html

consistent color range with a step of 1.5 °C. Besides, the reflected radiation was also considered, whenever necessary for the assessment of the combined cooling effects of greenery and shadow.

Table 1 Potential health-related risks according to heat index level. *Source* https://bigladdersoftware.com/epx/docs/9-6/engineering-reference/resilience-metrics.html

Heat Index in Celsius	Heat Index Level
Lower than 26.7 °C	Safe: no risk of heat hazard
26.7 °C - 32.2 °C	Caution: fatigue is possible with prolonged exposure and activity. Continuing activity could result in heat cramps.
32.2 °C - 39.4 °C	Extreme caution: heat cramps and heat exhaustion are possible. Continuing activity could result in heat stroke.
39.4 °C - 51.7 °C	Danger: heat cramps and heat exhaustion are likely; heat stroke is probable with continued activity.
Higher than 51.7 °C	Extreme danger: heat stroke is imminent.

3 Findings

3.1 Thermal Effects of Blue Infrastructure

Firstly, thermal comfort was investigated by assessing and comparing the variations between the heat index values calculated at different elevation levels of the two scenarios including the baseline model (BL) and the model with the added water body (W) (Fig. 6). On average, the water fountain helps to lower the heat index by 1 °C. The significant differences between the two scenarios were observed at $k = 0–2$, which means the cooling effects were most pronounced at the lower level. The highest variations in heat index were seen at 5:00–6:00 AM up to the pedestrian level ($k = 2$) with the highest value of 6.42 °C at 5:00 AM.

Following these preliminary results, a further examination of the potential health-related risks for pedestrians was taken for the first two scenarios (Fig. 7). On the hottest day, 91% of the time people must be extremely cautious when being outside, while during the remaining 9% time of the day, people were likely to experience heat cramps and exhaustion and even heat stroke. With the presence of water in the pocket park, the time with heat-related dangers was reduced to 4%, and the time when people must be with extreme caution was lowered to 52%. Interestingly, during 9% of the time of the day, people might remain outside without any concern about thermal risks, which was absent in the baseline model.

At the higher elevation levels in mid-afternoon (2:00 PM), the cooling effects of water were present, but only locally noticeable (Fig. 8). Up to 7.6 m ($k = 6$, or equal to the third floor), the air temperature was lower surrounding the water body. The difference in the temperature was observed with one color step, or around 1.5 °C on average. Remarkably, the cooling effects likely spread toward the prevailing wind direction (West-Southwest) (at 4.5 m high) and combined with the shadow of the existing trees for noticeable synergy effects (at 7.6 m high). The cooling effects were no longer visible at 11.0 m high.

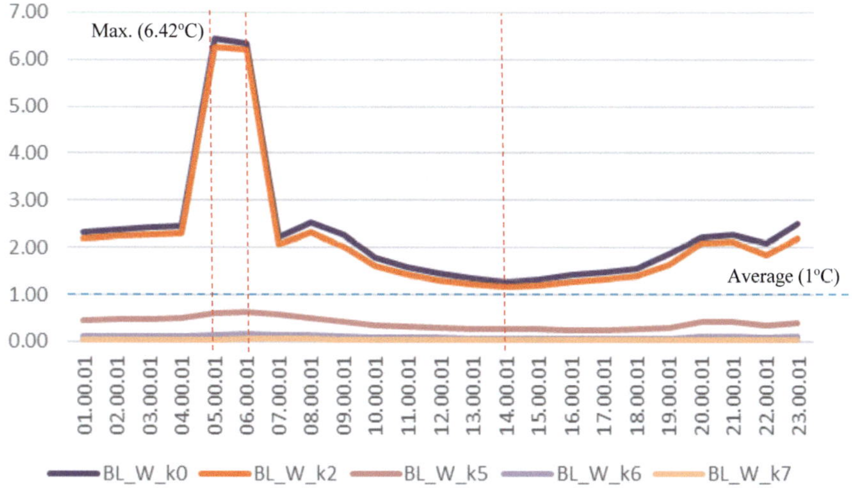

Fig. 6 Heat index differences compared between the baseline model and the scenario with the blue infrastructure

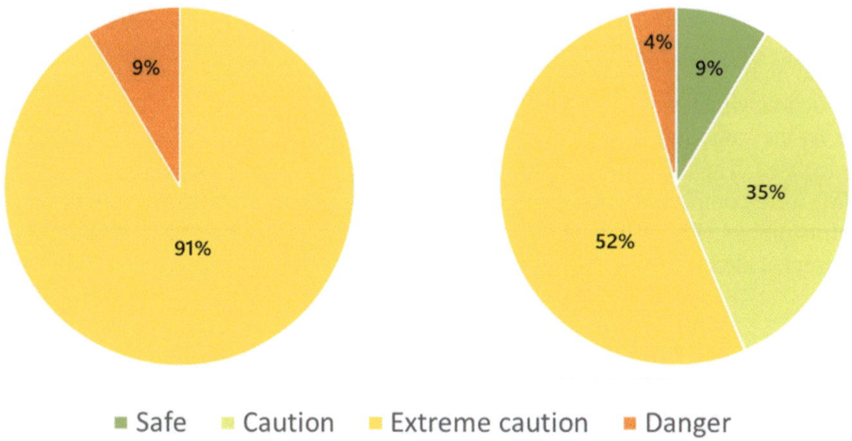

Fig. 7 Heat index assessment for the baseline model (left) and the second scenario with blue infrastructure (right) up to 1.5 m

3.2 *Effects of Green Infrastructure*

Similarly, by the differences in potential air temperatures, the thermal impacts of greenery were noticed (Fig. 9). Since plants were added to all outer walls, the cooling effects, which mainly took place along the green walls and above the green roofs, seemed to locate more widely throughout the site than those of the water body.

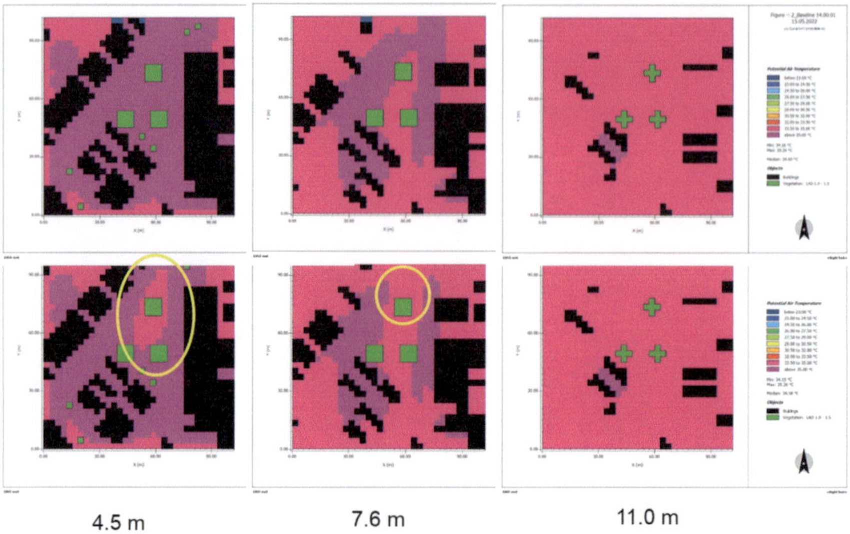

Fig. 8 Air temperature observed at 02:00 PM for the baseline model (top) and the second scenario with blue infrastructure (bottom) at the higher elevation levels

Based on the color differences, greenery proved to be beneficial for air temperature reduction, which was about 1.5 °C.

At the same time and elevation observed in Fig. 9, the radiation intensity was calculated. The reflected radiations from the Southwest (the sun's direction at 2:00 PM) were compared between the baseline model and the third scenario with green infrastructure (Fig. 10). The absolute differences between the two scenarios were presented along the green walls, above the green roofs, and in shadows.

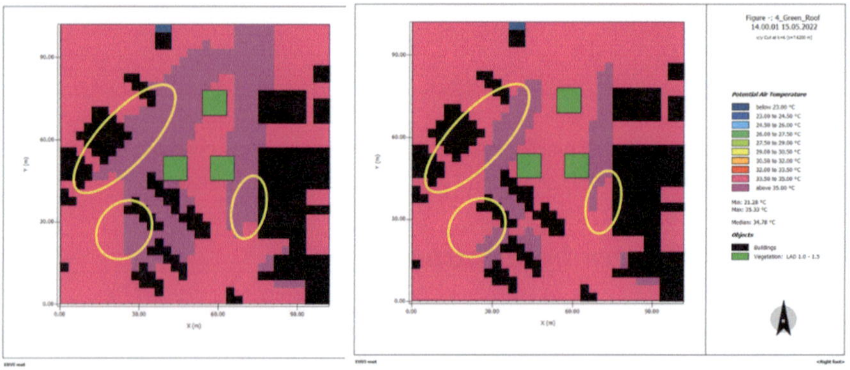

Fig. 9 Air temperature observed at 02:00 PM for the baseline model (left) and the third scenario with green infrastructure (right)

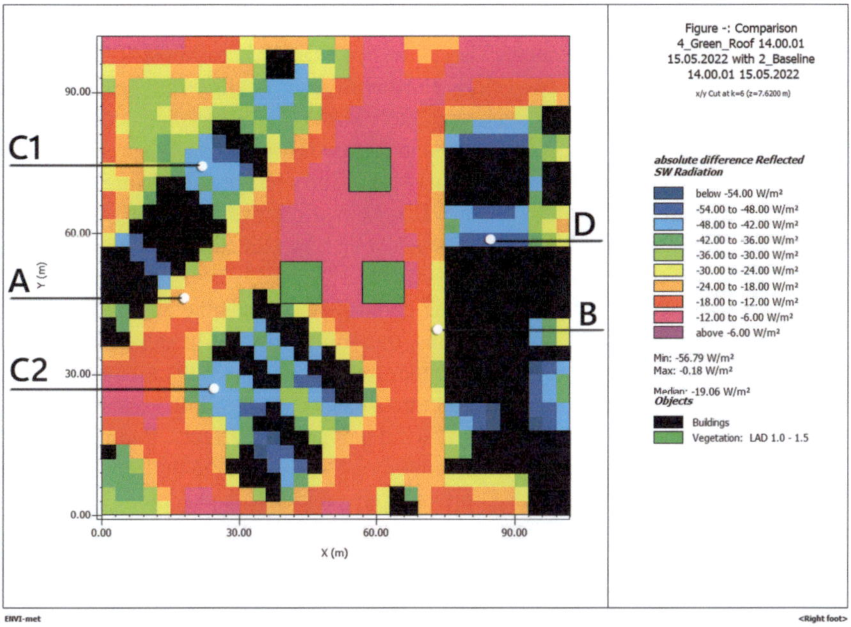

Fig. 10 The absolute differences between the third scenario with green infrastructure and the baseline model

At B, where the green walls were oriented towards the sun's direction, the reduction of radiation flux measured up to 30 W/m^2. At A, in combination with the shadow, the differences between the reflected radiation values ranged up to 42 W/m^2. The shadows between the buildings (C1, C2, and D) contributed to a visible synergy effect with the reduction of radiation flux measured up 48 W/m^2 at C1 and C2 (green roofs) to above 56.79 W/m^2 at D (along the green walls, above the green roofs and in the shadow).

4 Conclusion

In dense cities, the high urbanization rate can cause several urban climatic issues, including heat island effects. To investigate the potential solutions for the existing built environment using nature-based measures, a residential block in Ho Chi Minh City was taken as a case study in this paper using thermodynamic simulations and scenario analysis.

According to the findings, the heat index calculation provided proof of heat stress presence of more than 90% of the time during the hottest day. Through different scenarios, the thermodynamic simulation proved that thermal discomfort could be mitigated using blue and green infrastructure. Accordingly, heat stress was significantly reduced by the water feature. Therefore, pedestrians face almost 40% less time under hazardous temperature conditions. Furthermore, around 10% of the time can be considered non-harmful, which was completely absent in the baseline scenario. The cooling effects did not only take place around the water but also likely spread towards the wind with an air temperature reduction of 1.5 °C on average. Green infrastructure, which included green roofs and green walls, helped remove heat along and above the greenery surfaces, together with the shadow effects. Reflected radiation reduced up to 56.79 W/m^2 thanks to this synergy effect.

In conclusion, the paper confirmed the cooling effects of blue and green infrastructure on urban microclimate conditions, therefore lowering health-related risks. The numerical and spatial outcomes from the simulations also imply the significance of planning for blue and green infrastructure in urban design. For greater impacts, it is also important to take advantage of wind and shadow effects. Future research can take into account the contribution from on-site study, and the multiple effects of blue and green infrastructure such as urban biodiversification, psychological impact provision, air purification, and noise removal.

References

1. Huynh D (2015) The misuse of urban planning in Ho Chi Minh City. Habitat Int 48:11–19. https://doi.org/10.1016/j.habitatint.2015.03.007
2. Son N-T, Chen C-F, Chen C-R, Thanh B-X, Vuong T-H (2017) Assessment of urbanization and urban heat islands in Ho Chi Minh City, Vietnam using Landsat data. Sustain Cities Soc 30:150–161. https://doi.org/10.1016/j.scs.2017.01.009
3. Thi Van T, Duong Xuan Bao H (2010) Study of the Impact of Urban Development on Surface Temperature Using Remote Sensing in Ho Chi Minh City, Southern Vietnam: Study of the Impact of Urban Development on Surface Temperature. Geographical Research. 48, 86–96. https://doi.org/10.1111/j.1745-5871.2009.00607.x
4. Vu HN, Truong TH (2019) Management of Urban Environment in Ho Chi Minh City Facing Views of Climate Change. IJMRAP. 2:7–12
5. Doan Q-V, Kusaka H (2016) Numerical study on regional climate change due to the rapid urbanization of greater Ho Chi Minh City's metropolitan area over the past 20 years
6. Doan Q-V, Kusaka H, HO QB (2016) Impact of future urbanization on temperature and thermal comfort index in a developing tropical city: Ho Chi Minh City
7. Downes NK, Storch H, Moon K, Rujner H (2010) Urban Sustainability in Times of Changing Climate: The Case of Ho Chi Minh City, Vietnam. Presented at the 46th ISOCARP Congress
8. Moon K (2011) Determining Potential Areas for Mitigation of Climate Change Risks in Ho-Chi-Minh City
9. Dang HT, Pitts A (2020) Urban Morphology and Outdoor Microclimate around the "Shop-house" Dwellings in Ho Chi Minh City, Vietnam
10. Naiman RJ, Décamps H, McClain ME, Likens GE (2005) Biotic Functions of Riparia. In: Riparia. pp. 125–158. Elsevier
11. Anderson GB, Bell ML, Peng RD (2013) Methods to Calculate the Heat Index as an Exposure Metric in Environmental Health Research. Environ Health Perspect 121:1111–1119. https://doi.org/10.1289/ehp.1206273
12. enviadmin: ENVI-met A holistic microclimate model, https://envi-met.info/doku.php?id=filereference:output:atmosphere

Assessment of Cooling Effect by Urban Park Using a Multi-data Source Approach

Can Trong Nguyen, Amnat Chidthaisong, and Rungnapa Kaewthongrach

Abstract Urban park is a component of urban green infrastructures, essential in urban ecosystems because it delivers various benefits to urban dwellers, from tangible to intangible values. Urban park is widely accepted as a friendly adaptive solution in the context of warmer cities worldwide due to climate change and urbanization impacts. This research quantified the cooling effect of small and medium-sized parks in Bangkok Metropolis and identified primary controlling factors. An integrated approach was adopted to achieve the overarching objectives, including image interpretation from various data sources and Trends-Breakpoints Detection Analysis (TBDA). The research findings revealed that summer's most active cooling distance is around 100–200 meters. When the weather is cooler in winter, the cooling distance expands outward to + 400 and + 1,000 meters depending on park structures and neighboring backgrounds. A park in a region with denser vegetation should have a more moderate cooling effect. Increases in tree areas inside a park and their shape complexity can stimulate the cooling effect, especially in the cooler season. In contrast, the park's water surface only contributes to the cooling effect during the hotter period. The research outcomes are helpful for urban planners in heat mitigation strategies using urban green infrastructures.

Keywords Urban park · Urban green space · Cooling effect · Cooling distance · Park structure

C. T. Nguyen (✉) · A. Chidthaisong
The Joint Graduate School of Energy and Environment, King Mongkut's University of Technology Thonburi, Bangkok 10140, Thailand
e-mail: trongcan.ng@gmail.com

C. T. Nguyen
Faculty of Engineering, Prince of Songkla University (Hatyai Campus), Songkhla 90110, Thailand

C. T. Nguyen · R. Kaewthongrach
Space Technology Research Center, Geo-Informatics and Space Technology Development Agency (GISTDA), Chonburi 20230, Thailand

V. T. Ha et al. (eds.), *Proceedings of the 8th International Conference on Sustainable Urban Development*, Advances in 21st Century Human Settlements,
https://doi.org/10.1007/978-981-99-8003-1_9

1 Introduction

Many studies revealed that Southeast Asia (SEA) cities such as Kuala Lumpur (Malaysia), Bogor (Indonesia), Baguio (Philippines), Yangon (Myanmar), Ho Chi Minh City and Ha Noi (Vietnam), and Bangkok (Thailand) have experienced the warmer trend in land surface temperature (LST) and urban heat islands (UHI), about 0.03 °C to 0.92 °C per year [1, 4, 17, 20, 22, 23, 30]. Higher urban temperature is directly responsible for human thermal comfort and heat-related morbidity and mortality in the city [11, 18]. It also causes pressures on the economy and energy sector for cooling demands, especially during heat waves [16]. Moreover, urban temperature alterations always occur parallel with urbanization, which induces impervious surface extension and narrowing vegetation. These nexuses imply potential solutions for urban design strategies to reduce the harmful impacts of heat exacerbation based on the basic notion of maximizing urban green areas like public parks. In the context of urbanization and escalating climate change, the urban thermal environments are forecasted to be more severe for their inhabitants. Thus, the role of urban green spaces and public parks in mitigating the urban microclimate severity is more prominent and appreciated by managers and urban dwellers.

On the other hand, large parks frequently attract more attention because they are supposed to have diverse and more significant benefits than smaller green spaces. In a crowded city, most land budget is prioritized for residential areas, commercial purposes, and solid urban infrastructures rather than large patches of green spaces [15]. Planning extensive urban forests is challenging in most compact cities. Therefore, we intended to examine the cooling effect of small and medium-sized parks, potentially appropriate for urban greening plans. Additionally, the cooling distance is a principal parameter in the cooling models such as InVEST-Urban Cooling Model [7]. The cooling distance is diverse and fluctuates from park to park depending on many factors. When the cooling distance is specified, the model performance will be significantly improved. The factors controlling the cooling effect will substantially affect a prototype park design to apply to other current and future parks.

This research stands on exploiting diversely free-accessible data to investigate the cooling effect given by the public parks, in which Bangkok metropolis is an empirical study. The controlling elements moderating the cooling effect were then analyzed to identify impacts of the park's structure itself and external environments on the cooling effect, which are helpful information for urban planning toward heat mitigation strategies in the city.

2 Study Sites and Datasets

2.1 Study Sites

Five parks were selected along the urban-periurban gradient of Bangkok metropolis (Thailand), considering their popularity and usable area (Fig. 1). The most extensive park is Chatuchak Botanical Garden which comprises three adjacent parks of Wachira Benchtas, Queen Sirikit, and Chatuchak. Its usable area is about 110 hectares and it has become one of the essential urban green spaces in this compact city. It is followed by Suan Luang Rama IX (shortly Rama 9) with nearly 80 hectares. The following two parks are relatively similar regarding the usable area, Lumpini (57.6 hectares) and Serithai (56 hectares). However, they have significant disparities in location and landscape structures. Lumpini is supposed to be a "green asset" in a cramped neighboring area, while Serithai has its role in rainwater regulation in the periurban of eastern Bangkok. Finally, Thonburirom (*or Thonburi*, 10.058 hectares) is located on the city's west side, which plays a crucial role in aesthetics and its surrounding neighbors' environment.

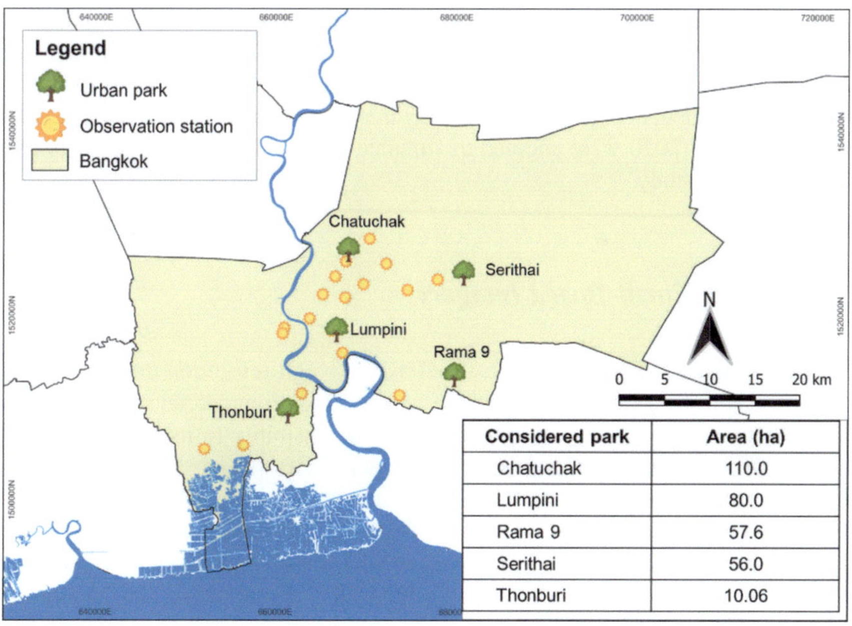

Fig. 1 Location of Bangkok, the considered parks, and air quality stations within Bangkok metropolis

2.2 Air Temperature

We obtained air temperature data from the Thailand Pollution Control Department (PCD), which includes 14 stations within the Bangkok metropolitan region. These stations are set up at roadside and ambient locations to hourly measure air temperatures. The temperature at each station was selected based on date and time criteria. In particular, we obtained the air temperature at 10:00 am for analysis because the initial test showed that LST and air temperature reached their highest correlation value at 10:00 am.

2.3 Landsat 8 Imagery

Landsat 8 (OLI/TIRS) surface reflectance is the primary data source to extract surface characteristics and simulate air temperature from these features. The land surface reflectance (LSR) data is atmospherically corrected using Landsat Surface Reflectance Code. The LSR, therefore, minimizes atmospheric influences, especially on temperature data, which is relatively susceptible to atmospheric and cloud conditions. We acquired images under satisfied weather conditions of clear sky or limited clouds covering the study sites and ambient regions. Seven scenes were captured at seven milestones from 2014 to 2016 for training and validating the air temperature predictive model. Whereas the clear-sky images acquired on 19/02/2020, 22/03/2020, and 17/11/2020 were adopted to simulate spatial air temperature for further analyzes in this study.

2.4 Google Earth-Based Imagery

The park's landscapes should be classified by very-high-resolution imagery (VHR) since its scale is frequently small. Yet, we cannot access commercial satellites in this study. Fortunately, Google Earth provides free accessibility to its VHR images at different times. Although Google Earth imagery (GEI) cannot offer diverse spectral information as a standard multispectral image, it can adjust the details to be observed. Thus, the GEI becomes a cost-effective VHR data source widely applied in many urban studies [12, 14]. We collected GEI at each park using the SASPlanet (v200606) tool. The pixel size of the collected GEI is approximately 30 centimeters per pixel edge.

2.5 Sentinel-2 Imagery

GEI is relatively approriate for mapping in detail land cover in a small-scale area. However, it shows limitations in a larger region when the detail of GEI becomes its drawbacks in terms of processing performance. More explicitly, there are potential noises from building shadows, colors of different roof materials, tree canopy structures, and even water surface waves. Therefore, we adopted moderate-high spatial resolution satellite imagery of Sentinel-2 (10 meters) to classify landscape information around the park from its border to 2 km. Sentinel-2 L2A images were collected with an acquisition strategy of the smallest difference in captured date of Landsat-8 images above. Two scenes were downloaded directly from the Sentinel Hub on 21/02/2020 (10:37) and 29/08/2020 (10:35). These images have a low cloud coverage rate of 0.47% and 5.81% for the image in February and August, respectively.

3 Methodology

3.1 Land Cover of the Park and Surrounding Areas

Landscapes of the parks were classified using Google Earth imagery. It comprises five land cover categories of wood tree, grassland, soil/pavement, buildings, and lake/pond. Firstly, the acquired images were reprojected before they were analyzed by object-oriented classification. The images were then overcome through a segmentation procedure, which groups nearby similar pixels together based on their similarities in spectral information, proximity, pattern, and so on. Segment Mean Shift (SMS) technique was adopted to analyze the GEI. Subsequently, the segmented images were classified by the ISODATA (Iterative Self-Organizing Data Analysis) unsupervised classifier. To cluster pixels into N user-defined groups, the algorithm randomly sets cluster centers and assigns pixels to clusters using minimum distance. The clusters will be merged or split based on the minimum distance among the cluster's centers. The progress is repeated until all pixels are precisely separated and the number of clusters reaches the user-defined number. In this study, the initial clusters of $N = 30$ were set for classification. The subclusters were combined altogether if they presented an identical land cover type when collated with image visualization.

The neighboring regions of each park were extracted by multispectral imagery of Sentinel-2, which grouped into major land cover categories such as impervious surfaces, vegetation, water bodies, and bare land. First, the multispectral bands (i.e., visible wavelength, Red-Edge, NIR, and SWIR) were consistently resampled pixel size of 10 meters. After that, we applied integration of principal component analysis (PCA) and ISODATA unsupervised classifier to retrieve the park's neighboring land cover. The initial cluster number was $N = 15$ because Sentinel-2 imagery is less detailed than GEI. Finally, the classified images were combined to generate a map based on general land cover types.

3.2 Air Temperature Estimation Using Machine Learning

Surface characteristics extraction. The surface characteristics were described by spectral indices, representing three primary land cover patterns of vegetation, impervious surfaces, and water bodies. Particularly, vegetation was depicted by the Normalized Difference Vegetation Index (NDVI) and Enhanced Vegetation Index (EVI). NDVI is a well-known index to identify live green foliage relating to photosynthetically active radiation (PAR) and Near Infrared (NIR) light [9, 24]. However, it shows limitations when vegetation becomes denser. EVI was proposed to overcome the NDVI limitations by adjusting the NDVI formula by blue light and constants [10]. Hence, we also obtained EVI to separate different vegetation canopies, which are expected to influence the air temperature. The next crucial land cover type is urban features, which mainly contribute to urban warming due to the thermal characteristics of urban materials. Therefore, we calculated two urban indices (i.e., Urban Index (UI) and Normalized Difference Built-up Index (NDBI)) to test which index better performs for air temperature estimation. Finally, the water surfaces were defined by the Modified Normalized Difference Water Index (MNDWI) – an optimal water index to locate water features in urban areas [28].

Land surface temperature retrieval. Land surface temperature (LST) was retrieved using a widely applied algorithm, which converts DNs values to LST by calibrating brightness temperature (T_B) using NVDI-based land surface emissivity (LSE) (Eq. 3) [3, 25, 27]. Firstly, vegetation fraction (FVC) was calculated by calibrating specific NDVI pixels by NDVI values of fully dense vegetation ($NDVI_V$) and completely bare soil surface ($NDVI_S$) (Eq. 1) (Carlson & Ripley, 1997). Then, LSE was estimated by an empirical equation using FVC for Landsat OLI/TIRs (Eq. 2) [21, 26].

$$\text{FVC} = ((\text{NDVI} - \text{NDVI}_S) / (\text{NDVI}_V - \text{NDVI}_S))^2 \tag{1}$$

$$\varepsilon = 0.00149 \times \text{FVC} + 0.985481 \tag{2}$$

$$T_S = (T_B/(1 + (\lambda T_B/\rho) \ln \varepsilon)) - 273.15 \tag{3}$$

where FVC is vegetation fraction; $NDVI_S$ and $NDVI_V$ are vegetation index of fully dense vegetation and bare soil, respectively; ε is land surface emissivity for Landsat OLI; TS the is the land surface temperature (°C); TB is brightness temperature in Kelvin; λ is the wavelength of emitted radiance (i.e., Landsat OLI is band 10, $\lambda_{B10} = 10.89$ μm); $\rho = hc/\sigma$, with $\rho = 1.438 \times 10^{-2}$ Mk.

Optimal distance determination. Air temperature at a particular location is regulated by its surrounding landscape rather than its land cover. Therefore, we should determine which distance that land cover mainly drives air temperature variations. Firstly, aggregated images of surface indices were generated with the number of pixels on each edge that belongs to an odd number subset, $F = \{3,5,7,\ldots, 65,67\}$.

These pixels correspond to distances from 90 to 2,100 meters from the stations. Next, the index values were extracted and compared to air temperature at 10:00 a.m. using the Pearson correlation coefficient. The distance of each index was noted when the correlation coefficient achieved the first highest value. The influential distance was subsequently determined as a general distance throughout all indicators. The effective distance is 750 meters based on our initial test.

Air temperature predictive model. LST is a critical variable in air temperature prediction models because it has the most significant relationship with other surface indicators [8, 19]. The best model was detected through a model performance test using cross-validation. Specifically, the simple model of only LST, its combination with each surface indicator, and two other Random Forest (RF) synthetic models were evaluated with iteration $N = 1000$. The most optimal model is determined when a model achieves a higher accuracy with fewer predictors. Finally, the optimally predictive model was applied to its contributors of spatial surface elements to simulate air temperature in entire study areas.

3.3 Analyzing Climate Regulation Effect

The influential distance of the park was analyzed by trend and breakpoint detection analysis using the greenbrown R package [5]. Theoretically, the greenbrown is performed on time series data to explore how the land phenology changes. This study assumed that air temperature variation every 100 meters until 2,100 meters is a node in time series data. As a result, the influential distance was detected at a breakpoint where air temperature significantly drops from the park site.

In addition to the land use, land cover (LULC) area, the landscape metrics, including Percentage of Landscape (PLAND), Aggregation Index (AI), and Landscape shape index (LSI), were computed for park structures and surrounding areas for each 100-m buffer zone [13]. Subsequently, we analyzed the correlation between air temperature difference (i.e., the gaps between near park zones and highly dense urban areas) and each landscape metric using the Pearson correlation coefficient.

4 Results and Discussions

4.1 Park's Land Cover Structure

Landscapes of the considered parks interpreted from GEI are shown in Fig. 2. Unlike the classic parks only occupied by vegetation and woodlands, these parks incorporate diverse landscapes of wood trees, lawns, and water surfaces (Fig. 2). These designs are based on modern design perspectives to provide diverse sceneries and take their environmental advantages. In addition to the difference in usable area, their structures

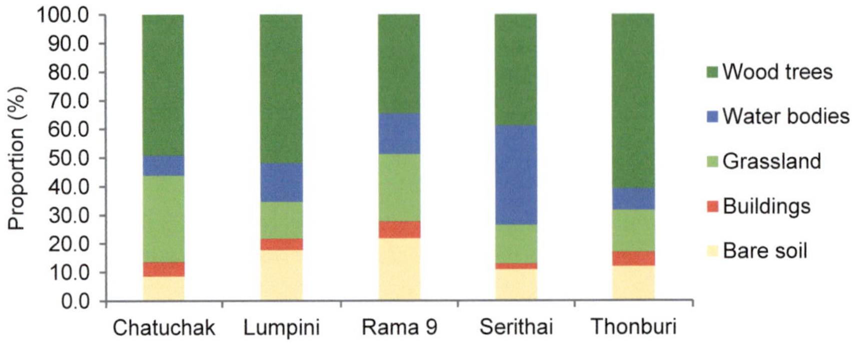

Fig. 2 Landscape proportion in the considered parks delineated from GEI

are relatively diverse. The pivotal object throughout the parks is vegetation, i.e., it is contributed by any green patch. The woodland proportion always maintains from 38.8% (Serithai) to 60.7% (Thonburi) (Fig. 2). Lumpini and Chatuchak also hold a relatively large area of the woodland, about 51.9% and 49.2%, respectively. Chatuchak Park uses 30.3% of the space for expansive lawns, the most extended lawn area among the parks. Regarding the contribution of water surfaces in the parks, Serithai is highlighted with a central lake, where the water surface area rate is up to 34.7%.

4.2 Spatial Air Temperature

The optimal distance for air temperature simulation was 750 meters. The influential distance was utilized as a buffer distance to extract surface indicators for model testing. Model accuracy for different RF-model combinations is shown in Table 1. Notably, the model with solely LST achieves relatively high accuracy, 0.91 ± 0.028. Adding one or more auxiliary data of surface indicators improves the accuracy of air temperature simulation. Among vegetation indices, the model contributed by EVI is more efficient than the NDVI model, with an accuracy is 0.94 versus 0.93, respectively. The models of the urban index are not much different in performance assessed by mean accuracy; nevertheless, the UI model reaches higher maximum accuracy implying potential efficiency compared to NDBI. The model with MNDWI contribution also obtained 0.94 ± 0.02. According to these analyzes, we proposed and tested an optimal model (M7), constituted by the favorable elements (i.e., LST, EVI, UI, and MNDWI). Its performance reached a high level of 0.96 ± 0.014 against the above models, with the highest value even achieved at 0.98. The model performance of all elements (M8) is not much distinctive from the M7 model, while the M7 model can simulate air temperature with fewer variables. As a result, the M7

Table 1 The accuracy obtained by cross-validation from different RF-model combinations

Model	Predictors	Average Accuracy	Std.
M1	LST	0.906	0.0285
M2	LST, NDBI	0.941	0.0185
M3	LST, EVI	0.932	0.0225
M4	LST, UI	0.942	0.0187
M5	LST, NDVI	0.924	0.0273
M6	LST, MNDWI	0.936	0.0208
M7	LST, EVI, UI, MNDWI	0.954	0.0150
M8	LST, NDBI, UI, NDVI, EVI, MNDWI	0.958	0.0136

model was considered as an ideal model for spatial air temperature estimation in Bangkok metropolis, $R^2 = 0.91$ and $RMSE = 0.89$. The spatial simulation for air temperature entirely in Bangkok on free cloud dates of 19/02/2020, 22/03/2020, and 17/11/2020 was obtained by applying the ideal M7 model.

4.3 *Spatiotemporal Distinctness in Cooling Effect Distance*

The average Ta within the park boundary extracted from the estimated air temperature is shown in Table 2. Overall, the park's air temperature increases from February to March; in November, it drops to values less than that in February. The highest temperatures throughout the months are at Chatuchak Park, while the lowest temperatures are held by Thonburirom Park (February and November) and Serithai (November). For example, we consider temperature variations between February and March, which shows that the most easily variated parks are Lumpini (ΔTa = 2.29 °C) and Thonburirom (ΔTa = 2.14 °C). On the contrary, Serithai and Rama 9 tend to be more stable in temperatures over time, especially Serithai ΔTa = 0.55 °C.

We investigated the cooling effect of each park using breakpoint analysis (Fig. 3). Firstly, the Ta value changes along the horizonal buffer zones from 0 to 2000 meters were assumed and analyzed as annual time series data. Then, the trend and breakpoints were tested to determine the statistically significant trends and breakpoints.

Table 2 Each park's average air temperature (°C) in February, March, and November 2020

Park	February	March	November
Chatuchak	27.678 ± 0.971	29.561 ± 0.879	27.417 ± 1.064
Lumpini	26.893 ± 0.166	29.180 ± 0.367	26.617 ± 0.496
Rama 9	27.087 ± 0.139	28.352 ± 0.522	26.577 ± 0.257
Serithai	27.659 ± 0.252	28.205 ± 0.487	27.989 ± 0.248
Thonburi	26.372 ± 0.035	28.513 ± 0.352	25.283 ± 0.001

The principles accepted for this analysis is the temperature at the park, which is controlled by cooling features such as wood trees, lawns, and water surfaces, being coolest in comparison to surrounding impervious surfaces; plus, the temperature gradually increases under the influences of dense urban areas until the temperature drops again when it reaches rural areas with extensive cooler surfaces and landscapes.

Chatuchak can control the air temperature around the park by about 600 ± 100 meters in November. The influential distance is narrowed from February to March, approximately 200 ± 100 meters (i.e., it can cool up to 400 meters) and 200 ± 100 meters. The Lumpini Park can also cool down its neighboring areas up to 600 ± 100 meters in winter. Yet, this distance is solely 200 meters in February.

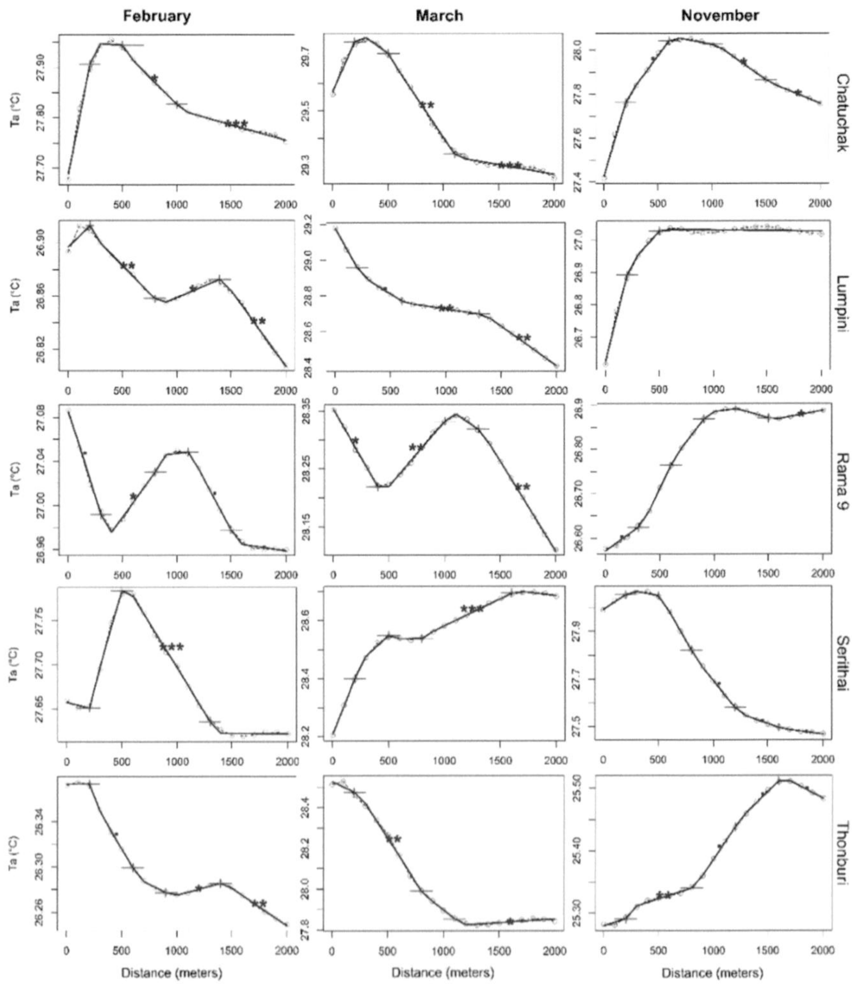

Fig. 3 Air temperature changes along the horizontal transect and corresponding changepoints (red nodes)

In the mid-summer, Lumpini Park does not show a cooling effect anymore to its surrounding areas.

Surprisingly, the biggest park of Rama 9 Park has no significant cooling effect through the analysis for both February and March. The cooling effect, however, significantly improves in the winter season. The areas around the park, about 900 ± 100 meters, are primarily affected, and the active distance can expand up to 1,200 meters. In Serithai Park, only the areas adjacent to the park (about 100 meters) can receive the cooling effect. The affected regions enlarged by 200 meters in November.

In contrast to other parks, i.e., the cooling effect weakens in the mid-summer, the Serithai Park shows a remarkable regulative capacity of 500 ± 100 meters. On the other hand, the smallest park of Thonburirom Park can reduce the air temperature by around 200 meters and 100 meters in February and March. However, these distances are insignificant. During the winter season, the active cooling distance achieves 1,600 ± 100 meters, the most extended distance among the considered parks in Bangkok.

4.4 Pivotal Factors Regulate Park Cooling Effect

The park's cooling effect fluctuates throughout the year depending on seasons, while its inside structures and surrounding environments remain relatively stable as a kind of evergreen vegetation. Our analyzes for air temperature variations within 100 meters and 500 meters, 100 meters represent a location cooled down by the park, and 500 meters is a place of denser urban areas (Fig. 4), revealing the parks' temperature and their cooling magnitude are closely associated with location and park structures. For instance, the temperature gap between the two mentioned locations in November (i.e., when all parks' cooling capacities tend to be consistent and clarified) decreases gradually in Chatuchak (0.367 °C), Lumpini (0.244 °C), Rama 9 (0.13 °C), Thonburi (0.042 °C), and Serithai (0.016 °C). The more significant the temperature gap, the more influential the park with a high cooling effect is. It means that Chatuchak and Lumpini have a higher cooling effect than other parks.

Figure 4 shows the background of the parks where they are located. Regarding urban density from 500 meters outwards, Chatuchak and Lumpini stand within the metropolitan areas with the urban proportion exceeding 50% from 300 meters. Similarly, Serithai and Thonburi are determined to be in periurban areas as the urban densities fluctuate around a threshold of 50%. Rama 9 is a suburban park since its surrounding urban density is approximately 35% within the first 1,500 meters from the park boundary. The correlation analysis of the park's neighboring areas influences the park's temperatures and cooling effect. The results concede that the park's cooling effect is assisted by outside vegetation. Increasing the shape complexity of vegetation patches and green areas is the most influential factor. In contrast, the cooling effect is weakened by increasing impervious surfaces.

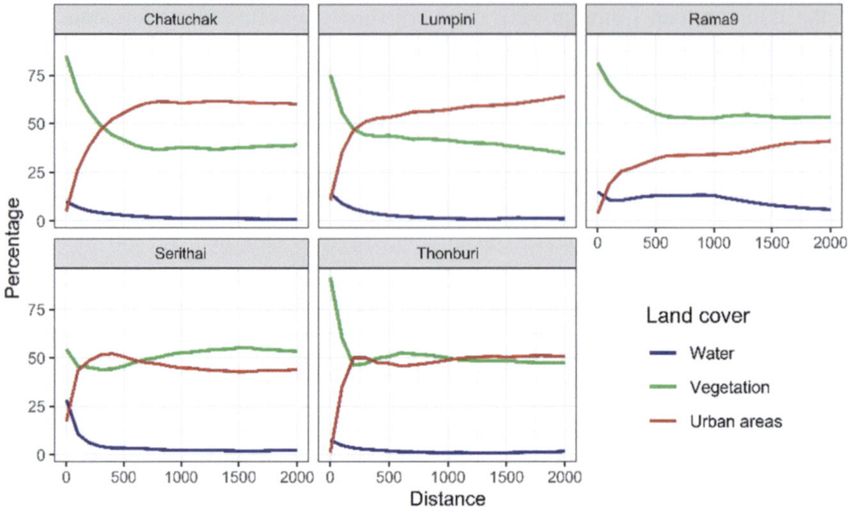

Fig. 4 Proportion of LULC from the park boundary outwards to neighboring areas

Regarding the park itself structures, it tends to be more influential in the cooling effect. The tree area is the most statistically significant element controlling cooling intensity (R = 0.97, P < 0.01). Moreover, the tree arrangement, i.e., green patches, are planted closely together into plates of complex shapes, which also governs to cool down park's neighboring areas. Besides, the water surface area inside the park in the form of a lake and wetland helps to improve the cooling effect outside the park.

The correlation analysis between the park characteristics, background patterns, and cooling intensity is shown in Fig. 5. The results indicate that both the outside and inside environments drive the cooling effect. In particular, the relationships are more significant in November against February and March. Outside vegetation substantially influences park cooling intensity, with an increase in vegetation proportion (PLAND) and its assembly into big patches (AI, LPI) being the most important determinants. In contrast, the compact impervious surfaces are represented by the spatial metrics of area, proportion, and aggregation, significantly weakening those effects encouraged by green spaces. Regarding the park's structures, the tree area is the most statistically significant factor influencing cooling intensity, notably in November ($R = 0.97$, $P \leq 0.01$). However, the influence is moderate in February and even non-existent in March. The tree arrangement, i.e., green patches, are planted tightly together to form plates of close and complicated shapes (i.e., AI and LSI), which also cools the park's surrounding regions. Furthermore, the park's water surface area, such as the lake and marsh, aids in improving the cooling effect outside the park, particularly during the hot season.

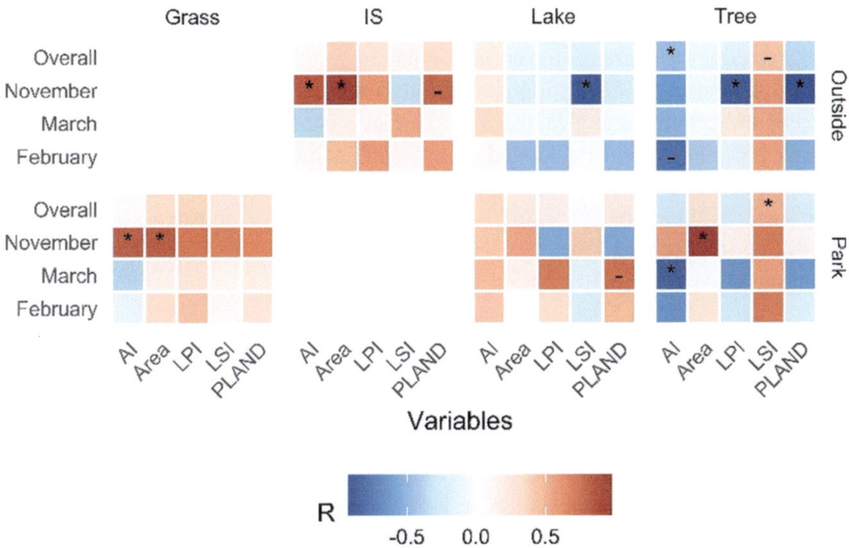

Fig. 5 Heatmap shows correlation coefficients between cooling effect and landscape metrics. Note: Symbols (*) and (-) indicate significance levels at ≤ 5% and ~ 10%. Grass: grassland; IS: impervious surfaces such as built-up, pavements, and roads; Lake: water surfaces; Tree: vegetation and woodland inside and outside the parks

4.5 *Implications for Urban Environment and Planning*

The considered parks in this study are ranked from small to medium size (10–110 hectares). Yet, the active cooling distance is smaller than about one-third of a 150-hectare park (200–300 meters) [6] and six times that of a 680-hectare large park (~1400 meters) [29]. Therefore, in urban green space design, small and medium-sized parks placed at the most beneficial distance from each other should be emphasized to maximize the limited urban land budget while ensuring open space accessibility for all inhabitants. Furthermore, a reasonable distribution of parks with other green infrastructures outside the parks disperses the concentrated pressures, improving the cooling effect.

Heat reduction is supported by combining blue and green spaces. Timber trees should be the primary focus. However, the water surface is also a supportive element. The integrated blue spaces (e.g., a tree-shaded lake and artificial wetland with appropriate aquatic plants) should be applied instead of the monotonic and extensive lake in urban planning. A lawn serves as a venue for outdoor leisure, such as picnics, assemblies, group activities, and a place to set up exercise equipments. At the same time, a cluster of wood trees significantly decreases air temperature. To minimize monotonic landscapes, the proportions of wood trees, lawns, and lakes within a park should be adjusted appropriately. In addition, the arrangement of trees should be considered, with individual trees being replaced with clusters of trees with intricate forms and edges to increase the cooling effect.

Though landscape integration is essential to enrich urban biodiversity and exploit different and beneficial angles, the water surface's cooling effect is prominent in hot weather [30]. The effect is reduced at nighttime due to heat release, especially for polluted water in the city. Therefore, in park design, the ratios of vegetated surfaces to water surfaces should be insightfully considered to diversify landscapes and optimize the cooling impact over time.

5 Conclusion

The cooling effect of public parks and the elements that influenced it were investigated. Using multiple data sources, we found the integrated landscapes in all parks with a high share of green and blue areas at various combinative ratios. Aside from that, there are differences in cooling distance, which vary according to the season. During the summer, the active cooling distance is usually between 100 and 200 meters. However, the cooling distance extends outward 400 meters and up to 1,000 meters, depending on park structures and the surrounding background. The tree area inside the park and the arrangement of green space patches are the most critical factors controlling climate regulation. Although the other neighboring elements are insignificant in statistics, they are worthily considered in urban planning strategies to mitigate urban heat islands, such as low urban density, high vegetation area, and vegetation shape complexity. Furthermore, the water surface provides an unsteady cooling effect both within and outside the park. As a result, using water surfaces as part of a heat mitigation plan should be carefully examined to achieve a more appropriate and sustainable strategy for urban heat reduction.

Acknowledgements This work was supported by the Petchra Pra Jom Klao scholarship of the King Mongkut's University of Technology Thonburi (no. 09/2562). We would like to show our gratitude to all the agencies that provided data for this study, including the Thailand Department of Pollution Control (PCD), United States Geological Survey (USGS), European Space Agency (ESA), and Google Earth.

References

1. Amanollahi J, Tzanis C, Firuz M, Makmom A (2016) Urban heat evolution in a tropical area utilizing Landsat imagery. Atmos Res 167:175–182. https://doi.org/10.1016/j.atmosres.2015.07.019
2. Carlson TC, Ripleya D (1997) On the relationship between NDVI, fractional vegetation cover, and leaf area index. Remote Sens Environ 62:241–252. https://doi.org/10.1016/S0034-4257(97)00104-1

3. Estoque RC, Murayama Y (2016) Quantifying landscape pattern and ecosystem service value changes in four rapidly urbanizing hill stations of Southeast Asia. Landscape Ecol 31(7), 1481–1507. https://doi.org/10.1007/s10980-016-0341-6
4. Estoque RC, Murayama Y (2017) Monitoring surface urban heat island formation in a tropical mountain city using Landsat data (1987–2015). ISPRS J Photogramm Remote Sens 133:18–29. https://doi.org/10.1016/j.isprsjprs.2017.09.008
5. Forkel M, Carvalhais N, Verbesselt J, Mahecha MD, Neigh CSR, Reichstein M (2013) Trend Change detection in NDVI time series: Effects of inter-annual variability and methodology. Remote Sensing 5(5):2113–2144. https://doi.org/10.3390/rs5052113
6. Hamada S, Ohta T (2010) Urban Forestry & Urban Greening Seasonal variations in the cooling effect of urban green areas on surrounding urban areas. Urban Forestry & Urban Greening 9(1):15–24. https://doi.org/10.1016/j.ufug.2009.10.002
7. Hamel P, Guerry AD, Polasky S, Han B, Douglass JA, Hamann M, Janke B, Kuiper JJ, Levrel H, Liu H, Lonsdorf E, McDonald RI, Nootenboom C, Ouyang Z, Remme RP, Sharp RP, Tardieu L, Viguié V, Xu D, Daily GC (2021) Mapping the benefits of nature in cities with the InVEST software. NPJ Urban Sustainability, 1(1). https://doi.org/10.1038/s42949-021-00027-9
8. Hereher ME, El Kenawy A (2020) Extrapolation of daily air temperatures of Egypt from MODIS LST data. Geocarto Int 0(0), 1–17. https://doi.org/10.1080/10106049.2020.1713229
9. Karnieli A, Bayasgalan M, Bayarjargal Y, Agam N, Khudulmur S, Tucker CJ (2006) Comments on the use of the Vegetation Health Index over Mongolia. Int J Remote Sens 27(10):2017–2024. https://doi.org/10.1080/01431160500121727
10. Liu HQ, Huete A (1995) Feedback based modification of the NDVI to minimize canopy background and atmospheric noise. IEEE Trans Geosci Remote Sens 33(2):457–465. https://doi.org/10.1109/36.377946
11. Lowe SA (2016) An energy and mortality impact assessment of the urban heat island in the US. Environ Impact Assess Rev 56:139–144. https://doi.org/10.1016/j.eiar.2015.10.004
12. Malarvizhi K, Kumar SV, Porchelvan P (2016) Use of High Resolution Google Earth Satellite Imagery in Landuse Map Preparation for Urban Related Applications. Procedia Technol 24:1835–1842. https://doi.org/10.1016/j.protcy.2016.05.231
13. McGarigal K, Cushman SA, Ene E (2012) FRAGSTATS v4: spatial pattern analysis program for categorical and continuous maps. Computer software program produced by the authors at the University of Massachusetts
14. Mering C, Baro J, Upegui E (2010) Retrieving urban areas on Google earth images: Application to towns of West Africa. Int J Remote Sens 31(22):5867–5877. https://doi.org/10.1080/01431161.2010.512311
15. Nguyen CT, Chidthaisong A (2022) Urban Green Space Inventory using Different Spatial Resolution Satellite Images: Practical notes in Bangkok. The 11th International Conference on Environmental Engineering, Science and Management (pp. 333–340). Environmental Engineering Association of Thailand (EEAT)
16. Nguyen CT, Diep NTH, Diem PK (2021) Factors affecting urban electricity consumption: a case study in the Bangkok Metropolitan Area using an integrated approach of earth observation data and data analysis. Environ Sci Pollut Res 28:12056–12066. https://doi.org/10.1007/s11356-020-09157-6
17. Nurwanda A, Honjo T (2019) The Prediction of City Expansion and Land Surface Temperature in Bogor City, Indonesia. Sustainable Cities and Society, 52(2020). https://doi.org/10.1016/j.scs.2019.101772
18. Santamouris M (2020) Recent progress on urban overheating and heat island research. Integrated assessment of the energy, environmental, vulnerability and health impact. Synergies with the global climate change. Energy and Buildings, 207, 109482. https://doi.org/10.1016/j.enbuild.2019.109482
19. Sohrabinia M, Zawar-Reza P, Rack W (2015) Spatio-temporal analysis of the relationship between LST from MODIS and air temperature in New Zealand. Theoret Appl Climatol 119(3–4):567–583. https://doi.org/10.1007/s00704-014-1106-2

20. Son NT, Chen C, Chen C, Thanh B, Vuong TH (2017) Assessment of urbanization and urban heat islands in Ho Chi Minh City, Vietnam using Landsat data. Sustain Cities Soc 30:150–161. https://doi.org/10.1016/j.scs.2017.01.009
21. Son NT, Thanh BX (2018) Decadal assessment of urban sprawl and its effects on local temperature using Landsat data in Cantho city. Vietnam. Sustainable Cities and Society 36(2018):81–91. https://doi.org/10.1016/j.scs.2017.10.010
22. Srivanit M, Hokao K (2012) Thermal Infrared Remote Sensing for Urban Climate and Environmental Studies: An Application for the City of Bangkok, Thailand. Journal of Architectural/Planning Research and Studies 9(1):83–100
23. Tran DX, Pla F, Latorre-Carmona P, Myint SW, Caetano M, Kieu HV (2017) Characterizing the relationship between land use land cover change and land surface temperature. ISPRS J Photogramm Remote Sens 124:119–132. https://doi.org/10.1016/j.isprsjprs.2017.01.001
24. Tucker CJ (1979) Red and photographic infrared linear combinations for monitoring vegetation. Remote Sens Environ 8(2):127–150. https://doi.org/10.1016/0034-4257(79)90013-0
25. USGS (2016) Landsat 8 (L8) Data Users Handbook (LSDS-1574 version 2.0). In USGS Landsat User Services. U.S. Geological Survey
26. Van De Griend AA, Owe M (1993) On the relationship between thermal emissivity and the normalized difference vegetation index for natural surfaces. Int J Remote Sens 14(6):1119–1131. https://doi.org/10.1080/01431169308904400
27. Weng Q, Lu D, Schubring J (2004) Estimation of land surface temperature-vegetation abundance relationship for urban heat island studies. Remote Sens Environ 89(4):467–483. https://doi.org/10.1016/j.rse.2003.11.005
28. Xu H (2006) Modification of normalised difference water index (NDWI) to enhance open water features in remotely sensed imagery. Int J Remote Sens 27(14):3025–3033. https://doi.org/10.1080/01431160600589179
29. Yan H, Wu F, Dong L (2018) Influence of a large urban park on the local urban thermal environment. Sci Total Environ 622–623:882–891. https://doi.org/10.1016/j.scitotenv.2017.11.327
30. Yi WC, Hu BKH, Myint SW, Feng CC, Chow WTL, Passy PF (2018) Patterns of land change and their potential impacts on land surface temperature change in Yangon, Myanmar. Sci Total Environ 643:738–750. https://doi.org/10.1016/j.scitotenv.2018.06.209

The Resilience of Structures in Times of Climate Change

Nikola Bisevac and Stefan Schäfer

Abstract The paper examines the structural interrelationships and resilient solutions commonly encountered in building construction projects during hazardous situations. The content of the paper will address design fundamentals and interrelationships of resilient building components. Further, the paper will provide insight into a module order, structural design, robustness, and building technology of the resilient structures. The final point of the paper will show the implications of resilient concepts, risk management, and determination of the hazards on human safety in times of climate change.

Keywords Resilience of constructions · Adaptability to climate change · Extreme engineering

1 The Concept of Resilience

1.1 Introduction

Resilience is the ability of systems to withstand disruptions. Resilient constructions remain fundamentally functional in crises. They are resistant to technical, environmental, and economic partial failures or disruptions through flexibility or redundancy of their systems, thus providing robust conditions for a long-term use [1].

Mutability of structures is understood as the ability of a system to adapt its construction actively and rapidly to temporally and unpredictably changing tasks from its own substance (adaptability) in conjunction with the ability to evolutionarily develop the structures in the face of temporally constant or longer-term predictably changing requirements from its own substance [2].

N. Bisevac (✉) · S. Schäfer
Institute for Constructive Design and Building Construction, Technical University of Darmstadt, Franziska-Braun-Straße 3, 64287 Darmstadt, Germany
e-mail: bisevac@kgbauko.tu-darmstadt.de

V. T. Ha et al. (eds.), *Proceedings of the 8th International Conference on Sustainable Urban Development*, Advances in 21st Century Human Settlements,
https://doi.org/10.1007/978-981-99-8003-1_10

There are many definitions and concepts of resilience of building structures and component connections. What all these different concepts have in common however, is that resilience is the ability or capacity of a system to maintain its functionality, or at least to regain it in the short term, even while facing unusual or unexpected stress situations. Resilience can refer to natural as well as technical systems [2].

In the following research approach, the resilience of a structure is to determine how susceptible to damage or resistant this structure is, when exposed to a natural event. It is largely determined by object characteristics, for example, the construction method, materials used, protective devices, etc. The risk can be reduced by improving the resistance of a structure to the effects of these natural hazards by regular maintenance (servicing, condition assessment, refurbishment), and by protective and damage mitigation measures before, during, and after the event.

Building structures and component connections are exposed to a wide variety of environmental effects. As a rule, these environmental effects on the building are adequately considered in the legal regulations. In addition to this, the approach of this paper should consider the extreme events, the intensities of which significantly exceed the already regulated impacts. The extreme events considered regarding wind, heavy rain, hail, snow, and floods are associated with greater hazards to people's health and risks of damage to property, from which higher requirements can be derived for the resistance of building structures [3]. The relevance of this topic is increasing due to the climate change that is already occurring and the associated increase in extreme weather events which, however, varies from region to region. Building structures and component connections must be adapted and prepared in this respect (adaptation).

The objective of this research is to improve the resistance of buildings and structures to current and future natural hazards at the site. This is to achieve, among other things

a protection of persons
a protection of material assets
the safeguarding of usability as well as the planned service life
the limitation of insurance needs
the compliance with the planned life cycle costs.

The evaluation of the resistance of the building structures and component connections to wind, heavy rain, hail, snow, and floods should be carried out taking into account the following (Fig. 1):

the nature and extent of current and future hazards at the site under consideration
resistance of the building to the concrete hazard. [4]

Fig. 1 Resilience of structures and environmental effects

2 Resilience in the Context of Extreme Weather Events

The concept of "resilience" receives special attention in the context of adaptation strategies to climate change and in particular dealing with extreme weather events and the resulting risks to building constructions and society through disruption or damage. Building resilience is in this context as a holistic response to new challenges, which goes beyond mere hazard prevention. Another characteristic of resilience is that the view is taken by adapting to future events. Thus, damage, processes, and experiences from the past must be analyzed and reflected upon. Building resilience refers to both physical structures as well as to systems. In the urban context resilience is an overarching concept, which covers all phases of dealing with an extreme event. In order to create resilient structures, adaptation processes must also be initiated in the built environment. A municipality must be aware of the fact that future events may be milder, but also stronger, than those experienced so far. This variability must be taken into account through a range of different scenarios. These scenarios must not only reflect the increasing frequency of extreme events. Also, other environmental changes that may affect the vulnerability of a city, its population, and its built environment in the long term should also be taken into account. Examples include land use changes, infrastructure replanning, and technological and demographic changes [5].

By consciously addressing the topic of extreme events in the context of urban resilience strengthening, the awareness and learning capacity of all actors involved should be increased. The planning view through the resilience enables a comprehensive perspective and a more comprehensive analysis scheme than, for example, in classical extreme event risk management. In summary, a resilient structure is so designed and managed in such a way that the impacts of extreme weather events are as low as possible.

The goal of building resilience is to be able to respond to challenges in an adaptive and anticipatory manner and, at the same time, to be able to derive learning and stabilization processes and thus adaptation options from past crises. Resilient

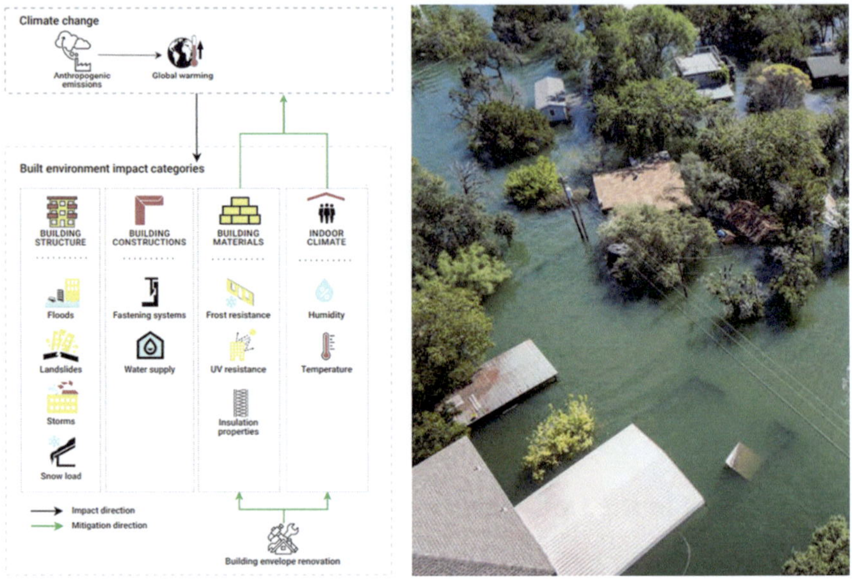

Fig. 2 Resilience, climate change, and built environment impact categories

construction can anticipate crises, absorb them, recover from them, and successfully adapt to the new circumstances in order to shape the structure in the future in a way that minimizes damage in the extreme event. In this context, resilience does not represent a desirable end-state. Rather, resilience can be understood as a dynamic process in which the reflection of past events and continuous learning processes by all actors leads to a transformation of built structures—accompanied by dynamic behavioral changes on the part of the actors, for example through improved preparedness and crisis management [6] (Fig. 2).

3 Main Principles of Resilience

Starting from a cyclical view (from one extreme event to the next), it remains that four main features are essential for the characterization of a resilient structure:

Ability to learn: In the aftermath of an event, it is important to learn from the crisis. This requires reflection with regard to patterns of damage and functional failures, among other things. Even communities that have not yet been affected by natural hazards can (and should) learn from the experience of other cities. The ability to learn in the sense of resilience also means, that actors recognize whether their objects/facilities are resilient or not. The ability to learn therefore also has a lot in common with raising awareness and with the development of institutional knowledge.

Adaptability: A resilient structure is characterized by its ability to adapt, e.g., by integrating knowledge gained from reflection of past (extreme) events into its planning and processes or in the participation/networking of the relevant actors among each other, in order to strengthen the protection and crisis management for the next event [7].

Robustness: Previous heavy precipitation events have produced enormous damage. Protecting people and critical facilities (and buildings as a whole) is therefore a central aspect of resilience. A resilient construction is able to withstand extreme events in such a way that damage is minimized.

Coping capacity: The analysis of risks and past damages makes it possible to take precautions. Due to the ubiquity of natural disasters and the many other uncertainties that are associated with, a total hundred percent protection against hazards is not possible. Therefore, not just resistance, but also to cope with the event accordingly that there are less serious failures and deformations. In order for this achievement, good organization and good crisis management are necessary, which includes both the local government (if necessary the district administration), the communal departments as well as the citizens and professionals [8] (Figs. 3 and 4).

1. Insufficient depth of foundation and soil located below the flood level.
2. Insufficient connection between the walls or absence of a frame structure to increase the stiffness.
3. Unsupported gable wall and exterior walls.

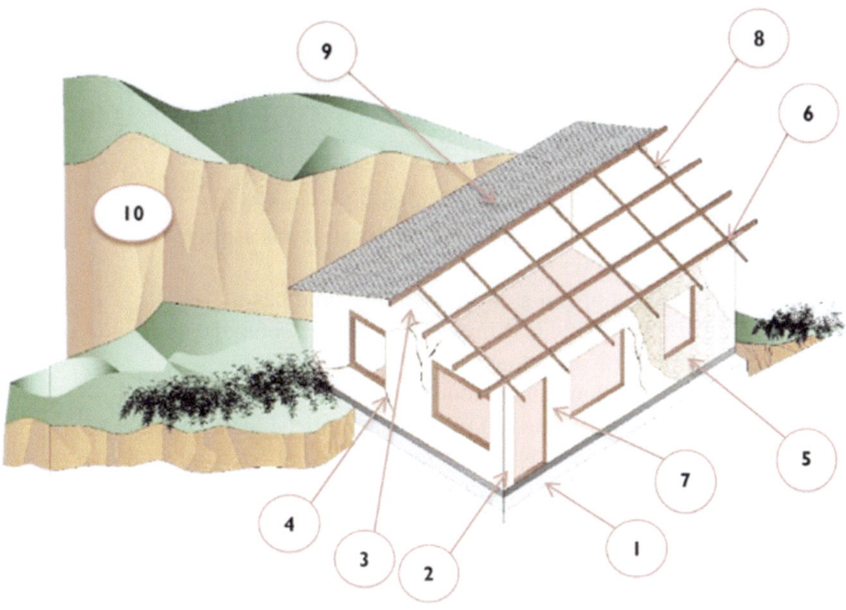

Fig. 3 Weaknesses of a non-engineered house

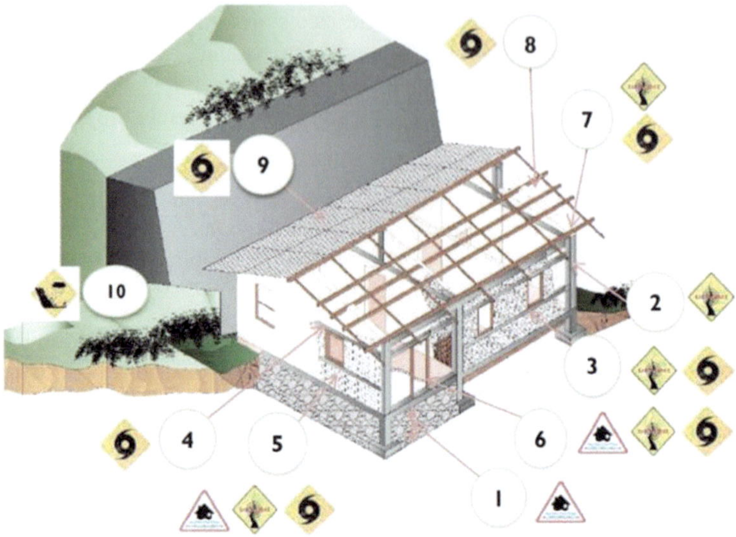

Fig. 4 Features of hazard resilient house

4. Diagonal cracking at the corners of the openings due to lack of sill and lintel beams.
5. Building material of inferior quality, lacking in required strength and performance.
6. Inadequate connection of the roof truss to the structure.
7. Inadequate anchorage of door and window frames frame to the structure.
8. Inadequate connection of gable to structure and the roof covering to the gable.
9. Inadequate anchorage of roof panels and/or roofing tiles to the roof framing.
10. Deterioration of existing soil conditions.

1. Increased depth of the foundation and raised ground level.
2. Truss construction with reinforced concrete columns supports.
3. Walls with adequate framing.
4. Bracing of walls at openings with lintel/threshold beams.
5. Use of good quality building materials in all the components used to provide the required strength and other properties.
6. Properly anchored door and window frames to the structure.
7. Properly erected and connected roof structure to the main structure.
8. Properly connected gable wall to structure and roof covering with gable wall.
9. Proper connection of roof covering to roof structure.
10. Minimal disturbance of the soil and supported cuts to preserve slopes.

4 Resilience of Components and Structures

4.1 Robustness as the Basis for Building Resilience

In common sense, the word "robust" is used in the context of insensitive or stable. It is assumed that small causes do not produce a large effect on robust persons or things. If this context is applied to construction, society expects robust structures whose damage or failure is limited to an extent that is in reasonable proportion to the cause. This means, for example, that local damage must not lead to failure of the entire structure. If this understanding is transferred to large buildings, they can only be considered robust if the failure of a load-bearing element does not lead to the collapse of the complete building. In this context, columns are the most critical load-bearing elements in building construction. They carry high loads and, due to their slenderness, are particularly vulnerable to impacts such as collision or explosion. Therefore, it is senseful to design large buildings also for the possible failure of a column. Columns occur in most cases in so-called skeleton buildings, the majority of which are made of reinforced concrete. This means, for example, that span widths or slab thicknesses do not occur in arbitrary orders of magnitude, but are limited to certain bandwidths. The failure of a column is a dynamic event, since the structure loses a load-bearing element within a short time. The dynamic load-bearing behavior plays an important role, since the possible failure time of a column is often less than one-tenth of the intrinsic period of the structure after failure. In the linear-elastic case, this means that the effects of a dynamic structural analysis can be up to twice as large as those of a static structural analysis. A structure failure is a rare event compared to storms or large snowfalls and can therefore be treated as an exceptional design situation. Similar to other exceptional design situations, e.g. earthquakes, the design values of the bearing resistances must therefore not be used for the verifications, but higher values, in case the characteristic bearing resistances, should be calculated. The correctness of this assumption must not be checked by means of a probabilistic analysis, but it can be assumed that the safety level for this hazard pattern is at an acceptable level. The structural analyses for structure failure could be performed for example using the finite element method, taking into account both geometric and material nonlinearities. In order to obtain realistic results from structural analyses, it is necessary to represent the load-bearing behavior of reinforced concrete as accurately as possible. In addition to the partially linear-elastic behavior of reinforced concrete, phenomena such as the flow of the reinforcing steel and the cracking or breaking of the concrete could also be simulated within the framework of the structural analysis. Shear force failure of the concrete and cracking of the reinforcing steel cannot be simulated directly in the structural analysis and therefore had to be excluded by a downstream design [9].

To verify the structural models set up, physical tests should be recalculated on three different load-bearing elements: a beam, a line-supported slab, and a point-supported slab. All three tests could be simulated with sufficient accuracy. Furthermore, it is important that a static structural analysis underestimates the effects after a column failure and that damping as well as distortion rate dependent material behavior can be neglected [10].

Comparative calculations should be performed between a dynamic, twofold nonlinear structural analysis and a simplified static, linear-elastic structural analysis. A static, linear-elastic structural analysis is only permissible if shear force failure of the slab can be excluded. In addition, the design after such a structural analysis must always be carried out using the design values of the load-bearing resistances. Although this type of structural analysis is less time-consuming than a dynamic, nonlinear structural analysis, the latter can still be more economical because significantly less reinforcement results from the design [11].

Overall, the load-bearing behavior of reinforced concrete skeletal structures after a building failure can be described as robust. Such structures can also withstand hazards without changing their geometry, but this requires an additional reinforcement of 5 to 20%. It is therefore reasonable to assume that structures of this type also exhibit similarly good load-bearing behavior under other extreme hazard conditions [11].

5 Strategies for the Resilience of the Structure

This section discusses five strategies that can be used to improve the resilience of structures. Not every strategy is evenly good for every structure. Therefore, it is the responsibility of the engineer and architect to decide which strategy or combination of strategies is appropriate for the building.

The first requirement serves to ensure sufficient deformation capacity of the structure. This means that even though the effects have been calculated using linear elasticity theory, for example, and the structure has been designed for these effects, it should not suddenly fail even if these effects occurred. In such a case, the structure should first deform strongly before it fails, in order that the upcoming danger can be recognized in time by the users. Therefore this is a strategy that increases the resilience of the structure without having a direct influence on the verification of the safety [12].

The strategy of continuity is not useful in all cases. A continuous beam may serve as an example. After the failure of one of its supports, the continuity of the bearing element can lead to the fact that even the double span can be bridged or that the entire bearing element fails. Consequently, this is not a good prescriptive rule, since the utility of this rule must always be verified by a structural engineer. Although the application of prescriptive rules is often easy for the planner, there are few rules that are beneficial for every structure. Therefore, there is a risk that the stubborn application of such rules may actually degrade the resilience of a structure. For this reason, engineers should always re-examine prescriptive rules to see if they make sense for each structure and, if necessary, see them only as ideas for forming an alternative load path.

The second strategy is the alternate load path (direct design). In the application of this strategy, it is assumed that one or more load-bearing elements suddenly fail. The reason for this failure is irrelevant and it is assumed that other parts of the structure are not damaged by the cause of this failure. The objective of the subsequent structural analysis and design is to show that the loads of the failed load-bearing element can be taken over by other parts of the structure. In other words, the structure is able to find another alternative load path after the failure of one of its elements without failing. Examples of such requirements are the hazard cases of a column failure in buildings with 5 to 15 stories or the failure of a cable in certain bridge structures. A disadvantage of this strategy is, that the structural analysis is often quite costly, since the dynamic and nonlinear structural behavior often cannot be neglected for such hazard patterns [13].

The third strategy is event control. The idea of this strategy is to keep an action away from a structure that would cause disproportionate damage. The aim is to mitigate the effect or prevent it from hitting the structure in the first place.

The fourth strategy is local reinforcement (direct design - local resistance). In this strategy, individual load-bearing elements are designed for increased actions. These effects can be determined, for example, for an assumed explosion or impact. When using this strategy, it is difficult to determine the necessary design level of the action. The reason for this is that any kind of impact cannot be predicted using statistical data. Furthermore, local reinforcements can also worsen the aesthetics of the structure. The advantages of this strategy compared to the alternative load path strategy are the lower design and construction costs, since both are limited to a local area of the structure.

The fifth strategy is compartmentalization. In this strategy, the structure is divided into individual sections. This is to prevent a local failure from spreading to the entire structure as a result of a chain reaction. The sections can be formed within the structure by joints or articulations. When joints are used however, it must be borne in mind that large forces can act on the joint during the failure of a section, but these must not lead to a failure of the intact sections. Overall, the strategy of sectioning can be considered to be easily implemented and cost-effective. But it should be emphasized at this point, that loss of life due to failure of a section of the structure must be accepted as a risk.

All five strategies aim to limit the potential failure of a structure to a point where it is reasonably proportionate to its cause. The goal of limiting a possible failure of the structure even further, in order that the serviceability is not limited, is only reasonable in a few cases. The reason is that serviceability can often only be achieved at great financial expense and with aesthetic restrictions [14].

6 Risk Management and Determination of the Hazard

6.1 *Resilience as a Result of the Risk Management*

Vulnerabilities and risk management focus on managing the potential risks posed by natural hazards. These risks can vary enormously depending on which building structure, population, or even study areas are considered. In order to be able to develop a successful hazard risk management, two levels of risk perception have to be distinguished: risk analysis and risk assessment. These two processes deal with different information [15].

In the risk analysis, the available information is analyzed exclusively objectively, aggregated or summarized, and classified among each other by means of fixed threshold values. Risk analysis is followed by risk assessment. This process is to be separated clearly from the analysis and contains a subjective evaluation based on the target system and planer decisions.

The most important prerequisite for strengthening the resilience of the structure is to identify potential hazard areas and vulnerabilities (especially critical infrastructures) during severe weather events in the neighborhood in order to subsequently determine the risk in the respective area under consideration by intersecting these data [16].

For example, heavy rainfall events can very quickly lead to rising water levels in water bodies, backups in the sewage system due to the sudden influx of water masses, as well as flooding and/or flash floods and thus to considerable hazards. In addition, there are influencing factors such as the topography, the degree of sealing, or the building density, which can affect the runoff dynamics and impoundment. They influence not only the flow velocity but also the amount of runoff water, which results in greater impoundments both in extent and depth. In cities and urban areas, vulnerability to heavy rainfall is particularly high as many people, goods and material assets, sensitive facilities, and buildings and infrastructure are found in a confined space.

A large number of buildings and infrastructures also exhibit a high degree of systemic criticality due to their enormous importance for the functionality of the entire urban system. Criticality can be seen here as the relative measure of the importance of an infrastructure in terms of the consequences that a disruption or functional failure has for the security of society's supply of important goods and services.

In the next step the risk analysis is followed by the risk assessment as an independent component, which includes in particular an evaluation of the various potentially affected objects of protection. This is a normative process in which the results of the risk analysis are evaluated by the municipality. Finally, the risk assessment serves as a basis for the formulation of priorities for action as well as for the prioritization of flood prevention and protection measures [17] (Fig. 5).

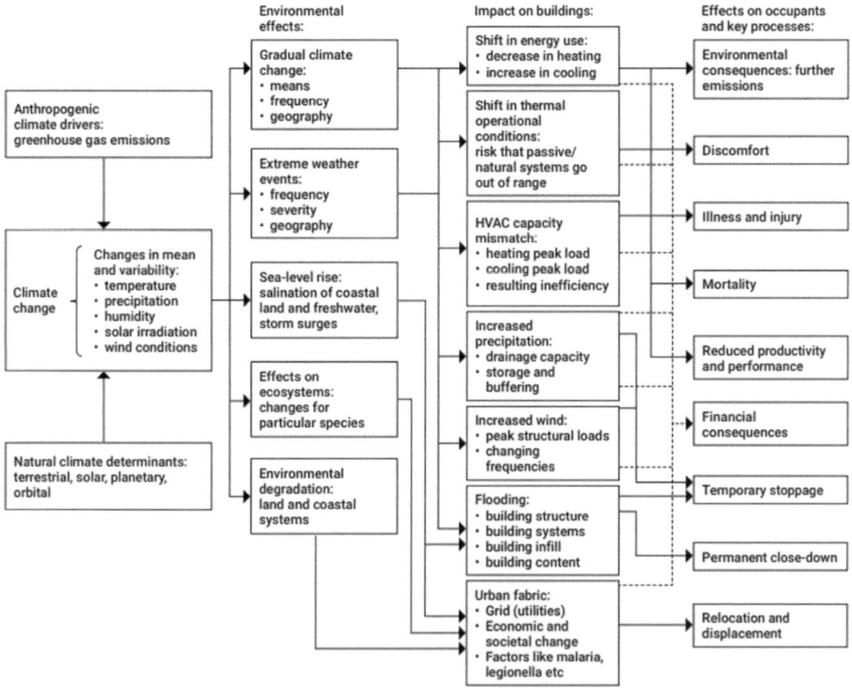

Fig. 5 Challenging impacts of a changing climate on the built environment

6.2 *Determination of the Hazard*

In order to determine a hazard or risk, analysis of the effects of extreme events in the urban context must be first drawn up in hazard maps. Hazard maps can be used to identify the areas of a municipality that could be for example flooded during a heavy rain event. They usually contain information on flood extension, flood depth, and/or flow paths and velocities for a specific design. The necessary modeling is usually contracted out to specialized engineering firms by cities with a less specialized administration and often a lack of human resources. Based on the determined flow paths, it is easy to show where the water flows into and through the settlement area on the surface. Depending on the flow velocity and the topographical conditions, these so-called runoff paths can also be used to assess a possible erosive hazard.

For a city-wide investigation of areas, structures, and infrastructures that are particularly affected by a hazard, meaningful threshold values are required to classify the risk. This is because it is often underestimated that even for example small flood depths of surface water can penetrate into ground-level or low-lying building openings, such as cellar shafts. In common practice, very different approaches and corresponding threshold values can already be found. The setting of these threshold values is influenced by various dimensions. For example, curbs usually have a height

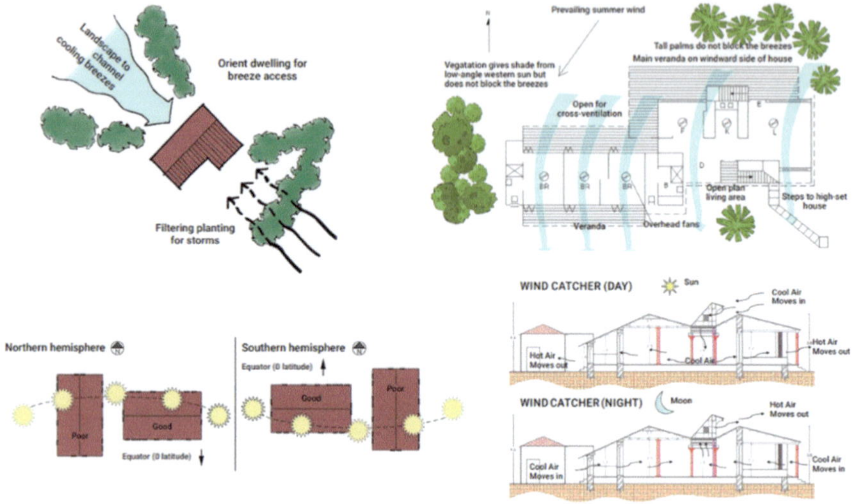

Fig. 6 Nature-based hazards and resilience

of 10–30 cm, multifunctional housings for power distribution and telecommunications technology usually have a height of 20–30 cm and conventional cars usually have a fording depth of approx. 40–50 cm. However, it is certain that from a water depth of 50 cm, significant material damage and, from 1 m, danger to human life must be expected. These examples can be used as an aid to approach these threshold values [18] (Fig. 6).

7 Case Studies

As climate change continues to pose significant threats to cities across Europe, building structures and their resilience to extreme weather events have become increasingly important. In this context, this paper will be exploring five case studies of urban development in Europe that showcase the resilience of building structures in the face of climate change. From innovative construction methods to the use of renewable energy sources, these case studies highlight different strategies that cities can adopt to mitigate the impact of climate change on their building structures. By examining these examples, we can gain insight into how cities can design and adapt their building structures to meet the challenges of a changing climate, while simultaneously providing safe, comfortable, and sustainable spaces for their inhabitants (Fig. 7).

Following case studies will be researched:

Coastal development and flood defense measures in northern Germany
Green roof initiative in Switzerland

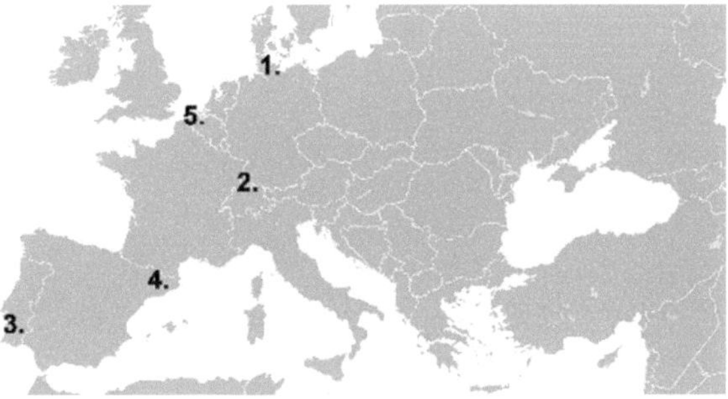

Fig. 7 Map of case studies

Water retention landscape in Portugal
Green infrastructure and biodiversity plan as a solution for heat island effect in Spain
Artificial dikes for flood protection and water filtration in Belgium.

7.1 Case Study 1: Coastal Development and Flood Defense Measures in Northern Germany

The coastal area of Timmendorfer Strand, situated along the Baltic Sea, is at risk of storm surges due to its location, which is less than 3 m above sea level. The current flood defence system, which is a natural beach ridge, was found to be insufficient to protect the population and the economic assets from the projected sea level rise of 45–65 cm in hundred years. However, the local community was hesitant to implement coastal flood defence measures due to the reliance on tourism and the fact that there had not been a catastrophic storm surge for 120 years [19].

To address this challenge, a socio-economic assessment was carried out to evaluate the potential damage and the need for improved coastal flood defence measures. Possible coastal defence measures were then discussed in stakeholder workshops under various sea level rise scenarios. The outcomes of these workshops were used to launch an ideas competition in the third step of the process, which enabled the community to participate actively in the decision-making process [20].

The final decision involved building a sheet pile wall integrated into the natural beach ridge, but lower than initially proposed, to maintain the scenic views and accessibility to the beach for tourists and locals. The success of the project can be attributed to the authorities' inclusive and participatory approach to decision-making, which took into account the values and concerns of the local community.

The success of the project in Timmendorfer Strand was due in part to the community's ability to finance some of the additional expenses. These expenses were necessary because the solution that was implemented had to be tailored to the needs of the tourist town. For example, glazed retention walls were constructed near cafes to preserve views of the ocean, and the beach promenade was rebuilt, combining architecture and landscaping.

Fig. 8 Timmendorfer Strand and coastal flood defence measures

Quantification of costs and benefits was performed for two different scenarios: one with a sea level rise of 0.35 m and another with a rise of 0.55 m. The analysis revealed that the estimated benefits outweigh the estimated costs of the measures, with the benefits ranging from four to eight times higher than the costs. The primary benefit of implementing the measures is the avoidance of damage from storm surges. Additionally, the implementation of the measures is expected to attract more tourists, resulting in substantial benefits [20] (Fig. 8).

7.2 Case Study 2: Green Roof Initiative in Switzerland

The city of Basel in Switzerland established a program to encourage the use of green roofs, which serve as both a mitigation measure to save energy and reduce emissions, as well as an adaptation by lowering indoor temperatures and absorbing rainwater to reduce flood risk. Green roofs also offer benefits for biodiversity and social aesthetics. Climate projections suggest that by the 2050s, the temperature in this region could increase by 2.1 °C in winter and 2.6 °C in summer, and extreme rainfall events are expected to become more frequent and severe [21]. The green roof initiative aimed to increase the coverage of green roofs through a combination of financial incentives and building regulations. In the early 1990s, Basel implemented a law supporting energy-saving measures, and green roofs were promoted through incentive programs funded by the Energy Saving Fund, which is financed by a 6% levy on energy bills for all customers in the canton of Basel. In 2002, an amendment to Basel's building and construction law required all new and renovated flat roofs to be greened and designed to improve biodiversity. As a result, Basel now has one of the highest per capita areas of green roofs in the world [22].

Installation of green roofs has become a common practice for building owners, who do not object to it. While green roofs are generally more expensive by 12–16% over their lifespan compared to conventional roofs, this is mainly due to higher initial costs, with maintenance expenses being similar to those of traditional roofs. As such,

Fig. 9 Green roofs as mitigation measure to save energy and reduce emissions

a 22% reduction in construction costs for green roofs through subsidy programs is deemed adequate to level the playing field for investors in green roofs and traditional roofs [23]. This makes green roofs not only an environmentally friendly option but also a financially viable one. Green roofs offer various direct advantages and co-benefits, such as reducing indoor temperatures (up to 6 °C) leading to energy savings, stormwater absorption and runoff delay, which lowers the risk of flooding in heavy rains, decreased temperature in densely populated areas, and provision of urban wildlife habitats and migration corridors, resulting in a more visually appealing urban landscape (Fig. 9).

7.3 Case Study 3: Water Retention Landscape in Portugal

Tamera is a farm spanning 155 hectares located in Alentejo, the most arid region of Portugal. This region has been plagued by soil erosion and desertification, which are likely to worsen with future climate change. However, Tamera has successfully countered these trends through the implementation of a water retention landscape (WRL) scheme, which has enabled the farm to achieve self-sufficiency in water and food. This has also increased Tamera's resilience to climate change and extreme events, such as droughts, water scarcity, and floods. The WRL system restores the full water cycle, whereby rain is retained by vegetation or water bodies and recharges the groundwater, eliminating rainwater runoff. Moreover, the retention area functions in place of the vulnerable humus layer, preventing landslides and floods through its high water-absorbing capacity. In Tamera, 28 lakes and retention spaces were constructed, and the water bodies' area increased from 0.65 hectares to roughly 8.35

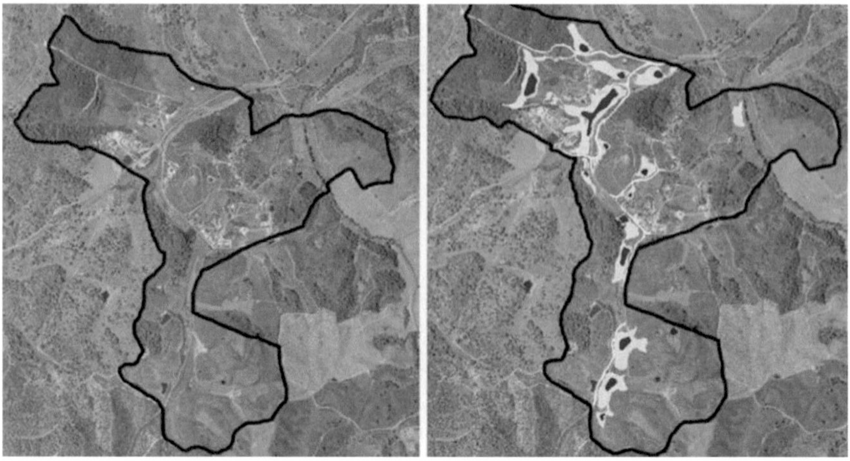

Fig. 10 Tamera ecovillage water retention landscape, Alentejo Portugal

hectares between 2007 and 2015. These interventions were integrated with other measures, including terrace construction, channel creation, and rotational grazing land [24].

As a result, Tamera is now equipped to manage even heavy, continuous rainfall. Located at the highest point of the valley, this expansive retention area utilizes gravity for irrigation within the catchment, without requiring additional energy for pumping. The direct advantages of this solution include decreased vulnerability to climate change and enhanced water management, such as reduced irrigation demands, better water quality, improved water availability, and stabilization of the groundwater table. There are also numerous co-benefits, such as increased carbon sequestration, greater agricultural productivity and diversity, recreational value of lakes, enhanced biodiversity, more visitors due to new water-based activities, and an improvement in the quality of life for local residents. For instance, the area of woodland has grown from 9.35 ha to 20.00 ha by converting natural grasslands. This has resulted in a yearly overall increase in carbon storage of 9.5% between 2006 and 2014. Additionally, it is noteworthy that the cost of water is expected to rise in the near future, making this type of resilient farming system very valuable [25] (Fig. 10).

7.4 Case Study 4: Green Infrastructure and Biodiversity Plan as a Solution for Heat Island Effect in Spain

The climate challenges facing Barcelona include rising temperatures, decreased rainfall, and an increased frequency of extreme weather events such as droughts and heatwaves. The high urban density of the city contributes to the heat island effect, which exacerbates these challenges. Barcelona has made a commitment to become a global model of a sustainable city in response to the challenges related to climate change. For many years, the city has placed a strong focus on planting and managing trees. Trees can mitigate the urban climate in two ways: by reflecting sunlight and transpiring through their leaves to lower air temperatures, and by providing shade to reduce surface temperatures and protect people from the sun. Additionally, trees can help prevent local flooding by reducing stormwater runoff. In addition to their climate-related benefits, city trees can also provide co-benefits such as removing air pollutants, storing carbon, reducing noise pollution, regulating humidity, and promoting ecological connectivity. Moreover, they can provide a habitat for urban biodiversity and create a pleasant urban landscape [26].

Barcelona implements the Green Infrastructure and Biodiversity Plan 2020 (BGIBP) to connect different parts of the city with green infrastructure. The Tree Master Plan for 2018–37 complements the BGIBP objectives and proposes measures to increase the number of trees and improve their resilience to climate change [27]. These measures include the use of resilient tree species, diversification of tree species, increased use of runoff water for irrigation, automatic irrigation systems, and water leak controls. Despite having a small amount of green space per person, Barcelona has more street trees than many European cities. The budget for the Tree Master Plan is EUR 9.8 million per year, with EUR 8.6 million per year already allocated to tree management. The remaining EUR 1.6 million per year will be invested in improving soil and water management [28] (Fig. 11).

Fig. 11 Green infrastructure and biodiversity plan in Barcelona

7.5 *Case Study 5: Artificial Dikes for Flood Protection and Water Filtration in Belgium*

The Hedwige and Prosper Polders are situated in a low-lying reclaimed land area, vulnerable to storm surges that threaten the Flanders coast, including Antwerp. The Sigma Plan, a comprehensive initiative to protect land along the Scheldt estuary, includes the Hedwige-Prosper Polder project. The project involves 'depoldering', the process of removing outer defences of the polders and opening up areas to the tides, while moving dike protection inland to provide room for water during tidal surges. The new dikes will protect the hinterland from flooding, and a network of creeks will be established in the polders to simulate the natural wetland. This project, combined with the adjacent Saeftinghe wetlands area, will create a large, cross-border, brackish intertidal area of approximately 4,200 hectares. The construction of the new inland dike on the Belgian side was completed in 2016, and work on the Hedwige Polder started in 2019 and took around three years to complete [29].

In addition to protecting against floods, the project is expected to bring about other advantages, such as enhancing water quality, expanding the natural wetland area, and providing recreational opportunities. However, the project could also result in losses, such as the loss of part of the cultural landscape for farming, and the removal of recreational opportunities, living space, and agricultural land. The majority of the project costs were covered by the Belgian Sigma Plan, as the project is vital to meeting Belgium's integrated water management objectives, including flood protection and shipping accessibility [30] (Fig. 12).

8 Summary

In many cities and municipalities, integrated concepts have been established as planning and control instruments for the entire city, with an integrated and participatory approach. The resilience concepts for areas of development funding should be supplemented with the following aspects regarding to:

- interaction with disaster control, disaster prevention, and health,
- socio-spatial monitoring and risk studies as a rule,
- temporary solutions and experimental spaces,
- measures to avoid, reduce, and adapt to existing and future risks,
- functional capacity, land reserves for resilience, and critical infrastructures.

This planning should be process-oriented, transparent, reversible, and open-ended, with goals to be agreed upon and evaluated between the stakeholders involved. To implement resilience strategies, a multidisciplinary body is needed for steering and coordination. In addition, existing processes for increasing resilience can be linked, for example in the context of smart city concepts or climate adaptation measures [31].

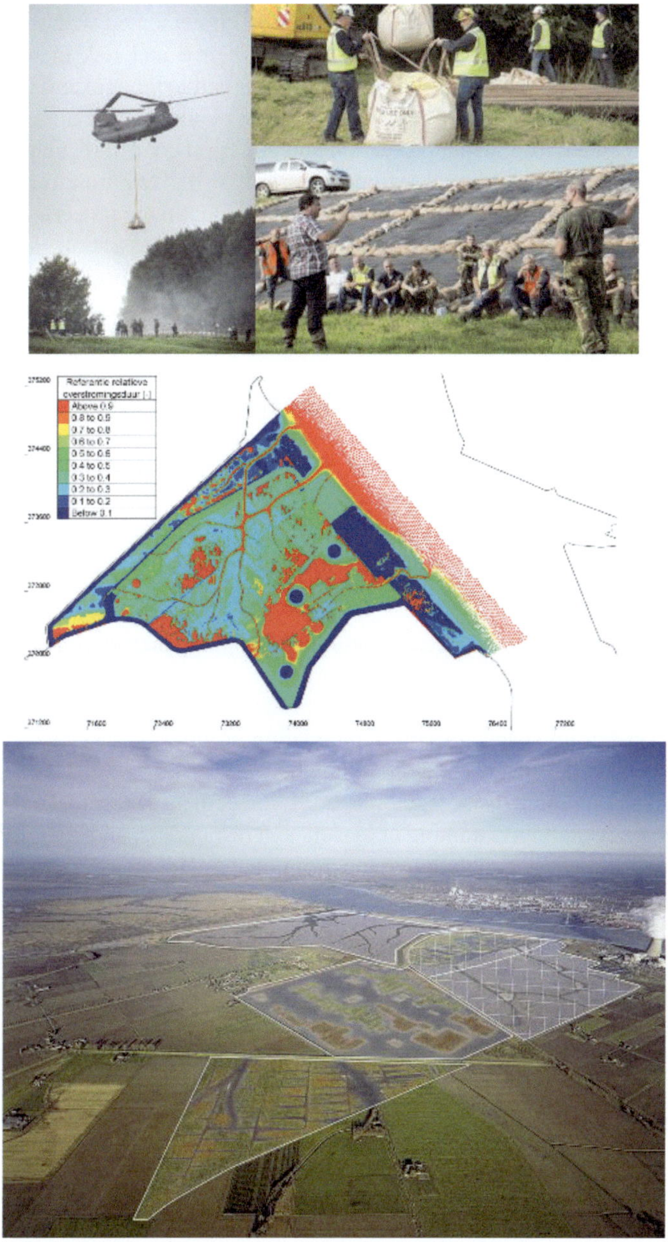

Fig. 12 Hedwige-prosper polder project

The goals for building resilience can largely be implemented with existing planning law. It is recommended that aspects of resilience and risk management be anchored as planning objectives. It should be examined to what extent further flexibilization and mixing objectives can be anchored in planning and building code law, in harmony with environmental protection and emission control regulations.

In addition to the citywide and regional perspective, this requires greater differentiation at the neighborhood level. According to their function as a steering instrument for the public sector, resilience concepts should also be used as a strategic framework for self-organization in neighborhoods. For this purpose, it is important to communicate desirable visions of the future into which concrete measures can be integrated. These should be developed in collaborative formats with the everyday experiences of all population groups [32].

Integrated risk and crisis management requires better linkages between prevention, preparedness, response, and recovery. This means taking greater account in the prevention of the lessons learned from crisis management and thinking about prevention in post-disaster reconstruction.

For the implementation of risk-informed building development, integrated risk and crisis management offers numerous tools and approaches that can contribute significantly to urban resilience [33]. These include:

- Conduct regular risk assessments, including mapping, analysis, and evaluation of local hazards, exposures, and vulnerabilities to increase risk awareness
- Ensure risk-informed planning and appropriate resources, taking into account the different needs and capacities of all population groups
- Learning from past events and rebuilding better
- Make decisions about the acceptability and management of residual risks, in terms of emergency planning, training of leaders and responders, and civic engagement
- Early warning of the population in the event of an incident and, where necessary, adaptation of local crisis management structures in order to be able to act across disciplines or localities
- Strengthening learning networks and exchange formats at the municipal level to optimize structures and processes.

References

1. Woods DD, Engineering R (2017) Concepts and precepts. CRC Press, United Kingdom
2. Shaw R (2014) Building resilient urban communities. Emerald Group Publishing Limited, United Kingdom
3. Resilience and risk: methods and application in environment, cyber and social domains. Netherlands, Springer Netherlands (2017)
4. Community resilience under the impact of urbanization and climate change: cases and experiences from Zimbabwe. Cameroon, Langaa RPCIG (2019)
5. Managing the risks of extreme events and disasters to advance climate change adaptation: special report of the intergovernmental panel on climate change. United States, Cambridge University Press (2012)

6. Goodman A, Mesa N (2021) Collaborating for climate resilience. United Kingdom, Taylor & Francis
7. De Florio V (2012) Innovations and approaches for resilient and adaptive systems. United States, Information Science Reference
8. Curtis DR (2016) Coping with crisis: the resilience and vulnerability of pre-industrial settlements. Taylor & Francis, N.p.
9. Knoll F, Vogel T (2009) Design for robustness. Switzerland, International Association for Bridge and Structural Engineering
10. Practical guide to structural robustness and disproportionate collapse in buildings. United Kingdom, Institution of Structural Engineers (2010)
11. Starossek U (2018) Progressive collapse of structures. ICE Publishing, A Division of Thomas Telford Limited, United Kingdom
12. Resilient structures and infrastructure. Germany, Springer Nature Singapore (2019)
13. Attoh-Okine NO, Engineering R (2016) Models and analysis. Cambridge University Press, United Kingdom
14. Uncertainty in mechanical engineering: Proceedings of the 4th international conference on uncertainty in mechanical engineering (ICUME 2021), June 7–8, 2021. Switzerland, Springer International Publishing (2021)
15. Alibašić H (2022) Strategic resilience and sustainability planning: management strategies for sustainable and climate-resilient communities and organizations. Springer International Publishing, Switzerland
16. Resilience M (2021) Adaptation and transformation in contexts of change. Oxford University Press, United States
17. Maintenance, safety, risk, management and life-cycle performance of bridges: proceedings of the ninth international conference on bridge maintenance, safety and management (IABMAS 2018), 9–13 July 2018, Melbourne, Australia. United States, CRC Press (2018)
18. Measuring Vulnerability To Natural Hazards: Towards Disaster Resilient Societies. Japan, TERI Press (2007)
19. Klimawandel in Deutschland: Entwicklung, Folgen, Risiken und Perspektiven. Deutschland, Springer Berlin Heidelberg (2016)
20. Gerstengarbe F-W, Welzer H (2013) Zwei Grad mehr in Deutschland: Wie der Klimawandel unseren Alltag verändern wird. Fischer E-Books, Deutschland
21. Green roof retrofit: building urban resilience. Deutschland, Wiley (2016)
22. Climate change adaptation manual: lessons learned from European and other industrialised countries. Vereinigtes Königreich, Taylor & Francis (2014)
23. Maldonado E (2014) Environmentally friendly cities: proceedings of Plea 1998, passive and low energy architecture, 1998, Lisbon, Portugal, June 1998. Vereinigtes Königreich, Taylor & Francis
24. Dregger L (2010) Tamera: ein Modell für die Zukunft. Verlag Meiga, Deutschland
25. Holzer S, Dregger L (2012) Desert or paradise: restoring endangered landscapes using water management, including lake and pond construction. Vereinigtes Königreich, Chelsea Green Publishing
26. The Sustainable City XIII. N.p., WIT Press (2019)
27. The green city and social injustice: 21 tales from North America and Europe. Vereinigtes Königreich, Taylor & Francis (2021)
28. Nordsee Und Klimawandel: Warnstufe Code Red. N.p., Satzweiss com (2021)
29. Gourbesville P (2022) Advances in hydroinformatics: models for complex and global water issues—Practices and expectations. Singapur, Springer Nature Singapore
30. Green infrastructure: current debates for policy, practice and implementation. Vereinigtes Königreich, Taylor & Francis (2018)
31. Building resilience to natural hazards in the context of climate change: knowledge integration, implementation and learning. Germany, Springer Fachmedien Wiesbaden (2021)

32. Principles for building resilience: sustaining ecosystem services in social-ecological systems. United Kingdom, Cambridge University Press (2015)
33. Martinez-Diaz L, Hill AC (2019) Building a resilient tomorrow: how to prepare for the coming climate disruption. United Kingdom, Oxford University Press

Evaluation of the Climate Adaptation Plans of Cities Against Heat Waves Using the Impact Logic of Municipal Action: A Study of Medium-Sized Cities in the "Warmest Climatic Regions" of Germany

Audrey Bourgoin

Abstract Urban populations are severely affected by prolonged periods of heat, whose frequency and severity increase with climate change, and which are escalated by urban spaces, as they combine concentrated systems (such as vehicles, AC systems, and industries) releasing heat into the urban environment, and a high degree of sealed urban structures, which intensifies heat island effects and hinders fresh air supply. Municipalities have limited resources to face escalating climate-related challenges and they need to find strategies to employ their resources to prepare structures and practices for the consequences of heat waves. The aim of this paper is therefore to present a methodology to evaluate the current adaptation plans of cities to warmer summers to adjust their planning practices (plans, instruments, financing) to counteract future climate threats. This research analyses current climate adaptation plans of medium-sized cities located in the "warmest climatic regions" of Germany, with regard to the diversity of instruments used, their implementation status, and the degree of detail of the developed strategies (e.g., using the "impact logic" of municipal action). The diversity of instruments used is assessed based on the distribution of the measures in the main categories "Grey infrastructure (urban and building structures)", "Green infrastructure", "Blue infrastructure" and "Complementary instruments", and their sub-categories. The implementation status of each measure is ranked on a four-level scale from "in planning" to "evaluable". The "impact logic" of municipal action is also a 4-level scale, which classes each strategy between "Input", "Output", "Outcome", and "Impact", whereas "Impact" represents the most conceptual level of planning with its goal and "input" constitutes the most concrete level, with a focus on the resource invested. The research intends to provide insights into the current climate adaptation planning practices, which have so far been rather conceptual and planning oriented. Finally, the research intends to expand

A. Bourgoin (✉)
Technische Universität Darmstadt, Franziska-Braun-Str. 7, 64287 Darmstadt, Germany
e-mail: bourgoin@geod.tu-darmstadt.de

V. T. Ha et al. (eds.), *Proceedings of the 8th International Conference on Sustainable Urban Development*, Advances in 21st Century Human Settlements,
https://doi.org/10.1007/978-981-99-8003-1_11

on the criticism that medium-sized cities must better define their measures to achieve previously abstract goals and implement (and monitor) the measures promptly.

Keywords Urban resilience · Cities · Climate adaptation · Heat · Germany

1 Introduction

1.1 Motivation

Climate change has led, in the last decades, to an increase in hot summers in Europe [7]. These already noticeable impacts are providing insight into the expected developments until the end of the century, as extreme weather events such as heatwaves are foreseen to increase along with the magnitude of global warming [35]. As climate change can no longer be totally prevented, it becomes increasingly necessary to plan and execute measures to adapt to future climatic conditions [21, 39].

Cities, their systems, and their population can be especially affected by prolonged periods of heat. Indeed, urban spaces combine concentrated systems (such as vehicles, AC systems, and industries) releasing heat into the urban environment, and a high degree of sealed urban structures, which intensifies heat island effects and hinders fresh air supply. All of this can lead to severe consequences for the health of the population.

Although some municipalities already recognized the need to take action on this topic, they have limited resources to face escalating climate-related challenges and they need to find strategies to best employ their resources to prepare structures and practices for the consequences of heat waves.

1.2 Aim and Structure of This Paper

The aim of this paper is therefore to present a methodology to evaluate the current adaptation plans of cities to warmer summers, in order to enable cities to later adjust their planning practices (plans, instruments, financing) to counteract future climate threats.

This paper begins in with an overview of the main impacts of climate change in Germany and in its "warmest climatic region". The effects of heatwaves on urban populations show the urgent need for urban climate adaptation measures (Chap. 2). In the following chapter, the basics of the evaluation theories for the assessment of urban climate adaptation plans will be presented: first, the requirements for the adaptation of a city to challenges are discussed, then the contribution that impact logic can make to the assessment of the effectiveness of municipal action will be explained, and finally, the categories of tools and measures to adapt cities to a warmer climate will be expanded upon. Chapter 4 presents the methodology used to conduct the different parts of the analysis: the diversity of instruments envisioned in the adaptation plan, their implementation status, and the degree of detail of the developed strategies in the different categories of the "impact logic" of municipal action will be evaluated. Chapter 5 is dedicated to the presentation of the analysis results of current adaptation plans of medium-sized cities located in the "warmest climatic regions" of Germany. In Chap. 6, the author will formulate recommendations, based on the results of the analysis. Afterward, the limits of the approach used will be discussed as well as the need for further data and research. Finally, Chap. 7 summarizes the main findings of this paper and concludes with an outlook.

2 Theoretical Framework: Impacts of Climate Change in Germany and Adaptation's Necessity

2.1 Impacts of Climate Change in Germany

The consequences of global warming are already noticeable in Germany, as since 1881, an increase in temperatures of 0.3 °C/decade during spring and 0.15 °C/decade in autumn has already been observed. The extreme values have also changed significantly: between 1881 and 2020, the daily minimum temperature and the daily maximum temperature each rose on average by about +0.5 °C/decade, which is twice as much as that of the daily mean temperature [7].

German climatic regions. The German Federal Environment Agency ("*Umweltbundesamt* (UBA)") evaluated the spatial patterns of 14 climate indicators with cluster analysis and defined seven homogeneous climatic regions[1] in Germany (see Fig. 1). Spaces clustered in the same climatic regions exhibit similar climatic conditions for the reference period (1971–2000), the middle of the century (2031–2060), and the end of the century (2071–2100).

2.2 Climate Change in Germany's "Warmest Climatic Region"

Germany's "warmest climatic region" comprises mainly parts of western Germany along the Rhine river as well as some areas in the east near the Spree river. It has currently the highest mean temperatures (2.0 °C in winter and 18.2 °C in summer)

[1] Climatic regions are relatively homogeneous, delimitable spaces in terms of climate, which were identified by means of a cluster analysis.

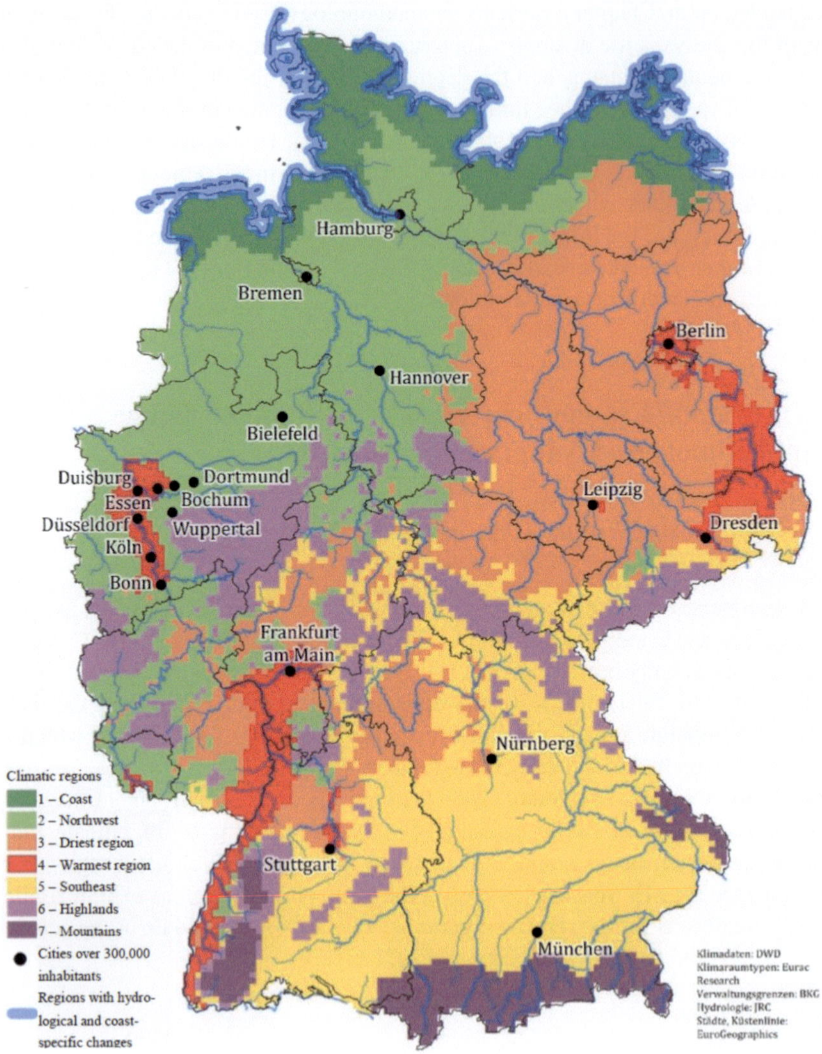

Fig. 1 Map of the seven climatic regions in Germany [39]. *Source* UBA

as well as the highest average of hot days[2] (9.2 days/year) and tropical nights[3] (0.5 nights/year) in Germany.[4]

Middle of the century. Current scenarios foresee that these areas will remain the warmest in Germany in the future as the greatest increase in hot days and tropical

[2] Days with maximum daytime temperature above 30 degrees Celsius.

[3] Nights with minimum temperature above 20 degrees Celsius.

[4] Mean values measured during the reference period (1971–2000).

nights is to be expected there. The 50th percentile of scenarios projections anticipates for the "middle of the century" in comparison to the reference period more than doubling of the mean number of hot days in this climatic region (+10.4 days/year) and more than a tenfold increase in the number of tropical nights (+5.0 nights/year). The increase of 11% in mean winter precipitation in the region for the "middle of the century" is in line with the average for other climate regions in Germany.

End of the century. Projections for the end of the century predict alarming increases in hot days (+27.8 days/year on average) and tropical nights (+19.1 nights/year on average). The 15th and 85th percentiles of the RCP8.5 scenario show a broad range in the possible values, which may vary from 20 hot days in the 15th percentile to a maximum of 56 hot days in the 85th percentile. An average increase of 20% in mean winter precipitation is to be expected, which is the highest increase after that of the "Southeast" region [38].

2.3 Impact of Climate Change and especially Heatwaves on the Urban Population

Climate change leads to significant impacts on human health, livelihoods, and key infrastructure, especially in more densely populated cities and regions [6, 21]. One of the most well-known consequences of global warming in cities is the increasing number of heatwaves and the formation of heat islands, which can greatly affect the health of the urban population.

Impact of climate change on cities. Climate change can lead to the following impacts on the built environment: damage to buildings, damage to infrastructure, damage to vegetation in settlements, changes in urban climate/heat islands, and affect the following elements of the urban environment: indoor climate, ground-level ozone, air quality (smog), hygienic situation in settlements, cooling/heating energy demand… [36]. The IPCC already reports an intensification of heatwaves in cities, which also aggravated air pollution events and hindered the functioning of key infrastructure. Economically and socially marginalized urban residents are among the most impacted [21]. The German Association of Cities (German: *"Deutscher Städtetag"*) also warns that risks for residents, infrastructure, or the urban greenery will continue to increase because of extreme summer temperatures, heavy droughts, and storms in Germany [6].

Urban climate and urban heat islands. In urban areas, the dense development and local emissions lead to the formation of special local climatic conditions forming what is called an "urban climate" [28, 41]. Urban heat islands (UHI) are typical features of the urban climate. They are characterized by a difference in air temperature between the warmer city and its cooler surrounding area. The difference can usually be up to about 10 °C in large cities [23]. The intensity of the UHI effect depends on the weather conditions and on the characteristics of the city. Its maximum is reached during cloud-free and windless nights. Factors such as the city geometry, the thermal properties of the building materials, the radiation properties of the surfaces, the energy balance, and anthropogenic heat releases contribute greatly to a warmer urban air temperature compared to the surrounding areas and thus to increased urban heat islands [8, 36].

Increasing urbanization contributes to an intensification of the heat island effect, particularly through increasing density and soil sealing in cities [15]. In addition, the expected increase in summer heat extremes may in turn increase the frequency and intensity of UHI in the future [17, 36].

The effects of urban heat islands are manifold. In the summer months, they increase the risk of heat stress for urban citizens. The high air temperature in cities during heatwaves also leads to increased use of cooling systems and air conditioning, which entails a high energy consumption and thus rising costs [8].

Direct heat-related risks for human health. Heatwaves and their associated effects can have extensive consequences for the population. Heat can directly affect the major organs of a person (brain, heart, lungs, kidneys…) and even negatively impact a pregnancy. Increasing temperatures and heatwaves can indeed induce "heat stress" for people, exposing their bodies to the risk of overheating, which can induce physiological stress, exacerbate illnesses, result in heat strokes, heat exhaustion and a higher risk of injuries and carries an increased risk of death [31, 40]. People with fragile health such as elderly people, people with pre-existing conditions, for example with troubles with the cardiovascular system, pregnant women, and small children are often particularly vulnerable to heat stress [8]. Outdoor workers and poor people can also be particularly exposed, making them vulnerable to excessive heat [40]. Heat can impact mental health too and increase mental health challenges, including anxiety and stress [21].

Heatwaves in Europe and Germany have repeatedly led to high mortality rates: between 1980 and 2020, about 130,000 fatalities from heatwave events were reported in European countries, which is more than 90% of the deaths caused by weather and climate-related extreme events during this period [12, 38].

Other indirect consequences of heat on people. In addition to the direct effects of heat on the human body, heat-related events can also negatively impact people: weather and climate extreme events can lead to trauma, and loss of livelihoods and cultures. Warm days and heatwaves also lead to the emission of wildfire smoke, atmospheric dust, and aeroallergens, which themselves can increase the risk of cardiovascular and respiratory distress [21]. The German Federal Environment Agency also lists additional side effects of global warming, such as: UV-related health damage (especially skin cancer), change in distribution and abundance change of potential vectors, etc. [37].

2.4 Necessity of Urban Climate Adaptation

Governmental and non-governmental organizations are united in their appreciation that "worldwide climate resilient development action is more urgent than previously assessed" [21]. The German Federal Environment Agency lists "Urban climate/heat island" and "indoor climate" among the fields with "very urgent needs for action", meaning that the climate risk is either presently already high or is estimated to be high for the middle or end of the century and that adaptation measures must already be started now, due to a medium or long adaptation period [39].

Urban resilience. It is the ability of an urban system to 1. "maintain or rapidly return to desired functions in the face of a disturbance" (i.e., the quality of "robustness"), and 2. "to adapt to change, and to quickly transform systems that limit current or future adaptive capacity" [29] (i.e., the quality of "adaptability"). It can play a central role in the mitigation of the risks arising from climate change. Robustness, as the ability to absorb crises or shocking events or to limit their adverse effects and consequences (i.e., reduce vulnerability[5]), cannot be fully achieved and maintained, as it is difficult to develop reliable predictions for complex dynamic events, with sometimes cumulative effects [25]. Therefore, adaptation—which is the ability to "adapt to the changed framework conditions (…) in order to permanently prevent or at least reduce adverse effects"—is also required to play a central role in reducing exposure & vulnerability to climate change [21, 25].

Adaptation. Adaptation can happen in anticipation of the risk or reactively. It can take the form of incremental changes and/ or transformational adaptation, where whole systems evolve [21]. There are currently quite big "adaptation gaps" between the current (or even the planned) adaptation level and the necessary goals for urban adaptation [5, 21]. The origin of these gaps lies in the weaknesses of the current adaptation practices: they are too often "fragmented, small in scale, incremental, sector-specific, designed to respond to current impacts or near-term risks, and focused more on planning rather than implementation" [21]. To "close this adaptation gap", the IPCC recommends going beyond the current too often short-term planning and instead developing long-term, concerted pathways as well as favorable conditions for ongoing adaptation to ensure timely and effective implementation [21].

Adaptation at the local level. According to Albrecht et al. [1], the regional and local levels are particularly well-placed to design and implement concrete measures for local climate change adaptation, especially as "there is no single recipe for the ideal adaptation", seeing as climate change impacts vary between regions. By implementing adaptation measures at an early stage, the dangers of climate change and the potential damage by more frequent and more severe extreme weather events can be minimized. Often, this requires structural measures that must first be integrated (with the appropriate lead time) into municipal planning processes. This makes urban and land-use planning a central instrument of climate adaptation, which, due to its long-term effectiveness, can point the way for the future climate resilience of municipalities. However, municipalities have limited resources to face escalating climate-related challenges and they need to find strategies to employ their resources to prepare structures and practices for the consequences of heat waves.

---

[5] Vulnerability is "the propensity or predisposition to be adversely affected" [21].

3 Basics of Urban Climate Adaptation

3.1 Requirements for the Adaptation of a City to Challenges

As outlined in the previous paragraphs, cities face significant challenges due to climate change. They also need to deal with additional challenges such as demographic and economic changes, digitalization, etc. Some of these problems have been known for a long time, some are new and many of them must be tackled under uncertain conditions.

Thus, cities need to plan and carry out far-reaching transformations, with future-oriented structures and functions, which can have far-reaching economic, social, and ecological effects on the complex system of cities. Therefore, careful impact analyses and assessments of the consequences of planning decisions as well as close cooperation between science and practice are needed to achieve a scientifically based, balanced, and far-sighted planning.

All these diverse requirements for cities must be taken into account in the planning strategies of the cities, which further need to be both long-term oriented and flexible enough to make place for short-term problem-solving. Urban planning should therefore develop strategies and implementation plans aiming for long-term sustainability, but also resilience to crises and disasters [26].

3.2 Measuring Municipal Action Using Impact Logic

Need to measure municipal action. Many cities want to improve their sustainability and/or their climate resilience or achieve some other goals, may they be self-imposed or coming from a higher level of government in a "top-down" logic. Therefore, cities develop diverse and sophisticated agendas, concepts as well as monitoring and management systems… However, there is too often a gap perceptible between, on the one hand, the enthusiasm of the concepts combined with the effects they want to achieve and, on the other hand, the activities undertaken, and resources available to reach these aims.

As the IPCC's calls are getting louder to end the period of orientation and begin a period of implementation and, above all, of action, it is necessary to provide cities with the means to measure the impact of their actions. Indeed, although numerous feasible[6] and effective[7] adaptation measures exist for urban systems, their feasibility and effectiveness are too often constrained by, among others, institutional and financial capacity, and depend on proper coordination and contextually appropriation of

[6] "Feasibility refers to the potential for a mitigation or adaptation option to be implemented" [21].

[7] "Effectiveness refers to the extent to which an adaptation option is anticipated or observed to reduce climate-related risk" [21].

a standard approach to the specific framework conditions of the city (e.g., physical, natural, and social infrastructure) [21].

Impact logic. Impact logic can provide cities with an effective tool to measure the results of their actions: it enables its users to systematically correlate and compare the planned impact goals with the resources necessary to achieve these goals and the services that the project provides. Thus, it facilitates control of the project's plausibility and feasibility. Impact logic has been used in the fields of project planning and evaluation since the 1970s and is thus a time-proven tool to represent and check how a project works or is supposed to work. An impact logic shows an "itinerary" for the project work and thus provides a basis for checking whether the project is (still) on the right track and is hence the foundation for impact-oriented project work and management [27].

There are different kinds of impact logics, known under different names, such as "programming logic", "theory of change", "impact chains", or "logical models". All of them aim to schematize the functioning of a project in a simplified form. In the following, the impact logic is described in the form of the so-called "logic model", which is one of the most widely used versions of the impact logic, as it can be used for most projects. Here, impact logic provides the frame to systematically identify the resources that go into an activity (*input*), the achievements that the activity produces (*output*), the effects of that activity on the target group (*outcome*), and lastly the effect for society as a whole (*impact*) [32]. In the following, we will describe the individual components in more detail and expand on how they relate to each other. A summarized representation of the impact logic of municipal action toward climate adaptation can subsequently be found in Fig. 2.

Inputs (resources). The inputs (resources) include all the means necessary to invest to successfully implement the project. These include first and foremost the staff and their working hours, the financial resources, the facilities, and the equipment needed for the implementation of the project. For realistic project planning and project expectations, all necessary resources must be listed here. It constitutes the foundation of the measure's effectiveness [27].

In the example of climate adaptation, the inputs can be the number of staff dedicated to the projects, the time they can devote to execute the requested projects, such as planting trees or conducting an information campaign on landscaped gardens and front yards, the financial investment the municipal consents to, to support its staff as well as the equipment available to the staff.

Outputs (services). The outputs comprise the (countable) offers, products, actions, first effects, or even services of a project, i.e., what a project does or offers ([27] ebd.). In our example of climate adaptation, the outputs can be the number of trees planted or the realization of the aforementioned information campaign.

The outputs can at times be directly related to the inputs (resources) used, which enables statements to be made about the efficient implementation of the project. In our example, it can be determined how much the project costs per planted tree or per resident reached.

From outputs to outcomes and impacts. Transforming outputs into outcomes and then impacts is crucial for the successful effectiveness of the project. Therefore,

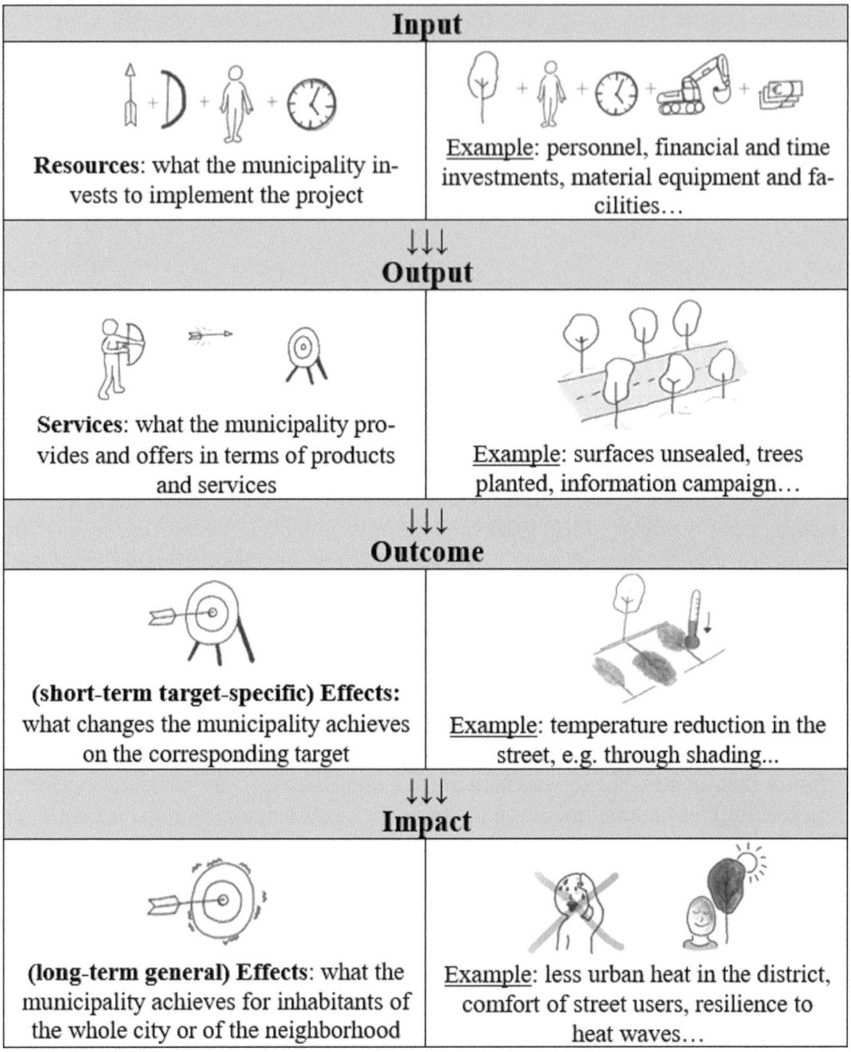

Fig. 2 Impact logic of the municipal action toward climate adaptation (own figure adapted from Peters [32])

making the difference between these levels is essential for impact-oriented project work, in order not to mistake one for the other and thus to miss the aim. The difference between outputs and impacts can be illustrated by the practical example of climate adaptation: climate adaptation project leaders and employees implemented their project with great commitment, they selected and then planted trees in the streets identified as particular places of heat urban island. But has this already had a positive effect (=outcome) on the street climate? The outputs are the prerequisite for a project to have an impact through its activities. However, the effects themselves only occur when positive changes occur in the selected streets, which ultimately enable them to influence the neighborhood or up to the whole city. A distinction can indeed be made between effects at the level of the target group (outcomes) and effects at the societal level (impacts) [27].

Outcomes (effects at the level of the local target). Outcomes are the direct effects of the project, or the changes the project aims to bring about, at the level of the local target (or target groups) [27].

Continuing with our example of measures for climate adaptation, after having planted the trees and/or conducted the information campaign on landscaped gardens and front yards, a temperature reduction should be observed in the street (*target*), compared to days with the same atmospheric conditions in the past.

Impacts (effects at the level of society). While outcomes are focused on the specific target of the project, impacts describe the desired changes at the "societal level". As it is not always meaningful, nor possible, to anticipate that inhabitants of the whole city will profit from the achieved local changes, it might often be more reasonable to consider the "larger society" to be the inhabitants of a neighborhood or of a district.

In our example of climate adaptation, the impacts could take the form of less urban heat in the district or better comfort for the users of the developed street.

The path from local outcome to societal impact is less clear than the previous relationships between the other elements of the impact analysis path. Indeed, some who benefit from the effects do not necessarily have to have come into direct contact with the organization's activities, e.g., passersby profiting from the street's ameliorated temperature may not have heard from the information campaign on landscaped gardens and front yards, which was targeted at residents of the street. Concurrently, it also means that the influence of the project on the achievement of some effects at the impact level is fuzzier than on the effects at the outcome level. This is because many other factors besides the project influence developments at the societal level. For instance, the quality of life of the street's passersby is not substantially improved, just because one street has become slightly more comfortable to walk through. Therefore, the project should only strive to contribute to the effects at the impact level, while focusing on the outcome level where the concrete effects that are to be achieved by the project take place (ebd [27]).

The creation of an impact logic path can be done in two directions: "from the impacts to the inputs" and "from the inputs to the impacts". The direction "from the impacts to the inputs" is better suited for impact-oriented project planning, as it focuses on the effects the project aims to achieve and plans the necessary resources accordingly. Projects starting their planning in the other direction risk limiting their aim by the limited resources they want to invest, thus staying in a "status quo" rather than having big ambitions. The planning direction "from inputs to impacts" can however be helpful in a second step to perform a "plausibility check" on the impact logic of the project [27].

3.3 Categories of Tools and Measures to Adapt Cities to a Warmer Climate

To adapt cities to a warmer climate, municipalities, associations of cities as well as some governmental branches supporting them in their tasks developed many measures, which they bundle into categories to give their public a better overview.

Categorization used by various stakeholders. The IPCC encourages policymakers to formulate responses for climate adaptation across physical, natural (ecosystem), and social infrastructures.

The C40 Cities Climate Leadership Group for instance proposes 6 categories of measures: "Green infrastructure", "Blue infrastructure", "Grey infrastructure", "Urban development", "Communication, and outreach" as well as "Policy" [4].

The German city of Herten divides its "Climate adaptation concept" into seven categories: "General measures", "Urban development and municipal planning", "Water management", "Biodiversity and nature conservation", "Human health, and social infrastructure", "Building and housing", as well as "Economy" [11].

The city of Landau in der Pfalz in its "Climate adaptation concept" opted for five categories: "Green and open space development", "Water-sensitive city", "Agriculture, forestry and nature conservation", "Planning and building", as well as "Communication, health, and social affairs" [14].

Thus, it is clear that there seem to be as many categorizations possibilities as there are of stakeholders dealing with the challenge of urban climate adaptation, but some topics are staying constant: the built infrastructure, the vegetation and nature (green infrastructure), and the topic of water. Other areas, such as the economy and communication, seem mostly to complete the three categories previously listed.

Categorization used in this article. As the Urban Heat Island (UHI) effect is due to a higher amount of sealed surfaces in cities, a lack of vegetation and a lack of cooling through vegetation and water bodies (see Sect. 2.3), it seems appropriate to measure the focus specially placed on these aspects by assigning them each their own category: "Grey infrastructure (urban & building structures)", "Green infrastructure, and "Blue infrastructure". Analysis of the concepts of various cities and organizations concur with these three categories.

As these three categories, through the main addressed by specialists, are not enough cover the whole spectrum of measures, a fourth category of "Complementary instruments" was adjoined to them, encompassing all additional measures accompanying and reinforcing other measures.

Grey infrastructure (urban & building structures). As increasing urbanization, particularly through increasing density and land sealing in cities, contributes to an intensification of the heat island effect [15], it was decided to assess the measures addressing the topic of "Urban development—density and air flow". Land sealing also contributes to the heat sensation outside, but it can be attenuated with shading (e.g., by man-made structures or vegetation), thus building the sub-category "heat outside". Increasing surface albedo through roofs, walls and street surfaces with high-reflectance materials can also reduce the consequences of urban heat waves, thus the inclusion of the topic of "albedo" in the category "grey infrastructure" [21]. Working on the design of "Building facades", through e.g., the use of natural thermo-regulating building materials, better insulation, facade greening, or cooling installations on the facade can also participate to increase the comfort of the population during warm days. Lastly, the "indoor climate" is also influenced by the urban outdoor climate and requires changes for building inner equipment (ventilation system…) [36].

Green infrastructure. It is internationally acknowledged that vegetation in settlements plays a central role in climate-friendly and climate-adapted settlement development. The German Federal Environment Agency advises the planting and care of vegetation providing "Cold air corridors" in the city to cool it, and the planting of "Trees", which cool their environment through evapotranspiration [36]. Both "Public spaces" and "Private spaces" can be specifically addressed, but some measures like the Reduction of gravel gardens or the installation of permeable paving can be promoted for both "Urban public & private areas". Lastly, it is important to consider "Climate-adapted fauna and flora", especially heat and drought-stress-tolerant plants and trees. Other vegetation-related measures, which are not part of the usual measures to adapt cities to climate change via its green infrastructure (e.g., measures related to allergens or agriculture), can be included in the sub-category "others".

Blue Infrastructure. Analyses of the main measures usually recommended enable to determine four main categories of water-related measures for urban climate adaptation.

- The measures directly contributing to a "Reduction of air temperature", like the creation and preservation of open and moving water areas or the implementation of integrated irrigation concepts for green spaces.
- The measures providing "Heat relief for people", like the installation of drinking water dispensers and drinking fountains in public spaces.
- The "rainwater"-oriented measures, like decentralized water retention, and infiltration areas.
- As well as measures for "flood protection".
- Other measures, such as the ones concerning the water quality were aggregated under the sub-category "others".

Complementary instruments. The category "complementary instruments" comprises a broad and very mixed range of instruments supporting the effects of the measures of the three categories. It was divided into the following seven sub-categories:

- "Communication and public relations": e.g., information campaign for citizens and cooling centers (also called "climate refuges", "cool spots", or "cool islands");
- "Public participation": e.g., cooperation exchange with companies, partnership systems between volunteers and vulnerable citizens, etc.

Table 1 Categories and sub-categories of instruments for urban adaptation to a warmer climate

Grey infrastructure	Green infrastructure	Blue infrastructure	Complementary instruments
• Urban development-density and air flow • Heat outside • Albedo • Building facades • Interior design in building	• Cold air corridors • Trees • Green spaces • Public spaces • Private spaces • Urban public & private areas • Climate-adapted fauna and flora • Others	• Reduction of the air temperature • Heat relief for people • Rainwater • Flood protection • Others	• Communication and public relations • Public participation • Resources • Regulatory instruments • Economic instruments • Planning instruments and concepts • Others

- "Resources" include what the city is willing to invest in climate adaptation-related projects, such as a climate manager or even a whole climate adaptation department, or explicitly listing measures for climate adaptation in the municipal budget.
- "Regulatory instruments" include among others climate-proof urban development and urban land use planning, (local) guidelines for climate design of buildings, and regulations for cool/green roofs.
- "Economic instruments" can be for instance tax rebate programs for rainwater retention or sustainable home cooling measures or subsidies for the installation of green/white roofs or facades on new and existing (residential or commercial) buildings or planting trees and greening courtyards…
- "Planning instruments and concepts" lastly include roader plans like heat action plans, planning information maps on urban climate, simulations of the climatic effects of urban development projects, etc.
- "Others" includes cooperation with other programs, statistics, and everything that could not be included in the six previous sub-categories.

Table 1 gives an overview of the four previously presented categories and their sub-categories of instruments for urban adaptation to a warmer climate, which will be employed in the analysis part of this article.

4 Methodology

This paper makes a qualitative analysis of the level of development of currently available climate adaptation plans and concepts of selected cities. Three aspects will be analyzed in the published concepts: the diversity of instruments used, their implementation status, and the degree of detail of the developed strategies, using the "impact logic" of municipal action.

4.1 Selection of the Cities Analyzed

As already discussed in Sects. 1 and 2, cities from the German climatic region "Warmest Region" are already particularly affected by heat days and heat waves and will remain so for "mid-century". However, German cities from other regions could experience similar climatic conditions by the end of the century [35]. Therefore, findings concerning the cities of the "warmest region" could provide valuable learning experiences for the other cities, which could be later confronted with the same kind of climatic challenges.

Medium-sized German cities (between 20,000 and 100,000 inhabitants) are commonly associated in German society with places of good quality of life, along with small towns, which make them most suitable to evaluate the efforts they invest to keep or even raise this "quality of life" along with the "profile of the city". Although research in Germany has been centered mostly on large cities in the last decades, medium-sized cities can also provide valuable insights for cities of other sizes. Small and medium-sized cities harbor certain weaknesses (e.g., scarcity and distance to material resources and personal resources) but also potentials (e.g., high degree of diversity with lower staffing levels, more flexible action…), which influence their way of dealing with planning tasks. Short distances, flexible administrative action with flatter hierarchies, and proximity to citizens are also positive attributes generally credited to medium-sized cities. They are also usually embedded in regional networks or even supra-regional cooperation strategies, which facilitates the transfer of knowledge. Professionals and researchers agree that experience gained by medium-sized cities in dealing with necessary, but voluntary tasks such as climate adaptation despite their intrinsic challenges can therefore provide useful or even exemplary insights for the administration of other cities, be their smaller or larger [3].

Therefore, it was decided to analyze the current climate adaptation plans and concepts of medium-sized cities located in the "warmest climatic regions" of Germany.

Germany has a bit more than 2000 cities, 586 of which are medium-sized cities. Out of these, 130 are in the "warmest climatic regions" of Germany and among them, 15 have (as of 1st September 2022) already published a climate adaptation concept. Moreover, 14 more cities are currently planning or working on a climate adaptation concept, which shows a marked interest in the topic on the part of the cities.

The climate adaptation plans and concepts of the 15 cities which already have one are the object of the analysis of this publication. Table 2 gives an overview of these cities in the different German federated states.

Table 2 List of the cities whose climate adaptation concepts were analyzed

Federated state	Cities analyzed (date of their plan)	Reference of the adaptation plan
Brandenburg (BB), 1 city:	Bernau bei Berlin (2020)	[22]
North Rhine-Westphalia (NRW), 6 cities:	Erkrath (2021)	[16]
	Herten (2021)	[11]
	Hürth (2018)	[9]
	Lohmar (2016)	[19]
	Meerbusch (2020)	[30]
	Siegburg (2018)	[20]
Baden-Württemberg (BW), 2 cities:	Bühl (2018)	[2]
	Ludwigsburg (2016)	[13]
Rhineland-Palatinate (RLP), 3 cities:	Landau in der Pfalz (2020)	[14]
	Speyer (2015)	[34]
	Worms (2016)	[18]
Hesse (HE), 3 cities:	Dietzenbach (2022)	[24]
	Hanau (2019) (district of Großauheim, focus on health)	[33]
	Rodgau (2019)	[10]

4.2 Diversity of Instruments

The diversity of instruments planned by each climate adaptation plan is evaluated based on the measures listed in the climate adaptation plan. For each analyzed climate adaptation plan, all the measures listed were recorded and allocated to the four main categories "Grey infrastructure (urban and building structures)", "Green infrastructure", "Blue infrastructure" and "Complementary instruments", and their subcategories of instruments listed in Table 1.

Thereupon the distribution of the measures in the four categories and their subcategories were analyzed, both for the whole group of cities, as well as for each city compared to the others.

4.3 Implementation Status

The implementation status of the measures in each climate adaptation plan is evaluated via the following four levels:

- **"1—In planning"**: for measures whose implementation has not yet started (or been communicated about).
- **"2—Test phase":** for measures being tested on a limited level, for instance as part of local pilot projects.

Table 3 The four levels of the implementation status of measures

1-In planning	2-Test phase	3-Ongoing	4-Evaluable
Planned measures, not yet started	Measure initially implemented only as part of pilot projects	Measure implemented city-wide	Measure completed or measure can already be assessed (e.g., for permanent measures)
e.g., planting of trees planned	e.g., planting of trees in neighborhood X	e.g., planting of trees city-wide	e.g., planting of trees Y years ago & so much survived

- **"3—Ongoing"**: for measures whose implementation is ongoing city-wide, but without results to date.
- **"4—Evaluable"**: for measures that can be assessed, either because they are completed or because their effects can already be evaluated if these measures are intended to extend over a longer period.

Table 3 provides an overview of the four levels. The evaluation of the implementation status of the measures was based on the information available online to the public (as of 1st September 2022): if there were somewhere mentions of some measures already in stages 2 to 4, it was transcribed in the evaluation of the implementation of the concerned city's adaptation plan. Should no information whatsoever be found about the implementation status of any of the measures, all measures of the city's adaptation plan will be noted as being still at stage "1-In planning", for lack of contradictory information.

4.4 Degree of Detail of the Developed Strategies

Lastly, the degree of detail of each climate adaptation plan is evaluated, based on the impact logic of the municipal action toward climate adaptation (see Fig. 2). Although there is a time delay between the different elements, it is important for the municipalities to clarify which resources they are ready to invest to reach their goals.

In "Impact" was listed everything concerning the overall aims of the city's plan, the long-term general effects (for inhabitants of the whole city or of the neighborhood), which are aimed for via the establishment of the climate adaptation plan.

The category "Outcome" contains all the direct effects which can be expected from the application of the climate adaptation plan or the changes the municipality aims to achieve concerning local situations.

"Output" is the category for all the services and products the municipality intends to provide with the enforcement of the climate adaptation concept.

Lastly, "Input" is the category where all information about the resources the municipality plans to invest is gathered: mostly personnel and financial investment, but also possibly a detailed description of the time investment necessary and of material equipment and facilities needed to realize the planned measures.

“Impact” represents the most conceptual level of planning with the very broad goals of the municipality, whereas “input” constitutes the most concrete level, with a focus on the resource invested. Therefore, the most precise description of the content should be found at the “input” level, whereas the vaguest formulation of the whole can be excepted at the “impact” level.

For each city, the elements concerning its climate adaptation plan in each of the four categories were recorded and the results were subsequently qualitatively analyzed.

5 Analysis Results

15 climate adaptation plans and concepts of medium-sized cities located in the “warmest climatic regions” of Germany were analyzed regarding their diversity of instruments used, the implementation status of their measures, and the degree of detail of the developed strategies.

5.1 Diversity of Instruments: Great Variations Between the Plans of the Different Cities

For each climate adaptation plan, all the measures listed were recorded and allocated to the four categories (Grey infrastructure, Green infrastructure, Blue infrastructure, and Complementary measures) and their subcategories of instruments listed in Table 1. Thereupon, it was analyzed which kind of measures and categories had the preference of the cities.

Overview of the categories. The first overview of the total number of measures shows a distinct willingness of the cities to tackle the challenge of climate change adaptation, as on average each city developed about 38 measures. A finer analysis shows that not all cities are equally prolific in the number of measures they developed, as the individual number of measures varies between 16 (Dietzenbach, in Hesse) and 105 (Bernau b. Berlin, in Brandenburg).

The four categories (Grey infrastructure, Green infrastructure, Blue infrastructure, and Complementary measures) are represented in every climate adaptation plan analyzed (except for blue infrastructure in Hürth), but not always to the same extend (see Fig. 3). Out of the 573 measures listed in the 15 analyzed plans, 45% (259 measures) belong to the very broad category of “Complementary measures”. A bit more than a quarter of all listed measures (149 measures) can be assigned to the category “Green infrastructure”, which is crucial for cooling the streets. Next in numbers (about 15% or 30 measures) are the measures related to the blue infrastructure. Measures concerning the grey infrastructure form the smaller of the four groups (13% or 75 measures), although it is crucial for cities to work on their built environment to counteract the worst effect on urban heat islands.

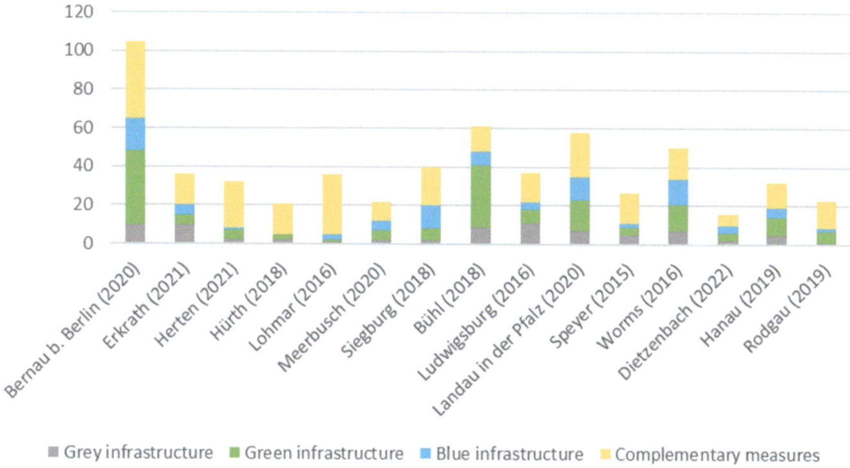

Fig. 3 Distribution of the measures of the 15 analyzed climate adaptation plans along the four categories "Grey", "Green", and "Blue" infrastructure, as well as "Complementary measures"

Some cities thus seem at the first glance to invest more energy in accompanying measures instead of in the more infrastructure-related measures, which can however have the most impact on making the cities more comfortable.

Grey infrastructure (urban & building structures). Out of the 75 measures planned in this category by the 15 cities, more than half concerned the "Building facades" and the topic of "Urban development—density and air". These two aspects are closely followed by the aspect of "Interior design in building". The topic "Albedo" was the least addressed (see Table 4).

A detailed analysis shows that no city developed measures in all the five sub-categories defined. 9 out of the 15 analyzed plans contained measures about the "interior design of the buildings", such as the use of natural thermo-regulating building materials (e.g., brick, limestone, or wood), better insulation of the building envelope, facade greening or shading or cooling installations on the facades. Noteworthy is also that three cities each developed 5 or 6 measures to address the topic of "Urban development—density and air", which includes measures such as the reduction of the

Table 4 Repartition of the measures of the 15 analyzed climate adaptation plans in the five sub-categories of "Grey infrastructure"

Grey infrastructure: 75 measures in total	
Building facades	23
Urban development—density and air	22
Interior design in building	19
Heat outside	8
Albedo	3

degree of sealing in new developments, loosened-up development with high proportions of green spaces and working on urban geometry or optimization of building orientation, height and the design of a building and building density, e.g., for air circulation. These measures are essential to ensure that cities allow fresh air to enter their quarters. The topic of the albedo was the most neglected as only two cities planned together three measures about it (see Fig. 4).

The results show that all the plans of the analyzed cities comprise some measures to mitigate urban warming via grey infrastructure, but it is also clear that some cities are more aware than others of the big potential to mitigate warming in cities through their action on urban development.

Green infrastructure. Out of the 149 measures planned by the 15 cities in this category, about a third concerned the "Green spaces", and a fifth the topic of "Trees". The five aspects "Cold air corridors", "Public spaces", "Climate-adapted fauna and flora", "Urban public & private areas" and "Others" total together half of the listed measures. The topic "Private spaces" was the least addressed (see Table 5).

A detailed analysis shows that two cities (Bernau bei Berlin and Bühl) planned each more than 30 measures, amounting together to nearly half (70 measures) of the total amount of the measures listed in the 15 analyzed climate adaptation plans. These two cities addressed respectively seven of the eight sub-categories (Bernau bei Berlin) and all the listed sub-categories (Bühl) in their adaptation plans. The city of Landau in der Pfalz also addressed all sub-categories of "Green infrastructure" measures in its climate adaptation plan (with 16 measures in total).

All cities addressed, however, the topic of "Green spaces" (with measures such as the preservation of parks and open green spaces or with climate-friendly development of urban green spaces), even the city of Lohmar for which it is its single measure in

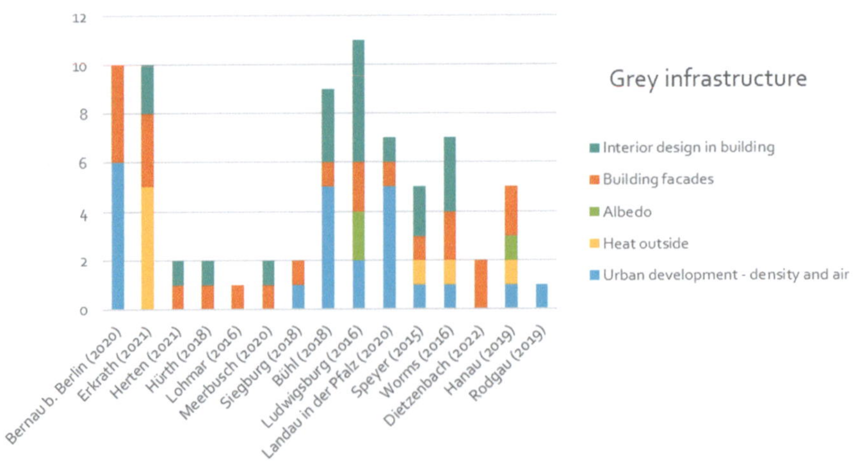

Fig. 4 Distribution of the measures of the 15 analyzed climate adaptation plans among the five sub-divisions of the category "Grey infrastructure (urban & building structures)"

Table 5 Repartition of the measures of the 15 analyzed climate adaptation plans in the eight sub-categories of "Green infrastructure"

Green infrastructure: 149 measures in total	
Green spaces	42
Trees	30
Others	20
Cold air corridors	15
Public spaces	15
Climate-adapted fauna and flora	14
Urban public & private areas	10
Private spaces	3

the category of "Green infrastructure". 10 cities out of the 15 took the topic "Climate-adapted fauna and flora" into account, and 9 out of the 15 analyzed plans contained measures about "trees", such as preserving the existing tree population, the addition of further climate change-tolerant trees and shrubs or the planning of green streets. The topic of "Cold air corridors" was also taken into account by eight cities, even if most of them only defined one or two measures in this field. Topics such as those of "Public spaces", "Private spaces", and "Urban public & private areas" were more sparingly addressed (see Fig. 5).

The results show great variation between the plans of the analyzed cities: with three cities having addressed all or all but one sub-categories of "Green infrastructure" measures, whereas four cities only addressed two categories and one city exhibited solely a single measure. It thus shows that some cities are much more aware than

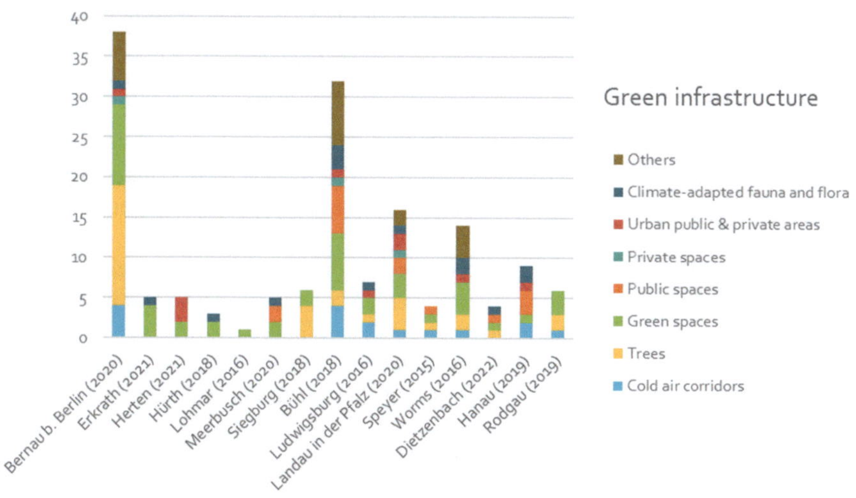

Fig. 5 Distribution of the measures of the 15 analyzed climate adaptation plans among the eight sub-divisions of the category "Green infrastructure"

Table 6 Repartition of the measures of the 15 analyzed climate adaptation plans in the five sub-categories of "Blue infrastructure"

Blue infrastructure: 90 measures in total	
Rainwater	28
Heat relief for people	24
Flood protection	20
Reduction of air temperature	16
Others	2

others of the wide number of possibilities offered to them by measures in the category of "Green infrastructure".

Blue infrastructure. Out of the 90 measures planned by the 15 cities in this category, about a third concerned "Rainwater" and nearly as much the topic of "Heat relief for people". Afterward came the topics of "Flood protection" and of "Reduction of air temperature" (see Table 6).

It is striking to note that one city (Hürth) did not address the category of "Blue infrastructure" at all in its climate adaptation plan. In contrast, the city of Worms included measures in all five sub-divisions of the category and two cities (Bernau bei Berlin and Bühl) addressed four topics. The topic of "Rainwater", which is the most mentioned by all measures in total, was addressed by 12 out of the 15 analyzed plans. The topic of "Heat relief for people" is a bit less represented, as it is included in 9 of the analyzed plans. There are measures concerning the topic of "Flood protection" in 7 out of the 15 analyzed plans. The topic of "Reduction of air temperature" is by contrast only addressed by four cities (Bernau b. Berlin, Bühl, Ludwigsburg, and Worms), one of which (Bernau b. Berlin) concentrates 75% of all recorded measures on the topic (see Fig. 6).

The results show here too great variation between the plans of the analyzed cities: three cities have addressed all or all but one sub-category of "Blue infrastructure" measures, whereas three cities only addressed one sub-category and one city did not address the category "Blue infrastructure" at all. It thus shows that some cities are much more aware than others of the impact measures in the category of "Blue infrastructure" can have on the comfort of their urban spaces.

Complementary instruments. Out of the 259 measures planned by the 15 cities in this category, a bit more than a third concerned "Communication and public relations". About half as many measures can be clustered under the sub-category "Planning instruments and concepts". Afterward came the topics of "Public participation", "Resources", and "Regulatory instruments". The topics "Economic instruments" and "Others" were the least addressed (see Table 7).

A detailed analysis shows that all 15 cities abundantly scoop into the possibilities of "Complementary measures", as all of them plan measures in at least four of the seven defined sub-categories. The cities planned between 6 and 40 "complementary measures", with an average of 18. Two cities even have measures in all the sub-categories and three cities have measures in all but one sub-category.

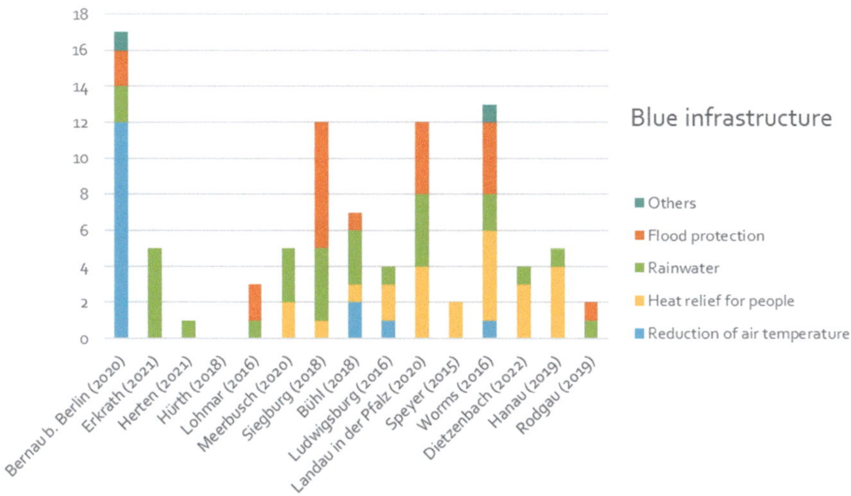

Fig. 6 Distribution of the measures of the 15 analyzed climate adaptation plans among the five sub-divisions of the category "Blue infrastructure"

Table 7 Repartition of the measures of the 15 analyzed climate adaptation plans in the seven sub-categories of "Complementary measures"

Complementary measures: 259 measures in total	
Communication and public relations	92
Planning instruments and concepts	47
Public participation	32
Resources	27
Regulatory instruments	27
Economic instruments	17
Other	17

Both sub-categories "Communication and public relations" and "Planning instruments and concepts" are represented in all the climate adaptation plans analyzed. It is reasonable, as the population needs to be informed about and included in the actions and plans of the local government such as heat action plans, planning information maps on urban climate, simulations of the climatic effects of urban development projects. The other sub-categories are quite well addressed by most of the cities (see Fig. 7).

The results show that all the plans of the analyzed cities comprise a bountiful amount of "Complementary measures", most of all "Communication and public relations" and "Planning instruments and concepts". Other sub-kinds of "Complementary measures" are also well taken into account.

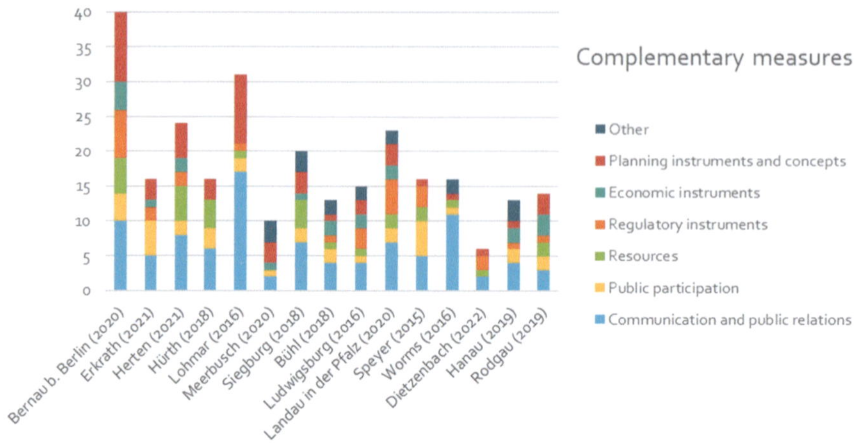

Fig. 7 Distribution of the measures of the 15 analyzed climate adaptation plans among the seven sub-divisions of the category "Complementary measures"

5.2 *Implementation Status: Good Effort to Implement Some of the Planned Measures*

For each climate adaptation plan, the general implementation status of the plan was based on the collated records publicly available of the four-level implementation status of the measures listed in the adaptation plan.

Table 8 provides an overview of the status of the measures listed in each adaptation plan. A check mark in the columns "2—Test phase", "3—Ongoing" or "4—Evaluable" means that at least one measure in the whole plan could be assigned to the corresponding category. A red cross means that no measure could be found at the corresponding stage, based on the information publicly available on 1st September 2022.

Looking at the information summarized in Table 8 it seems clear that most of the analyzed cities are making efforts to implement the measures foreseen in their climate adaptation plans. Only for the city of Landau in der Pfalz, which published its plan in 2020 could no information be found about the implementation status of the measures listed in its plan.

Otherwise, 13 of the cities provided information showing that they are already implementing some of their planned measures in some form (either in a test phase or city-wide) and even six cities could boast of having evaluable results about some of their measures.

This quick analysis thus shows already a trend that some cities with a few years old climate adaptation plans have already implemented some measures, but it is unclear to which extend and there seems to be no official monitoring report measuring the effects of these measures. For a more precise analysis in the absence of monitoring reports, interviews would be needed to clarify with each city the real implementation status of each measure and thus of the whole plan.

Table 8 Overview of the implementation status of the measures in the 15 analyzed cities' climate adaptation plans

Cities (classed by year of their climate adaptation plan)	1 – In planning	2 – Test phase	3 – Ongoing	4 – Evaluable
Speyer (2015)	✓	✗	✓	✓
Lohmar (2016)	✓	✗	✓	✗
Ludwigsburg (2016)	✓	✗	✓	✗
Worms (2016)	✓	✓	✗	✗
Bühl (2018)	✓	✓	✓	✓
Hürth (2018)	✓	✓	✓	✗
Siegburg (2018)	✓	✗	✗	✓
Hanau (2019)	✓	✓	✓	✓
Rodgau (2019)	✓	✗	✓	✓
Bernau b. Berlin (2020)	✓	✓	✗	✗
Meerbusch (2020)	✓	✓	✓	✗
Landau in der Pfalz (2020)	✓	✗	✗	✗
Erkrath (2021)	✓	✗	✓	✗
Herten (2021)	✓	✓	✓	✗
Dietzenbach (2022)	✓	✗	✓	✓
	14 / 15	**7 / 15**	**12 / 15**	**6 / 15**

5.3 *Strategies Unequally Developed in the Different Categories of the Impact Logic of the Municipal Action*

For each city, the elements concerning its climate adaptation plan in each of the four categories of the impact logic of the municipal action ("Impact", "Outcome", "Output", and "Input") were recorded and the main results of the qualitative analysis are presented here.

The first observation derived from the compilation of information is that not every city provides some kind of information about each category of the impact logic of the municipal action, and when the information is provided, the degree of details provided varies greatly. Thus, it shows that cities are not aware of the underlying mechanisms of the effects of their measures.

"Impact". The category of "impact" is the one for "mission statements" and "long-term goals". The aims of 13 of the 15 analyzed cities can be clustered in this category, such as:

- Adapting the city to the consequences of climate change (Meerbusch, Lohmar, Rodgau)
- Long-term reduction of vulnerability (against heat waves or heavy rainfall events) and increase of resilience (Bernau b. Berlin, Erkrath, Herten, Meerbusch, Rodgau)
- Ensuring quality of life and healthy living conditions for present and future generations (Lohmar, Ludwigsburg, Bühl, Worms, Landau in der Pfalz, Hanau)
- Anchoring climate protection and climate adaptation requirements in urban land use planning (Rodgau)

- Permanent participatory & ecosystem-based adaptation planning (Bernau b. Berlin)
- Structural and institutional adaptations of ecosystem management (Bernau b. Berlin)
- Ecologically-functioning ecosystems (Bernau b. Berlin, Landau in der Pfalz, Landau in der Pfalz)
- Exemplary function in responsible and climate-conscious action (Hürth, Speyer, Rodgau)
- 3 pillars of sustainability: social, economy, and ecology (Landau in der Pfalz)

For two cities (Siegburg and Dietzenbach), no real information could be found at the "impact" level. One of the measures the city of Siegburg defined in its plan was to "define or further develop the energy and climate policy model and goals". The city of Dietzenbach does not give any information about the overall aims of the city's climate adaptation plan.

"Outcome". In the category "Outcome" are collected all the direct effects that the cities expect from the application of their climate adaptation plan, or the changes the municipalities aim to achieve concerning local situations. Analysis of the content of the 15 cities' climate adaptation plan shows that they do not always clearly distinguish the concrete "outcomes" expected from the "impacts" hoped for, or from the measures themselves. The city of Bühl provides the best example of "outcomes", which show concrete expectations from their adaptation plan by defining 12 central objectives for the thematic areas of living and residential environment, health, and landscape and recreation, among whose are for instance:

- Preservation, development, and creation of areas for thermal relief in the settlement area.
- Preservation and expansion of areas for thermal relief in outdoor areas
- Safeguarding and development of climate-adapted open spaces in outdoor areas
- Reduction of air-hygienic pollution
- Preservation and development of CO2-storing land uses
- Climate-adapted design of buildings.

The city of Bernau bei Berlin also defines such outcomes, even if they are more "ecosystem oriented" and less "urban spaces oriented".

The "core objectives" or "sub-goals" of the cities of Erkrath, Herten, Hürth, Meerbusch, Ludwigsburg, Rodgau, and Hanau are mostly a slightly more detailed version of the "impact" aimed for by the whole process.

Some cities (e.g., Speyer, Siegburg, Landau in der Pfalz) prefer to define "fields of action" they expect their measures to influence, which can be either topic-specific like "Urban planning and green spaces", "Economy", "Tourism", "Health", "Mobility" and "Flood, or related to cross-sectoral domains, like "communication/public relations/education".

For two cities (Worms and Dietzenbach), no real information could be found at the "outcome" level. The city of Worms concentrated on their measures and drew up a catalog of measures and priorities with 48 measures but did not define their expected outcomes. The city of Dictzenbach does not give any information about the outcome expected from the measures in its climate adaptation plan.

"**Output**". All services and products that the municipalities intend to offer with the implementation of their climate adaptation concepts can be found in this category. In most cases, the "outputs" can be derived directly from the results of the single measures planned in the climate adaptation plans, for instance:

- The areas of sealed ground newly greened
- The newly shaded facades (through trees, façade greening, or blinds or awnings) to reduce direct irradiation and promote climate-friendly buildings
- The deciduous trees planted
- The installed temporary and/or mobile solutions such as mobile greenery for heat prevention in the inner cities
- The existing (near-natural) water bodies and wetlands are preserved
- The new water installations with moving water or spray mist in areas subjected to very high thermal loads
- The promotion campaign & competition for near-natural garden design among private households
- The newly created funding programs and financial incentives, to promote roof and façade greening (private, industry, and commerce).

As each of the 15 cities defined concrete measures (the one analyzed in part 5.1), each of them defined concomitantly the outputs they are expecting from the measures. The cities which planned the most measures were mathematically the ones with the bigger number of "outputs" that can be expected from their climate adaptation plan.

"**Input**". The "Input" category, which is the one about the resources (e.g. personnel, financial, etc.) the municipality plans to invest, is along with the category "outcome", one of the most unequal ones among the cities: some cities provide indeed very clear descriptions of the resources needed for the realization of the measures, in terms of financial and personal planning, whereas others give near to no indication on how the implementation of the measures will be concretely possible.

Financial planning indeed often simply refers to funding programs, mostly used to develop the climate adaptation plan, but is rarely clear whether these same programs will help finance the foreseen measures of the plans (Erkrath, Speyer, Hanau). For the city of Lohmar, no information about any kind of financing was found.

Concerning personal planning, some cities (Bernau b. Berlin, Lohmar) focus mainly on project groups, representatives from the municipality, and department managers of the municipality and tend to forget to take into account the people needed to implement the measures decided. Some cities (e.g., Hürth, Siegburg) list also the personnel costs per measure or for the controlling of measures and reporting. Lastly, some cities (Herten, Meerbusch, Ludwigsburg, Bühl, Worms, Rodgau, Dietzenbach) created detailed "Profile sheets" for their measures, with for instance information concerning:

- Who is responsible and/or which actors (department) are implicated in the realization of the measure?

- What are the financing and funding possibilities for the implementation of the measure?
- What will the implementation cost? Including personnel costs, direct implementation costs, public relations, and follow-up costs, if applicable.
- What is the duration of the measure? Preparation period, implementation period, and duration of the measure itself.

These are exactly the kind of information one needs, to check the feasibility of the impact logic of the municipal climate adaptation plans, as without this information it cannot be checked if the cities have the means available to execute everything written in their adaptation plan or if the project of the adaptation of the city will go no further than the elaboration of the plan.

Evaluation of the degree of detail of the elements of the impact logic of the municipal action. The above analysis shows that cities have little to no trouble defining "mission statements" and "long-term goals" belonging to the category of the wider "impact" they hope from their climate adaptation plan.

The category "outcome" is a bit more difficult to elaborate on for the cities, as they seem to have trouble distinguishing between the concrete "outcomes" expected, and the broader "impacts" hoped for or between the "outcomes" and the measures themselves. Two cities among the 15 analyzed, however, still managed to formulate clear "outcomes", which should result from the implementation of the measures.

The level of the "outputs" is in turn much easier for the cities to define as it comprises mostly the results of the single measures planned in the climate adaptation plans.

Lastly, the "input" category is, after the category "outcome", the second the most unequal among the cities. Some cities give created detailed "Profile sheets" for their measures, which contain very clear descriptions of the resources needed for the realization of the measures, especially in terms of financial and personal planning, whereas other cities give near to no indication about how the implementation of the measures will be concretely possible.

It is therefore clearly apparent that the degree of detail of the 15 cities' climate adaptation plans is very varied, with some having no trouble providing a precise description of the content of some category of the impact logic of the municipal action, whereas others seem to struggle to express the steps of the impact logic necessary for them to reach their goals.

6 Recommendations and Limits of the Approach

This paper presented a methodology to evaluate the current adaptation plans of cities to warmer summers. Based on the results presented in Chap. 5, the following recommendations can be formulated to help cities adjust their planning practices.

6.1 Recommendations

The analysis of the elements of the impact logic of the municipal action in the climate adaptation plan of the 15 selected cities shows that cities have difficulties distinguishing between the concrete "outcomes" expected, and the broader "impacts" hoped for from the plans. Therefore, cities should, in addition to their already defined "mission statements" and "long-term goals" (belonging to the category "impact"), also formulate the concrete "outcomes" or direct (local) effects that they expect from the application of the measures of their climate adaptation plan, or the changes they aim to achieve concerning local situations. Impacts can indeed only result from the aggregation of all outcomes at the city-wide level, and it is unrealistic to expect single measures to affect the whole city if one does not first take into account the local effect of these measures.

Further, the analysis of the diversity of instruments used in the climate adaptation plans of the 15 analyzed cities showed clearly that cities seem to invest more energy in "complementary instruments" instead of in the more infrastructure-related measures (be they grey, green, or blue), although these kinds of measures are the most adequate to make the cities more comfortable for future climatic conditions.

Detailed analysis revealed that the measures of many cities' adaptation plans are limited to a few sub-categories of measures. Thus, they are missing out on the potential of some measures, which they could exploit if their plans were more diverse. Therefore, cities should be encouraged to use the full range of instruments available and to take advantage of the wide number of possibilities offered by all categories and sub-categories of measures. They should use the full potential of the possible fields of action and develop measures in each category or sub-category. They especially should develop as many or more measures in the first three categories ("Grey infrastructure", "Green infrastructure", and "Blue infrastructure") as in the last ("Complementary measures").

Moreover, cities should create a "profile sheet" for each measure or group of measures, with a precise description of the measure and information about for instance the implementation cost of the measure (including personal cost, direct implementation costs, public relations, and follow-up costs, if applicable), the financing of it, the means to be made available to carry out the measure, the time needed as well as all the necessary conditions and the actors involved (responsible persons or departments for the planning and for the implementation). Cities should take care of promoting integrated approaches, where different departments cooperate, to foster cooperation, to avoid conflict and to ensure the prompt achievement of the goals.

Furthermore, cities should establish a monitoring system, to follow the implementation of each measure and thus of the whole of the plan. Such a system would indeed enable to assess the efforts already made and clearly show those still necessary.

For five or more years old plans and concepts, it might also be worth it to adapt the plan or concept to the current knowledge on climate change and on the state-of-the-art measures as well as the possibly changed local conditions, like political constellations and aims or finances…

In addition, local and national stakeholders should be involved as broadly (meaning as many stakeholders as possible involved in as many aspects as meaningful) and as deeply (meaning stakeholders involved at a detailed level) as possible. Regionals and federal governments should also encourage the cities to meet the needed preparation to adapt themselves to climate change, especially by adapting their own agendas, by lowering the barriers to implement some measures, and by financing adaptation planning and implementation. They should also encourage networking between the cities for them to exchange best practices.

Cities, which are just beginning to draw up and implement their climate adaptation plans and/or whose financial situation is too unstable to implement all the foreseen measures, should focus on "no-regret measures", i.e., measures that make sense, regardless of the degree of warming. Such measures indeed minimize maladaptation[8] and keep options open, while ensuring already some benefits for the city and its society [21].

6.2 Limits of the Approach and Transferability

This paper proposes a methodology to evaluate the current adaptation plans of cities to warmer summers, based on the analysis of the diversity of instruments envisioned in the adaptation plan, their implementation status, and the degree of detail of the developed strategies in the different categories of the "impact logic" of municipal action.

The focus was placed here on the adaptation to warmer summers, independently of concomitant challenges such as climate protection or increased drought or heavy precipitations. A thorough analysis of climate adaptation plans should take all effects of climate change into account, which was not possible here. Therefore, it is possible that the categories and sub-categories of measures used to evaluate their diversity ("Grey infrastructure", "Green infrastructure", "Blue infrastructure" and "Complementary instruments", with their sub-categories) in this article might not all be suitable for every city, and it may be possible that some sub-categories cannot be used for some specific cities. If other effects of climate change are taken into account, then other categories and sub-categories of measures would need to be taken into account.

The analysis of the implementation status of the measures was based on the information available online to the public, which is often not complete and not always

[8] "Maladaptation refers to actions that may lead to increased risk of adverse climate-related outcomes, including via increased greenhouse gas emissions, increased or shifted vulnerability to climate change (…) Most often, maladaptation is an unintended consequence" [21].

up to date but enables to see which measures are important for the cities to report about. For a more precise analysis of the real implementation status of the measures, cities should be contacted and directly asked about the real implementation status of each measure and thus of the whole plan.

Concerning the impact logic of the municipal action toward climate adaptation, some limitations need to be taken into account, like the unavoidable time lag between the different elements, and the diverse paths from input to impact—which can be fraught with loss—which means, that when analyzing a project at a given time, the input is often not yet reflected in impact. However, despite the previously listed limitations, impact logic still provides a very valuable overview of the various activities undertaken as well as clues to answering the questions of whether these are sufficient and what current and future activities should cities focus on [32].

Moreover, not all outcomes and impacts can be expressed in figures: some qualitative information and subjective perceptions also play important roles, like the feeling of the population of being prepared for heatwaves, although they cannot be accurately measured due to the lack of uniform and efficient measurement concepts [32].

6.3 Need for Further Data and Research

Research and cooperation with the municipalities are therefore needed to support them in identifying and developing best practices, which proved effective in other cities with similar conditions. "No regret measures", which will be useful regardless of the degree of warming, should also be further explored, as they can encourage cities to actively implement their plans. Research should also focus on the contrary of such measures and identify which measures are more susceptible to lead to maladaptation on how to avoid it. Consequently, there is a need for methods to evaluate the potential cost–benefit analysis of measures and which conditions need to be taken into account.

Concerning the implementation of the measures, municipalities also need some support on how best to create "profile sheets" for their measures, and which information should be included in the "profile sheets" to best facilitate the implementation and the checking of the feasibility of the impact logic.

Lastly, cooperation between research and municipalities is also necessary, to develop efficient monitoring reports, which best show the path already covered and clearly point out the distance still to be covered and the room for improvement in this crucial challenge for cities in the coming century.

However, it is widely known that data availability is often insufficient, especially at the small-scale level of (small and medium) cities. Therefore, monitoring reports should, where possible, concentrate on available information instead of asking cities with limited resources to set up new data collection for purposes of local monitoring processes. Yet, some indicators require new individual assessments and some of them can only be operationalized with considerable financial and human effort. There, researchers should meet with practitioners like cities and regions, which themselves should cooperate to identify the most efficient ways to acquire the necessary data or to identify which more easily available data can replace the desired information.

7 Conclusion

Research shows that municipalities are aware of the challenges of climate adaptation that await them, and they are trying to actively tackle the issue. However, cities have difficulties distinguishing between the concrete "outcomes" expected, and the broader "impacts" hoped for from the plans. Therefore, cities should work on also formulating concrete "outcomes", i.e. direct (local) effects that they expect from the application of the measures of their climate adaptation plan, as impacts at the societal level can only result from the aggregation of all outcomes at the city-wide level.

Moreover, the analyzed climate adaptation plans of the selected medium-sized cities located in the "warmest climatic regions" of Germany showed that cities have a large choice of categories of measures to choose from to adapt themselves to a warmer climate and that they already use a significant part of this broad possibilities. In all but one case, the analyzed plans indeed contained measures in the four defined categories "Grey infrastructure", "Green infrastructure", "Blue infrastructure", and "Complementary measures". But many measures consist of accompanying measures, instead of more infrastructure-related measures. Cities should therefore increase their effort in this direction. They should also develop a detailed "profile sheet" for each measure and closely monitor their implementation progress and impact.

Research is therefore needed to support the municipalities to identify and develop best practices, to identify "no regret measures", to avoid maladaptation, and to know which information is necessary to create useful "profile sheets". Cooperation between research and municipalities is also necessary, to develop efficient monitoring reports, which best show the path already covered and clearly point out the distance still to be covered and the room for improvement in this crucial challenge for cities in the coming decades.

Although cities have been more or less consciously adapting to changing conditions for centuries, the speed of the current climate change requires them to now make decided conscious adaptation efforts, and where efforts have already begun, to scale up and speed up their endeavors. Researchers and practitioners need to work hand in hand to identify possible challenges and solutions early on, but as some lessons can only be "learned by doing", it is crucial that knowledge gained "on the field" is relayed back to theoreticians but also quickly and efficiently spread around to other practitioners.

References

1. Albrecht J, Schanze J, Klimmer L, Bartel S (2018) Klimaanpassung im Raumordnungs-, Städtebau- und Umweltfachplanungsrecht sowie im Recht der kommunalen Daseinsvorsorge. Grundlagen, aktuelle Entwicklungen und Perspektiven. Climate adaptation in regional planning, urban development, and environmental planning law as well as in the law of municipal services of general interest. Fundamentals, current developments, and perspectives. Umweltbundesamt (UBA), Dessau-Roßlau
2. Bäumer C, Schuster L, Hage G (2018) Klimawandel und Klimaanpassung: Zukunftsfähige Gesundheits-, Lebensumfeld- und Erholungsvorsorge in der VVG Bühl-Ottersweier. Climate change and climate adaptation: sustainable health, living environment and recreation provision in the VVG Bühl-Ottersweier. LUBW Landesanstalt für Umwelt, Messungen und Naturschutz Baden-Württemberg (ed.) KLIMOPASS-Berichte. Karlsruhe: LUBW Landesanstalt für Umwelt, Messungen und Naturschutz Baden-Württemberg
3. Baumgart S, Flacke J, Grüger C, Lütke P, Rüdiger A. (Ed.) (2004) Klein- und Mittelstädte – Verkleinerte Blaupausen der Großstadt? Dokumentation des Expertenkolloquiums am 29. April 2004 an der Universität Dortmund (SRPapers, Nr. 1). Small and Medium-Sized Towns - Downsized Blueprints of the Big City? Documentation of the expert colloquium on 29 April 2004 at the University of Dortmund. Dortmund
4. C40 Cities Climate Leadership Group, Inc (2021) Urban Cooling Toolbox
5. Chen L, Shi J (2016) Analysis and predication of urban water security: a case study of Chengdu City, China. IOP Conf. Series Earth Environ. Sci. 39:1755–1315
6. Deutscher Städtetag (Ed.) (2019) Anpassung an den Klimawandel in den Städten – Forderungen, Hinweise, und Anregungen. Adaptation to climate change in cities - demands, hints, and suggestions. Berlin and Cologne
7. Deutscher Wetterdienst (DWD) (2022a) Zeitreihen und Trends. Time series and trends. https://www.dwd.de/DE/leistungen/zeitreihen/zeitreihen.html Last accessed on 23.09.2022
8. Deutscher Wetterdienst (DWD) (2022b) Stadtklima – die städtische Wärmeinsel. City climate - the urban heat island. https://www.dwd.de/DE/klimaumwelt/klimaforschung/klimawirk/stadtpl/projekt_waermeinseln/projekt_waermeinseln_node.html Last accessed on 23.09.2022
9. Energielenker Beratungs GmbH (2018) Integriertes Klimaschutz- und Klimaanpassungskonzept der Stadt Hürth. Integrated climate protection and climate adaptation concept of the city of Hürth. Hürth
10. Energielenker Beratungs GmbH (2019) Integrierter Klimaschutz und Klimaanpassung. Konzept der Stadt Rodgau. Integrated climate protection and climate adaptation. Concept of the City of Rodgau. Rodgau
11. Energielenker Beratungs GmbH (2021) Klimaanpassungskonzept für die Stadt Herten. Abschlussbericht 2021. Climate adaptation concept for the city of Herten. Final report 2021. Herten
12. European Environment Agency (EEA) (2022): Economic losses and fatalities from weather- and climate-related events in Europe. In: EEA Briefing 21/2021. Luxembourg. https://www.eea.europa.eu/publications/economic-losses-and-fatalities-from Last accessed: 23.09.2022
13. faktorgruen Freie Landschaftsarchitekten BDLA (2016) Strategisches Fachkonzept Klimaanpassung (KliK). Strategic concept for climate adaptation (KliK). Ludwigsburg
14. faktorgruen Freie Landschaftsarchitekten BDLA (2020) Stadt Landau in der Pfalz. Klimaanpassungskonzept. Endbericht. City of Landau in der Pfalz. Climate adaptation concept. Final report. Landau in der Pfalz
15. Früh, B, Koßmann, M, Roos, M. (2011) Frankfurt am Main im Klimawandel. Eine Untersuchung zur städtischen Wärmebelastung. Frankfurt am Main in climate change. An investigation of the urban heat load. Berichte des Deutschen Wetterdienstes 237. Deutscher Wetterdienst (DWD), Offenbach am Main
16. GEO-NET Umweltconsulting GmbH, MUST Städtebau and Dr. Pecher AG (2021): Klimaanpassungskonzept der Stadt Erkrath. Teil A: Klimaanpassungsstrategie. Climate adaptation concept of the city of Erkrath. Part A: Climate adaptation strategy. Erkrath

17. Hackenbruch, J. (2018): Anpassungsrelevante Klimaänderungen für städtische Baustrukturen und Wohnquartiere. Adaptation-relevant climate changes for urban building structures and residential areas. Wissenschaftliche Berichte des Instituts für Meteorologie und Klimaforschung des Karlsruher Instituts für Technologie 77. Kottmeier, Christoph, Karlsruhe.
18. Hofstetter, W, Jurisch, K, Lieser, R, Reinholz, K. (2016): KLAK – Worms handelt! Konzept zur Anpassung an den Klimawandel. KLAK - Worms acts! Concept for adaptation to climate change. Worms
19. Infrastruktur & Umwelt Prof. Böhm und Partner (2016): Klimaschutzteilkonzepte zur Anpassung an den Klimawandel für die Stadt Lohmar, die Gemeinde Much und die Gemeinde Ruppichteroth. Abschlussbericht – Kurzfassung. Climate protection sub-concepts for adaptation to climate change for the city of Lohmar, the municipality of Much and the municipality of Ruppichteroth. Final report - abridged version. Lohmar, Much and Ruppichteroth.
20. Infrastruktur & Umwelt Prof. Böhm und Partner (2018) Integriertes Klimaschutz- und Klimaanpassungskonzept für die Kreisstadt Siegburg – Endbericht. Integrated climate protection and climate adaptation concept for the town of Siegburg - final report. Siegburg
21. IPCC (2022) Climate Change 2022: Impacts, Adaptation and Vulnerability. Contribution of Working Group II to the Sixth Assessment Report of the Intergovernmental Panel on Climate Change [H.-O. Pörtner, D.C. Roberts, M. Tignor, E.S. Poloczanska, K. Mintenbeck, A. Alegría, M. Craig, S. Langsdorf, S. Löschke, V. Möller, A. Okem, B. Rama (eds.)]. Cambridge University Press. Cambridge University Press, Cambridge, UK and New York, NY, USA, 3056 pp., doi:https://doi.org/10.1017/9781009325844
22. Jankowiak, J, Brinkmann, J, Hirschfeld, S. (2020): Strategie zur ökosystembasierten Anpassung an den Klimawandel in Bernau bei Berlin. Anpassung an den Klimawandel in Bernau mittels ökosystembasierter und partizipativer räumlicher Planung. Strategy for ecosystem-based adaptation to climate change in Bernau bei Berlin. Adaptation to climate change in Bernau using ecosystem-based and participatory spatial planning. Bernau bei Berlin
23. Ketterer C, Matzarakis A (2014) Human-biometeorological assessment of heat stress reduction by replanning measures in Stuttgart, Germany. Landsc Urban Plan 122:78–88. https://doi.org/10.1016/j.landur-bplan.2013.11.003
24. Klimakommunen Hessen (2022): Hessen aktiv: Die Klima-Kommunen. Aktionsplan für die Kreisstadt Dietzenbach. *Hesse active: The climate municipalities. Action plan for the district town of Dietzenbach.* Dietzenbach.
25. Kötter, T, Weiß, D, Heyn, T, Grade, J, Lennartz, G. (2018): Stresstest Stadt – wie resilient sind unsere Städte? Unsicherheiten der Stadtentwicklung identifizieren, analysieren und bewerten. Stress test city - how resilient are our cities? Identify, analyse and evaluate uncertainties in urban development (Bundesinstitut für Bau- Stadt und Raumforschung (BBSR) of the Bundesamt für Bauwesen und Raumordnung (BBR), Ed.), Bonn
26. Kötter T (2022) Zur Bedeutung der Planungswissenschaften für die Planungspraxis. On the importance of planning sciences for planning practice. Flächenmanagement und Bodenordnung (fub) 84(1), 25–34
27. Kurz B, Kubek D (2013) Kursbuch Wirkung. Das Praxishandbuch für alle, die Gutes noch besser tun wollen. Coursebook Effectiveness. The practical handbook for all those who want to do good even better. Berlin. http://www.bertelsmann-stiftung.de/fileadmin/files/BSt/Publikationen/GrauePublikationen/GP_PHINEO_Kursbuch_Wirkung.pdf
28. Landsberg HE (1981) The Urban Climate. Academic Press, New York
29. Meerow S, Newell JP, Stults M (2016) Defining urban resilience: A review. Landsc Urban Plan 147:38–49. https://doi.org/10.1016/j.landurbplan.2015.11.011
30. MUST Städtebau and Energielenker Beratungs GmbH (2020) Klimafolgenanpassung in Meerbusch. Konzept zum Umgang mit den unvermeidbaren Folgen des Klimawandels im Meerbuscher Stadtgebiet. Climate Impact Adaptation in Meerbusch. Concept for dealing with the unavoidable consequences of climate change in the Meerbusch urban area. Meerbusch: Stadt Meerbusch
31. National Institute for Occupational Health and Safety (NIOSH) (2020) Heat Stress. Available online: https://www.cdc.gov/niosh/topics/heatstress/default.html Last accessed: 23.09.2022

32. Peters O (2021) Nachhaltigkeit im Kontext städtebaulicher Entwicklung - Kommunen auf dem Weg zu mehr Wirkungsorientierung. Sustainability in the context of urban development - municipalities on the way to more impact orientation. Flächenmanagement und Bodenordnung (fub) 83(5), 203–209
33. ProjektStadt (2019) Anpassung an den Klimawandel in Großauheim. Fokus Gesundheit. Adapting to climate change in Großauheim. Focus on health. Hanau: Stadt Hanau
34. Stadtverwaltung Speyer - Stadtentwicklung (2015): Klimawandel Speyer folgen. Integriertes Konzept und innovative Strategien zur Anpassung. Abschlussdokumentation 2012 – 2015. Follow climate change in Speyer. Integrated concept and innovative strategies for adaptation. Final documentation 2012 - 2015. Speyer: Stadtverwaltung Speyer, Stadtentwicklung
35. Umweltbundesamt (UBA) (Ed.) (2021a) Klimawirkungs- und Risikoanalyse 2021 für Deutschland – Teilbericht 1: Grundlagen. Climate Impact and Risk Analysis 2021 for Germany - Subreport 1: Fundamentals Kahlenborn W, Linsenmeier M, Porst L, Voß M, Dorsch L, Lacombe S et al Dessau-Roßlau
36. Umweltbundesamt (UBA) (Ed.) (2021b) Klimawirkungs- und Risikoanalyse 2021 für Deutschland – Teilbericht 4: Klimarisiken im Cluster Infrastruktur. Climate Impact and Risk Analysis 2021 for Germany - Sub-report 4: Climate Risks in the Infrastructure Cluster. Voß M, Kahlenborn W, Porst L, Fritsch U, Renner K, Zebisch, M. et al Dessau-Roßlau
37. Umweltbundesamt (UBA) (Ed.) (2021c) Klimawirkungs- und Risikoanalyse 2021 für Deutschland – Teilbericht 5: Risiken und Anpassung in den Clustern Wirtschaft und Gesundheit. Climate Impact and Risk Analysis 2021 for Germany - Sub-report 5: Risks and Adaptation in the Economy and Health Clusters. Voß M, Kahlenborn W, Porst L, Fritsch U, Renner K, Zebisch, M. et al Dessau-Roßlau
38. Umweltbundesamt (UBA) (Ed.) (2021d) Klimawirkungs- und Risikoanalyse 2021 fürDeutschland – Teilbericht 6: Integrierte Auswertung – Klimarisiken, Handlungserfordernisse und Forschungsbedarfe. Climate Impact and Risk Analysis 2021 forGermany - Substudy Report 6: Integrated Evaluation - Climate Risks, Action Requirements and Research Needs. Voß M, Kahlenborn W, Porst L, Fritsch U, Renner K, Zebisch, M. et al Dessau-Roßlau
39. Umweltbundesamt (UBA) (Ed.) (2021e) Klimawirkungs- und Risikoanalyse 2021 für Deutschland – Kurzfassung. Climate Impact and Risk Analysis 2021 for Germany - Executive Summary. Voß M, Kahlenborn W, Porst L, Fritsch U, Renner K, Zebisch, M. et al Dessau-Roßlau
40. World Health Organization (WHO) (2018) Heat and Health. https://www.who.int/news-room/fact-sheets/detail/climate-change-heat-and-health Last accessed: 23.09.2022
41. World Meteorological Organization (WMO) (Ed.) (1996) Climate and Urban Development, Genf.

An Accessibility of New Urban Area's Residents to Social Infrastructure in Hue City. Case Study: An Cuu and My Thuong New Urban Area

Duyen Thi My Thieu

Abstract Vietnam's urbanization process has been characterized by a rapid increase in the quantity and scale of new urban areas (NUAs). Aside from major cities such as Hanoi and Ho Chi Minh City, Hue city has experienced significant change through the expanded urban area [1, 2]. It is largely shaped by the development of new residential areas, particularly NUA projects. Furthermore, the development of the corresponding infrastructure system can be regarded as a critical condition of the NUAs in specific and of the city in general. In this case, social infrastructure plays an important role in providing and meeting people's basic daily needs [3, 4]. In practice, however, are they given effective distribution? The study's focus on the accessibility of NUAs residents to social infrastructure is on assessing the type of facilities and distance. Data is collected from the actual distribution of social infrastructures and NUAs in Hue city and combined processing and visualizing by using Google Earth and GIS. The study specifically looks into (1) the coverage of basic types of social infrastructure and (2) people's accessible distance to infrastructure (non-motorized modes of transportation). This study can also be viewed as a foundation for future research into perfecting the criteria for assessing accessibility in urban planning.

Keywords New urban area · Accessibility · Social infrastructures

1 Introduction

NUA (also called New Town, or Khu đô thị mới in Vietnamese) is recognized for its significant proliferation in Southeast Asia, and Vietnam is no exception [1–4]. After the Reform (Doi Moi) in 1986, the urbanization by market economy implementation and industrial transformation was accelerated throughout the country. The NUA definition was officially issued in Decree No. 52/1999/ND-CP. In the very first stage of urbanization, the 1990s, NUA form was only largely invested in the two largest

D. T. M. Thieu (✉)
Sustainable Urban Development, Ho Chi Minh City, Vietnam
e-mail: duyen.thieu@stud.tu-darmstadt.de

V. T. Ha et al. (eds.), *Proceedings of the 8th International Conference on Sustainable Urban Development*, Advances in 21st Century Human Settlements,
https://doi.org/10.1007/978-981-99-8003-1_12

cities, Hanoi and Ho Chi Minh. It then, from the 2000s, has spread to many provinces of Vietnam mostly by transforming from agricultural land to urban land [5–7].

Located in the coastal line, the Complex of Hue monuments recognized by UNESCO as a World Cultural Heritage in 1993 and considered one of the central cities in Vietnam's middle would constitute a thrust for Hue's development, the city, however, gets in stuck between conversation and accelerating urbanization. Though the long-term development since being a historical capital of the Nguyen Dynasty in the 1800s–1900a, the center of Hue is mainly located along two sides of the Huong river. Since the orientation of the expanded area to the south of the Huong river in 1999, which was officially issued in 2021. The upgraded city's rank 1 in 2005, Hue city has more prosperous development. Most of the new urban development area is land converted from agricultural land, some of which are the area to be merged into Hue city in 2021 (Fig. 1). Hue City developed general planning in a context quite similar to Hanoi which carries out conservation work and constructs modern urban areas based on city expansion in the direction of a "heritage city" (đô thị di sản). Therefore, it can be said that although Hue city has a fairly early history of urban development, its development speed is quite slow compared to the development process in time and neighboring cities of the same rank.

Along with the government's recent efforts in urban development and extension of Hue city, modern urban projects have been focused on addressing inner-city challenges as well as preparing for future development needs. To expedite urbanization, NUA development in both quantity and size is the preferred option. On the one hand, these initiatives help to accommodate a variety of housing needs and improve quality of life, all of which contribute to the modernization of Hue. On the other hand, the challenges typically encountered in the compensation process, spatial and social segregation, rapid development, and mismatches in planning management or capital balance can all have impacts on the qualitative development of NUA projects.

Infrastructure is referred to as an "integral component of the urban fabric" [8]. In terms of "macro-area", it is defined to cover four areas: health, education, culture, and environment following Torrisi [9]. It refers not only to the physical infrastructure, which is obviously a visible component of the city, but also to a variety of facilities and services ranging from nonsocial to SI. According to Williams and Pocock [10], the social infrastructure is broadly defined and comprises "soft" services that support the social environment in general as well as "hard" facilities. Furthermore, social infrastructure is defined as networks of amenities, institutions, and locations with various roles [8]. Similarly, "SI is comprised of the facilities, places, services, and networks that promote the quality of life and well-being of our communities," with these facilities run by governments and other service providers [11, p. 388]. Specifically, commercial spaces are also included in SI [8, 11].

SI are listed by various scholars. According to Brown and Barber [12, p. 100], they defined SI including "schools, health centers, leisure and recreation facilities, libraries, community centers, religious facilities, local shops, open spaces, transport and utility services and emergency services". More specifically, Klinenberg [13,

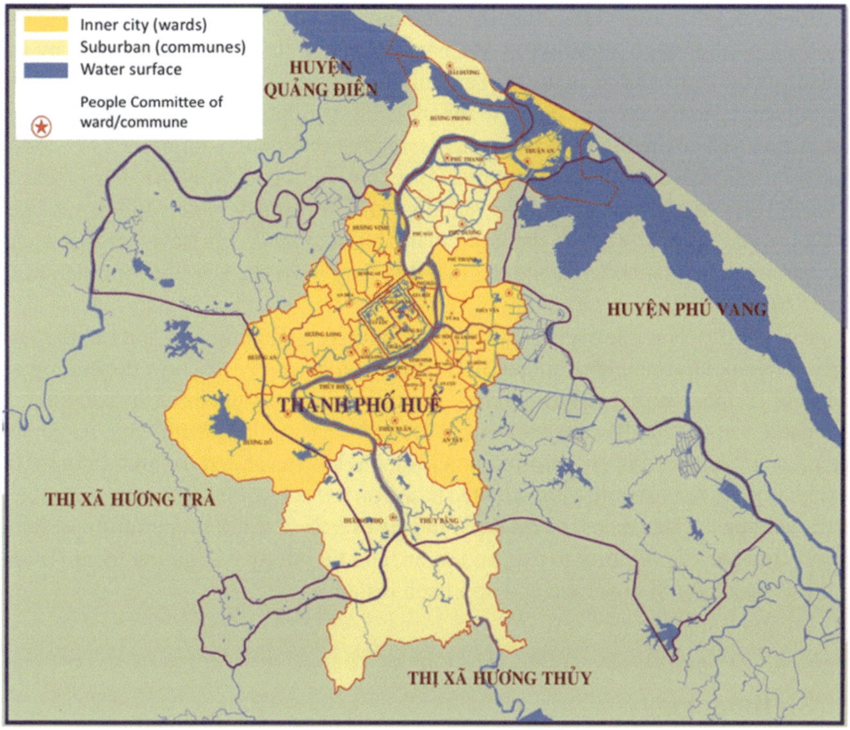

Fig. 1 The new administrative boundary of Hue city (From 1st July 2021). *Source* http://huecity.gov.vn

p. 33] illustrated in his research "Public institutions, such as libraries, schools, playgrounds, parks, athletic fields, and swimming pools are vital parts of the social infrastructure. So too are sidewalks, courtyards, community gardens, and other spaces that invite people into the public realm. Community organizations, including churches and civic associations, act as SIs when they have an established physical space where people can assemble, as do regularly scheduled markets for food, furniture, clothing, art, and other consumer goods". Therefore, it is clear that SIs are kinds of facilities necessary for cities [8], social identity contribution [14], and increasing social comfort and to act on the economic productivity [15].

The importance of physical infrastructure in urban development has been noted elsewhere in Vietnam [16, 17], mostly through road and power networks. Satisfactory infrastructure has also become a criterion for urban appraisal and ranking [18]. However, unlike technical infrastructure, SI investment is frequently established after the formation of an urban or residential area [19], and the distribution of SIs is impacted by a variety of other circumstances rather than being determined. People's efficiency and accessibility are critical aspects. All of these have an impact on people's comfort and quality of life.

However, the reality shows that the planning and investment in the SI system has not been paid enough attention in Vietnam. That can first be demonstrated by the lack of specific provisions in the SI planning in the legal system. It will be easy to see the difference in the level of detail in the planning regulations between the technical infrastructure and the SI, such as standards, regulations, explanations, etc. Secondly, based on the reasons for the limitation of economic conditions, SI investment is often developed after the urban or residential area is formed, which affects the lives of the people [15]. Moreover, there are many NUAs that do not have enough social infrastructure as planned, despite the fact that they have been in use for a period of time [20].

Accessibility is a relatively new problem in urban planning in Vietnam, particularly in residential neighborhoods. It is defined in a variety of ways depending on the field of interest of each researcher, but they all have something in common. It is regarded in the small-scale residential area as the simplicity with which a person may reach various sorts of facilities, and it should be considered for non-automobile residents [21]. Accessibility is evaluated based on a variety of factors such as distance (various types of distances, or multiple means of access if the scope of the study is large), time, an opportunity to use, and so on, and this study will go into detail of the analysis based on SI coverage and approach distance.

2 Methodology

A desk review is used to gather data from secondary sources. Secondary data was used to establish the broad NUA development background of Hue city as well as detailed planning information for the case study. Furthermore, based on associated previous research on the assessment of SI accessibility in general and specific facilities, develop a suitable procedure that can be represented using ArcGIS software.

In June and July 2021, observation and semi-structured interviews were carried out. The author collected, categorised, and recorded the SI kinds in the form of actual location coordinates. Eight interviews were conducted with two groups of people, including residents of An Cuu and My Thuong NUA. All these persons have lived in NUAs for at least three years and were chosen at random from the park and streets. In addition, online interviews with members of local administrations and the Office of the People's Committee of Thua Thien Hue Province were held (Fig. 2).

ArcGIS software was used as the primary tool to map the distribution of SIs and calculate distances according to the study's objectives.

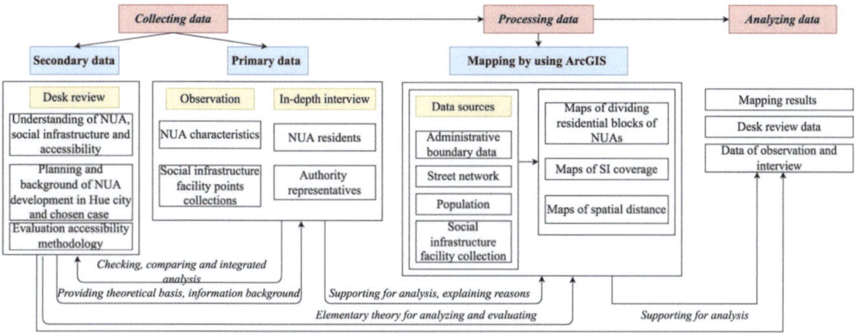

Fig. 2 Process of methods application. *Source* Author

2.1 SI Classification and Accessible Factors

(1) SI classification

SI is a broad notion that should be considered as a whole, comprising the system of facilities and services, with the objective of meeting inhabitants' requirements and improving their quality of life. The components of the SI are dynamically determined based on the methods of various fields. Furthermore, each country's classification is based on its own traits and conditions.

In this research, SI types were defined based on three scopes: serving daily needs, public and private sector, and distribution within 1000 m radius of the NUA centroid. First, "daily needs" are mentioned to emphasize the necessity and frequency (or urgency) of people's use of SI types [16]. This determination varies from country to country, and it is also influenced by other factors, such as population scale (or population unit). The second element is updated on the diversity of investors and operators of SI types from both the public sector and private sector. Typical examples are retail outlets or schools. Finally, a distance of 1,000 m was chosen for the distribution of basic SI facilities/services [22]. Besides, in the context of urban planning in general and urban residential areas in particular, the study also identifies SI types based on Vietnam's Building Code (2021) and Construction Law (2014).

The SI categories mentioned in this study were eligible from three selected scopes and legal contexts, particularly the population size of a hierarchical housing unit in Vietnam (in this case, NUA). SI contains (1) healthcare which should be a medical station and belongs to the ward-level administrative unit, (2) education which is required for a kindergarten, primary school, and/or low-secondary school depending on the size of the population and surrounding distribution, (3) park/public green space, (4) cultural and sports center which are defined as a gathering place for community meetings and activities, as well as a site for sports, and (5) commerce which be defined as a place where people able to come and use various services or buy products by free or payment [13], hence, in the local context, it should include market

Table 1 Types of SI

	Types of chosen SI
Healthcare	Healthcare unit
Education	Kindergarten
	Primary school
	Low-secondary school (optional)
Cultural-Sport center	Sport/training ground
	Cultural/community house
Commerce	Market
	Grocery store
Park	Park

Source Author

(or informal market—street market), mini-market, and grocery shop. These specific SI types are listed in the table below (see Table 1).

(2) **Accessible factors**

According to relevant studies [16, 23–25], it is important to identify three basic variables before implementing accessibility assessment modeling. These are origin, destination, and distance.

- Origin places are defined as areas where population groups with the ability to use utilities can be found. Origins are indicative of a chosen NUA and residential blocks within that NUA that are users or have a need to access SI amenities in this study. There are a NUA's centroid point and residential block's centroids (see Fig. 3).
- Destinations represent locations of specific SI kinds based on the SI scopes section which are the utilities and services that residents from the original locations have access to. These destinations are in point format.

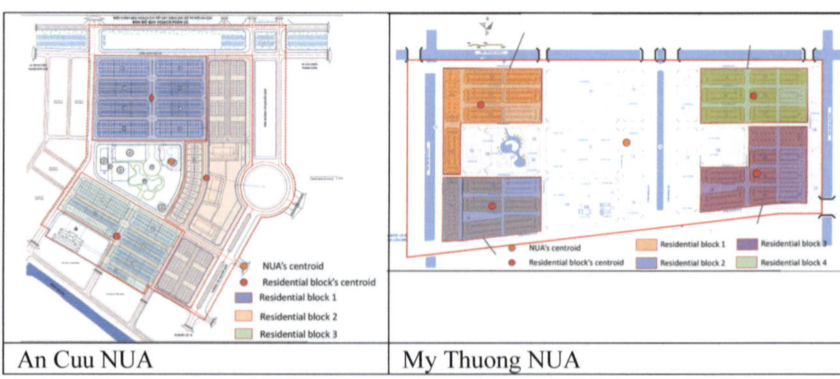

Fig. 3 The origin places. *Source* Author

– Distance is defined as the distance between two points, one of which is the origin and the other is the destination. To assess spatial accessibility, the study measured a network distance which is the actual trip distance determined based on the existing traffic infrastructure.

2.2 Evaluation Accessibility Stages

Based on the research results of [16, 24], this study was conducted to evaluate the accessibility of NUA residents to SIs through 5 steps (see Table 2).

– In addition to the An Cuu NUA's boundaries, a specific area surrounding the NUA must be defined. A radius of 1 m kilometer could be a suitable choice for spatial extending based on the argument of a favorable distance to approach the underlying SIs [22].
– The factors required for quantification and accessibility assessment are identified in the second stage. Origins, destinations, and the distance between the two points above are variables. As a result of the foregoing factors, the study solely looked at accessibility within walking distance.
– The coverage of SIs is measured using ArcGIS software's buffer analysis which figures out how many and what kind of SIs are in a given radius [26, 27].
– The distance between each point of origin and destination is used to determine spatial accessibility. The second type of distance, often known as network distance, is measured by using the traffic system. The network analysis tool in ArcGIS is used.
– The final step of the study is to evaluate the accessibility of the residents of the NUAs to the SIs based on the shortest distance determination.

Table 2 Stages of accessibility

Stage	
(1) Define research scale	Area of 1 km from the centroid of the NUA (polygon)
(2) Define accessibility variations	• Origins (point) • Destinations (point) • Distance between these points (line)
(3) Measure the coverage of SI	• Number of SI • Types of SI
(4) Compute distances	Network distance
(5) Evaluation access distance	Nearest distance

Source Author

3 Results and Discussion

3.1 Context of NUA

The general planning of the An Van Duong project was officially promulgated in 2005 under Thua Thien Hue province's Decision No. 1577/QD-UBND. The study area for this project is 2,150 ha, with 1,700 ha authorized in the 2005 master plan and Zone E approved in 2018. The present population status at the time of the planning decision was 11,900 people, with a projected population of 60,000 people by 2020. The project aims to create a modern NUA with a variety of housing types while also serving as the province's new administrative center, sports center, commercial facilities, and entertainment facilities.

An Van Duong NUA (Zone A, B, C, D) and zone E accomplished comprehensive design and construction of 112 projects covering an area of about 891.71 ha between 2005 and 2020. The state government invests in 52 projects, the majority of which are technical infrastructure, and there are six schools (3 primary schools, two junior high schools, and one high school); private capital invests in 49 projects (501.9 ha). Although there are 18 new urban/residential projects totaling 337.89 ha, only 8 are actively under development and in use.

(1) **An Cuu NUA**

The An Cuu project, which is part of the An Van Duong NUA, is one of numerous NUAs in Zone A which spans 32.2 ha. An Cuu NUA was approved in 2006, one year after the An Van Duong project was approved, under Decision No. 56/QD-UBND Thua Thien Hue Province (2006) which serves the high-quality housing needs of the city in the future. It was designed on agricultural land and is located in An Dong ward (Thuy An commune before 2007). An Cuu NUA is one of the few early approved NUA projects on the scale of An Van Duong NUA and within Hue city in general.

The revised An Cuu NUA planning 2019 includes numerous adjustments, mostly in terms of building types and larger housing plots. Sports facilities are diverse in terms of public services; however, the construction overlaps the park's territory and significantly diminishes the park's actual space. Furthermore, NUA does not intend any schools, according to the initial approved plan. After then, an extra secondary school was erected on the land that the investor had given to the state. However, when it comes to social infrastructure needs for residential units, the educational infrastructure is not really appropriate since there are not enough kindergarten and primary schools in the official approval compared to NUA's population. In addition, except for the shophouse areas that were approved in 2019, commercial facilities providing people's everyday needs, such as markets or supermarkets, are not included in the plan.

(2) **My Thuong NUA**

My Thuong NUA has a total area of 622,443 m^2 and is expected to have a population of 6,916 people was approved by Thua Thien Hue province's Decision No.1886/

QD-UBND in 2005 and partially adjusted by Decision No.1399/QD-UBND 2018. It was planned with the intention of meeting housing and service demands.

In terms of planning, My Thuong NUA's social infrastructure is more diversely developed than An Cuu NUA due to its larger scale in both area and people. For example, the area for parks, markets, primary school, etc., however, kindergarten is not mentioned in the planning map. Although the project was formally completed in 2008, there are still many incomplete components, particularly social amenities, at the time of the research survey in 2021.

3.2 Coverage of SIs

Based on the classification method and unified understanding of SI types mentioned in the above classification, the author identifies the types of SI according to the survey results in the study area. It is necessary to repeat the determination method based on the radius of 1 km from the center of NUAs. Facilities both in the public sector and private sector are also included.

It is important to note which sorts of SI are found within and outside of the An Cuu NUA's planned region. An Cuu NUA has six SI facilities, however, they only belong to three of the five SI classes initially designated by the Vietnam Building Code (2021). Educational facilities (1 private kindergarten and 1 middle school), a sports facility (3), and a park are among them (1). The number of SIs surrounding the An Cuu NUA area is often more diverse, particularly within the commercial group. In addition, there are preschools and primary schools in the immediate vicinity. Within this radius, a formal market trade facility, such as a market or supermarket, is absent in terms of the basic SIs that must be met for an accommodation unit. Some residents of this NUA, according to in-depth interview data, say they have to travel further if they want to buy food at markets or supermarkets because street markets or grocery stores can only accommodate one few types of goods (see Table 3).

Thirteen SI facilities are documented within a one-kilometer radius of the centroid of My Thuong NUA. It contains four facilities located inside the My Thuong NUA area, including two football grounds, three grocery stores, and a park. Furthermore, similar to An Cuu NUA, the majority of commerce is in the private sector, and educational facilities are located outside of the NUA. In addition, within the radius, there is no healthcare unit. This NUA has fewer SI facilities than An Cuu NUA, and the SI type is incomplete in comparison to the basic SI types required for the day-to-day needs of the originally recognized residents. Several SI facilities in My Thuong NUA were approved in 2005, but have yet to be built. Hence, they are not included in the assessment of resident accessibility because this study prefers to look at access to current SI amenities (see Table 4).

Table 3 Coverage of SIs within 1 km of An Cuu NUA centroid

Types of SI		Public	Private
Healthcare	Healthcare unit	An Dong	
Education	Kindergarten	An Dong (LM)	Thao Nguyen Xanh*
		An Dong (LQĐ)	Bup Sen
			Phuoc Van
	Elementary school	An Dong 1 (1)	
		An Dong 1 (2)	
		An Dong 2	
	Low-secondary school (optional)	Dang Van Ngu*	
Cultural-Sport center	Sport/training ground	An Cuu swimming pool*	Tennis courts*
			An Cuu football ground*
	Cultural center	Community house Zone 6	
		Community house Zone 2	
Commerce	Market		Street market
	Grocery store		Grocery store 1
			Grocery store 2
			Grocery store 3
			Grocery store 4
			Grocery store 5
			Grocery store 6
			Grocery store 7
			Grocery store 8
			Grocery store 9
			Grocery store 10
			Grocery store 11
			Grocery store 12
Park	Park	An Cuu*	
	Pocket park	Nearby residential area	
		Riverside	
		Nearby The Manor residential area	

Noted * SI within An Cuu NUA
Source Author

Table 4 Coverage of SIs within 1 km of My Thuong NUA

Types of SI		Public	Private
Healthcare	Healthcare unit		
Education	Kindergarten	Phu My	Newstar
	Elementary school	Phu Thuong 1	International school
	Low-secondary school (optional)		International school
Cultural-Sport center	Sport/training ground	Foodball ground*	Informal football ground*
	Cultural center	Duong Mong village	
Commerce	Market		Informal market
	Grocery store		Grocery store 1
			Grocery store 2*
			Grocery store 3*
			Grocery store 4*
	Park		
Park	Pocket park	Pocket park	

Noted * SI within My Thuong NUA
Source Author

3.3 *Accessible Distance*

Distance is used to measure the spatial accessibility of households to SIs for daily use, particularly Network distance within the scope of this research. The Network distance—actual travel distance is measured using the road system that residents take from their location. It is determined from the centroids of residential blocks to the SIs. The distance is computed for pedestrians, hence, factors such as traffic density and road type are not taken into account in this research. It is classified into three categories: less than 500 m (friendly walkability distance), 501 to less than 800 m (walkability distance), and more than 800 m (un-walkability distance).

(1) **An Cuu NUA**

Residents of residential block 1 often have easier access to many types of SI than residents of the other two blocks. The number of SI facilities distant from block 2 over 800 m is the highest. For example, if within a distance of fewer than 500 m of block 1, there are 11 facilities in terms of Network distance. Meanwhile, residents of block 2 can only reach 5 SI by network distance. In contrast, it has 14 SI facilities that people need to travel over 800 m to reach.

Residents of each block can travel a different distance, and the SI that represents each group of SIs reached is also varied. Because of their position inside An Cuu NUA, private sports facilities, parks, and preschools are generally accessible. However, it is essential to drive a considerable distance to reach a public preschool or primary school. The actual accessible distance to the informal market is more than 500 m for all three blocks. Finally, the medical station is always the sort of SI that must progressively go as little as 1223 m to reach its destination (for block 3).

The shortest distance to contact the statistical author is provided in the table below based on measured distances from the residential blocks to the SI facilities. People may easily access private kindergarten, lower-secondary school, sports center, park, and grocery stores; nevertheless, the distance of more than 800 m to primary school, market, and the medical station is deemed un-walkable (see Table 5).

(2) **My Thuong NUA**

In all blocks, the number of SI utilities that require homeowners to walk more than 800 m to obtain always accounts for more than half of the total. According to Network distance, residential block 4 has 13/16 utilities more than 800 m distant. Meanwhile, in residential blocks 2 and 3, the number of SI utilities required to go beyond 800 m is 9 SIs, which is less than in block 4. The common feature of the apartment blocks is the simple walking access via network to sport center facilities and grocery stores, which are less than 500 m away. However, block 2 has more private kindergartens while block 3 has a pocket park. In the meanwhile, the healthcare unit is always the utility with the furthest reach.

It is evident that there is a distinction between residential blocks 1, 2 and 3, 4. For example, although inhabitants of blocks 1 and 2 must travel an average distance of 1,553 m and 1,228 m, respectively, to attend the public preschool, residents of blocks 3 and 4 can do so in 573 m and 861 m, respectively. The same may be said for amenities like private primary schools, cultural centers, informal markets, and parks. However, for institutions such as private kindergartens and public elementary schools, blocks 1 and 2 are closer together.

Table 5 The shortest distance to SIs of three residential blocks of An Cuu NUA

Types of SI			Network distance		
			Block 1	Block 2	Block 3
Healthcare	Healthcare unit		1,312	1,448	1,223
Education	Kindergarten	Private sector	263	378	216
		Public sector	962	1,340	1,199
	Elementary school		864	1,116	825
	Low-secondary school (optional)		545	543	200
Cultural-Sport center	Sport/training ground		131	186	304
	Cultural center		611	717	828
Commerce	Market (informal)		550	840	699
	Grocery store		250	527	593
Park	Park		407	288	72

Source Author

Table 6 The shortest distance to SIs of three residential blocks of My Thuong NUA

Types of SI			Network distance			
			Block 1	Block 2	Block 3	Block 4
Healthcare	Healthcare unit		2,830	2,506	1,728	2,016
Education	Kindergarten	Private sector	582	274	1,069	1,283
		Public sector	1,553	1,228	573	861
	Elementary school	Private sector	951	643	1,438	1,651
		Public sector	912	610	1,415	1,637
	Low-secondary school (optional)		1,284	976	712	1,000
Cultural-Sport center	Sport/training ground		320	234	126	339
	Cultural center		1,489	1,163	563	851
Commerce	Market (informal)		1,674	1,351	572	860
	Grocery store		39	49	185	470
Park	Pocket park		1,338	1,015	439	190

Source Author

The below table illustrates the shortest distance to access each type of daily SI of 4 residential blocks (see Table 6). The distance from the centroid of residential blocks to the SIs differs among SI types and from block to block when assessing accessibility. The accessibility of residential blocks 2 and 3 is superior to that of the other two residential blocks. This is explained by the number of SIs that these blocks can reach within walking distances, which are fewer than 500, and 501–800 m.

4 Conclusion

The development of Hue city's NUAs plays an important role in the construction and development of the city in particular, and Thua Thien Hue province in general, in order to implement the orientation of modern urban development planning alongside conservation, and plan for the central municipality. An Van Duong NUA, in particular, is planned to become a new urban center of the city in the future, in addition to the existing center's conservation purpose. Despite the fact that projects on technological infrastructure, resettlement housing, NUAs, and so on have been designed and implemented over a period of more than 15 years, however, when compared to the anticipated progress plan from 2005 to 2020, the actual outcomes have highlighted some of the limits in planning calculation and duration planning in terms of climate change, investment management, and social demand forecasting [28]. Furthermore, the project plans are out of date in order to meet current needs and regulations on urban growth, particularly for several projects produced during this period.

The fact that SIs have not received appropriate attention and concentrated on development, in particular, can be considered an issue. Firstly, while housing conditions are well invested in and developed, the development of SIs is not given the attention it deserves and remains limited, many SI facilities were in use later, or even not completed, however, according to Article 35 of the Government's Decree No.11/2013/ND-CP, the construction of SI works is to be completed on time in accordance with housing projects. Secondly, in the large area of the An Van Duong, An Cuu, and My Thuong NUA projects planned alongside certain other social infrastructure, several residential projects are already occupied, but the majority of the SI (of An Van Duong) have not yet been built. Thirdly, the NUAs do not have easy access to SIs such as healthcare, public education, and basic commerce (here defined as commercial facilities for daily buying needs such as markets, supermarkets, and so on). As a result, NUA residents must drive further to find amenities that match their needs. Motorized vehicles are favoured over walking over longer distances. This not only impacts NUA inhabitants' utility use and quality of life, but it also places some strain on the infrastructure of the existing development area.

The current state of SIs in NUAs only serves a portion of the needs of residents. This may be seen in the number and type of SIs now held within each NUA, as well as the breakdown of SIs by day-to-day needs. According to the classification of SIs based on daily use demands, NUA residents can generally only access up to 4/5 types of SI, excluding healthcare. Residents of An Cuu NUA, on the other hand, have more options for the types of SI they can use due to the nearby existing inner city location. They have easier access to a range of sports facilities and parks within NUA, however, residents of My Thuong NUA have fewer options due to an incomplete planned SI system. Commercial services have not existed in the An Cuu NUA area. Residents can only reach nearby grocery stores within a reasonable distance, but not the official market. My Thuong NUA has three grocery stores which are located within the NUA, however not all inhabitants have access to them. Residents of the NUA also have fewer options in terms of commercial establishments. Educational institutions are subject to the same constraints.

The establishment of NUAs in Vietnam differs from the assumptions regarding spatial and social fragmentation established by many academics about NUA development [1, 2, 29]. Instead, it depicts segregation and integration coexisting. Furthermore, existing urban planning in Vietnam does not totally segregate NUA projects from former surrounding villages [25]. These are similar to Hue's NUA development master plan. This statement is partly explained in terms of the distribution and SIs approach of new spatial products. It is specifically explained practically through the SIs designed and structured in the general plan of An Van Duong NUA, or the "exchanges" between the NUAs and the surrounding region, such as cultural (modern–traditional), commercial, public space, and so on. Based on this, NUA citizens' access to SI amenities should be examined not just within the NUA but also outside of the NUA. It does, however, necessitate establishing a reasonable distance for access to essential SIs, which is normally done on foot or by bicycle.

This study only looked at spatial accessibility to SIs for NUA residents, rather than a variety of other parameters. It is indicated by the coverage of basic SIs within given radius limits, as well as the representative measured distance between SI and residential sites. As a result, the study has some limitations in terms of measuring and assessing accessibility in general.

- Firstly, the approach distances to the SI sites are determined relative to the centroids of the residential blocks. As a result, measured distance measurements are just approximate and representative. It does not represent the exact worth of any housing plot.
- Secondly, network distance is the shortest distance people must travel to get the desired sort of SI, as assessed by Hue's existing road system. However, the elements influencing inhabitants' walking ability/effectiveness have not been considered. For example, the width of walkways, trees, and so on.
- Finally, the many types of SI found in the study, both within and outside of the NUAs, necessitate an assessment of inhabitants' everyday access to basic types of SI. However, the population density and capacity of SIs outside the NUA have not been carefully examined when determining residents' accessibility.

Developing a set of evaluation criteria for many aspects included in the accessibility assessment can be considered a useful research step in the future. In which the criteria are applied flexibly based on developed general principles and local realities. This proposal promotes project planning in particular, as well as the effective development of the SI system, with a focus on residential development.

References

1. Douglass M, Huang L (2007) Globalizing the city in Southeast Asia: Utopia on the urban edge—The case of Phu My Hung. Saigon 3(2):43
2. Fauveaud G (2016) Residential enclosure, power and relationality: rethinking sociopolitical relations in Southeast Asian Cities: residential enclosure, power and relationality. Int J Urban Reg Res 40(4):849–865
3. Harms E (2015) Porous enclaves: blurred boundaries and incomplete exclusion in South East Asian Cities. South East Asia Research 23(2):151–167
4. Labbé D, Boudreau J-A (2011) Understanding the causes of urban fragmentation in Hanoi: the case of new urban areas. Int Dev Plan Rev 33(3):273–291
5. Labbé D (2010) Facing the urban transition in Hanoi: Recent urban planning issues and initiatives. INRS-Urbanisation, culture et société. https://central.bac-lac.gc.ca/.item?id=HanoiUrbanization&op=pdf&app=Library
6. Luan TD (2014) Living in "New Urban Areas": towards sustainable urban communities in Hanoi. Vietnam 333–344
7. Minh NQ, Tung TM (2017) A preliminary study on the development of new urban areas in Hanoi over the past ten years and their Interactions with surrounding villages—A review from Van Quan new town 11(6):8
8. Latham A, Layton J (2019) Social infrastructure and the public life of cities: studying urban sociality and public spaces. Geogr Compass 13(7):e12444
9. Torrisi G (2009) Public infrastructure: definition, classification and measurement issues
10. Williams P, Pocock B (2010) Building 'community' for different stages of life: physical and social infrastructure in master planned communities. Commun Work Fam 13(1):71–87
11. Australian Government: An Assessment of Australia's Future Infrastructure needs (2019)

12. Brown J, Barber A (2012) Social infrastructure and sustainable urban communities. In: Proceedings of the institution of civil engineers. Eng Sustain 165(1):99–110
13. Klinenberg E (2018) Palaces for the people. How social infrastruture can help fight inequality, polarization, and the decline of Civic life. The United States by Crown
14. Hansen NM (1965) The structure and determinants of local public investment expenditures. Rev Econ Stat 150–162
15. Le V. Urban areas lack social infrastructure: The burden overwhelms localities (Khu đô thị thiếu công trình hạ tầng xã hội: Gánh nặng "đè" các địa phương). https://dantri.com.vn/bat-dong-san/khu-do-thi-thieu-cong-trinh-ha-tang-xa-hoi-ganh-nang-de-cac-dia-phuong-20200208232002593.htm. Accessed 29 May 2023
16. World Bank (2020) Vietnam's urbanization at a crossroads: embarking on an efficient, inclusive, and resilient pathway, Washington, DC 20433. https://documents1.worldbank.org/curated/en/856201604684234125/pdf/Main-Report.pdf
17. Adenwala A, Turner S (2020) Small city politics in the Global South: state imaginaries and everyday realities of a frontier city in northern Vietnam. Urban Geogr 1–22
18. Coulthart A, Quang N, Henry JS (2016) Urban development strategy: meeting the challenges of rapid urbanization and the transition to a market oriented economy. Vietnam: World Bank in Vietnam
19. Nguyen XH (2017) Social infrastructure planning—Foundation for sustainable urban structure. Archit Mag Vietnam Asoc Archit
20. Le V. Lack of social infrastructure in urban areas: "Finding a cure for incurable diseases" (Thiếu công trình hạ tầng xã hội khu đô thị: Tìm thuốc chữa bệnh nan y). https://kinhtedothi.vn/thieu-cong-trinh-ha-tang-xa-hoi-khu-do-thi-tim-thuoc-chua-benh-nan-y.html. Accessed 29 May 2023
21. Talen E (2003) Neighborhoods as service providers: a methodology for evaluating pedestrian access. Environ Plann B Plann Des 30(2):181–200
22. Barton H, Grant M, Mitcham C, Tsourou C (2009) Healthy urban planning in European cities. Health Promot Int 24(Supplement 1):i91–i99
23. Hoang A, Apparicio P, Pham T-T-H (2019) The provision and accessibility to parks in Ho Chi Minh City: disparities along the urban core—Periphery axis. Urban Science 3(1):37
24. La Rosa D (2014) Accessibility to greenspaces: GIS based indicators for sustainable planning in a dense urban context. Ecol Ind 42:122–134
25. Yhee H, Kim S, Kang S (2021) GIS-based evaluation method for accessibility of social infrastructure facilities. Appl Sci 11(12):5581
26. Donkin AJ, Dowler EA, Stevenson SJ, Turner SA (2000) Mapping access to food in a deprived area: the development of price and availability indices. Public Health Nutr 3(1):31–38
27. Witten K, Pearce J, Day P (2011) Neighbourhood destination accessibility index: a GIS tool for measuring infrastructure support for neighbourhood physical activity. Environ Plan A: Econ Space 43(1):205–223
28. Tran P, Tran A, Nguyen T, Dang L (2017) Assessment of urban planning and urban development in An Van Duong, Hue city. Inst Soc Environ Trans-Int
29. Labbé D, Boudreau J-A (2015) Local integration experiments in the new urban areas of Hanoi. South East Asia Res 23(2):245–262

Improving Urban Industry Through Cultural Landscape Analysis Framework: A Size Study of the Riverside of Ha Thanh-Quy Nhon City-Central-Vietnam

Quang Viet Pham, Anh Hoang, and Tuyet Thi Anh Cu

Abstract Urban identity is a combination of relationships formed from many attributes of tangible elements (terrain, climate, landscape) and intangible elements (tradition, custom, culture, society) that create the unique characteristics of the city. Along with the trend of global integration and urbanization, the current Vietnamese urban architecture, especially the traditional urban spaces, is being gradually lost because of the influences of new architecture imported from the other countries. In terms of management, the choice of development and preservation of identity is always a difficult problem when taking into account the economic and social development of the city. It is necessary to include the goal of harmonious development of urban identity in the planning and design process. Through synthesizing and surveying local cultural values that have been evaluated and analyzed to distill some outstanding tangible and intangible values, and at the same time recreate urban public spaces to regenerate landscape at Ha Thanh river, in Dieu Tri area-Quy Nhon city-Central region-Vietnam.

Keywords Cultural landscape · Identity · Behavioral culture · Ha Thanh river

1 Introduction

In the current urban functions, public space is increasingly used by people, becoming a place to attract the community with diverse and rich exchange activities. The current use of public space has the role of improving people's lives [1]. Public space also creates a familiar sense of location, one that represents societal perspectives and illustrates harmony amongst groups of people that have various goals and points of view, including politics. As a result, creating public spaces with identity is essential

Q. V. Pham (✉) · A. Hoang · T. T. A. Cu
Faculty of Architecture, Thu Dau Mot University, Thu Dau Mot, Vietnam
e-mail: quangpv@tdmu.edu.vn

V. T. Ha et al. (eds.), *Proceedings of the 8th International Conference on Sustainable Urban Development*, Advances in 21st Century Human Settlements,
https://doi.org/10.1007/978-981-99-8003-1_13

to the process of urban growth [2]. Urban identity is one of the key elements in urban studies that helps define the standard of urban life and motivates individuals to take care of and feel accountable for their local urban environment. During urban development, environmental and social forces combine to form urban identity. Urban identity is expressed by 3 elements: natural, cultural, and artificial of the city. The identification component will be identified via the process of describing and talking with locals about what a location genuinely looks like and what major elements influence each person's own perception of the area. Therefore, "identity" is a term that is very related to the living time of residents, the longer people live in the city, the more they feel attached and loved to the city [3]. Perspectives on urban identity are seen through different social, cultural, and economic perspectives. The primary focus an urban public space must have meaning and memory for its users [2]. The process of industrialization has taken place more and more strongly since the 2nd and 3rd technological revolutions have significantly changed the face of urban organization, leading to problems that need to be solved in space urban construction. However, keeping urban identity, especially in the modern era, is a challenge for urban planners, designers, and urban researchers.

Most urban challenges nowadays are centered on the urbanization and globalization processes. These two issues become a challenge in the connection between elements of natural landscape, artificial landscape, culture, and people's perception in urban development. The conventional structure of urban functions serves as a reflection of urbanization, and building projects seldom take into account the compatibility of cultural traits and the surrounding environment, which has an impact on the process of developing urban area identity [4]. People now have a more significant role in modifying the organizational structure of urban space and enhancing the natural landscape thanks to advancements in science and technology. However, the natural landscape is being pushed farther and farther away from the urban environment by the traditional urban spatial organization structure for all locations. As a result, the natural environment progressively starts to have less of an impact on urban development. Globalization changes the urban face towards civilization and modernity in a flat world. The loss of traditional cultural identity, the loss of place identification, and the rise in ecological disasters are the drawbacks of globalization, nevertheless [5]. The process of urban planning and design is stereotyped and offers building design solutions as an isolated object placed in the landscape space, while public spaces only have functional meaning. In the book "Lost Space: Theories of Urban Design" Trancik argues that the current urban space development does not bring people a sense of belonging due to the absence of cultural values in the form of public space organization, and at the same time losing the traditional character of the city. The significance of traditional cultural values in contemporary public spaces relies on how individuals view their surroundings. In the urban built environment, the absence of natural landscape elements, the adaptation of people to the environment and the decline in the perception of place, are losing the urban identity. Place construction is based on the relationship of tangible, social, and historical factors of the place, to show the cohesion of the material and the perception of the local people [6]. Therefore, the benefits of creating a place will help to form a cultural space and a livable society. At that time, the urban public space will give people a sense of familiarity and modernity. Thereby, the cultural and natural environment characteristics should be considered and integrated into the urban development stages. The study of cultural landscape aims to assess the interaction between tangible and intangible cultural elements. The change of these two elements in space and time serves as a scientific basis in urban design projects, through providing a multi-structural design framework, which is called "framework for analyzing cultural landscape". This analytical framework's main goal is to pinpoint the cultural components of locations and conserve them through the reproduction of cultural spaces in urban public areas. In urban studies, it should be noted that any urban landscape (including UNESCO sites, monuments, and historic locations) can be regarded as a cultural landscape.

The study is set up as follows: Sects. 2.1 and 2.2 respectively describe the main representative features of the local cultural identity as well as the main features of the components in the cultural landscape. In Sect. 2.3, relate elements of local identity and cultural landscape in the framework of cultural landscape analysis to identify urban morphology. Then, the study conducts a qualitative discussion of each functional area in the cultural landscape analysis framework to extract the necessary information from the analytical framework. From there, in Part 3, research on the selection of public space in Quy Nhon-Binh Dinh city aims to build a cultural space through the framework of analyzing the cultural landscape, to contribute to improving the urban identity.

2 Research Methods

The study begins by highlighting the traditional cultural factors as the basis for the process of site identification. From there, determine the meanings, functions, and forms of traditional cultural elements, to come up with a design to renovate the riverside landscape space with traditional cultural meanings in Quy Nhon urban area.

2.1 Components of Cultural Identity

The organization of the landscape architecture area during the urban development process has its own qualities and not all aspects share the same identity. Until its subject ceases to exist, identity is a factor that cannot be lost. However, the factors that form the identity have changed in the development process, these factors have historical characteristics of the current and future development. Therefore, identity evolution in urban areas is seen as subject through genetic means. An expression of the concept of urban identity can be made from the combined perception of urban formation factors at a location (e.g., streets, squares, buildings, public spaces, urban decoration, and sculpture) (Fig. 1).

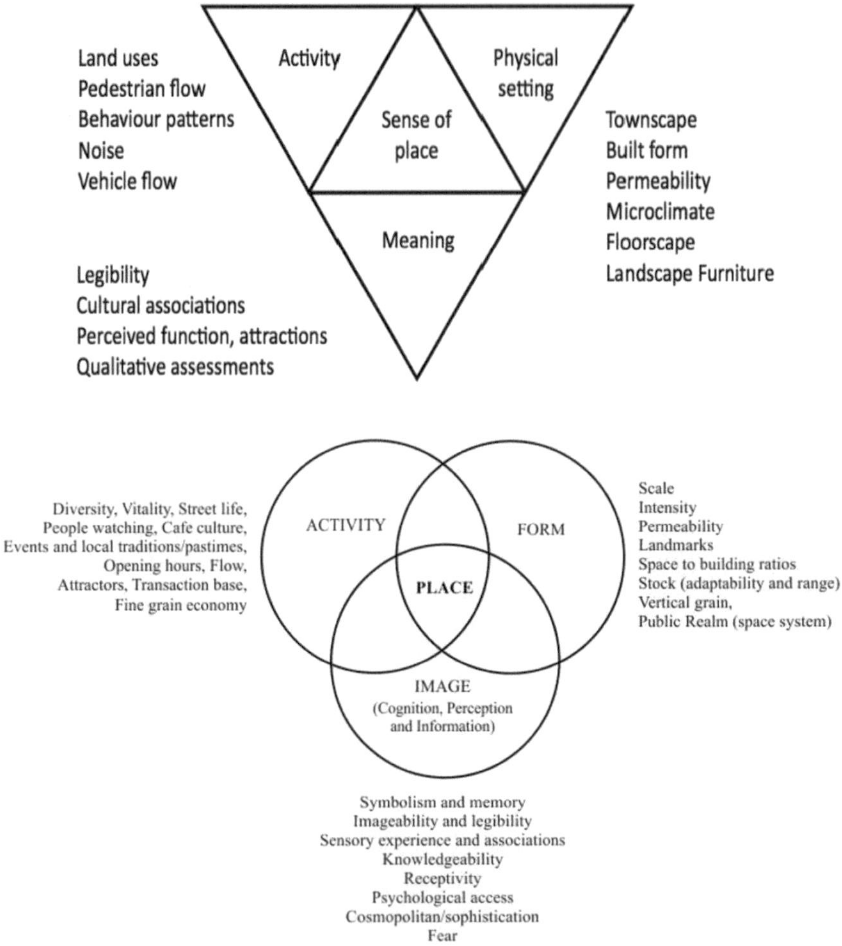

Fig. 1 Descriptive components of place identity suggested by **a** Punter [9], and **b** Montgomery [10]

According to Relph,[1] identity is the basic relationship in our daily life, this relationship refers to the similarity and long-term unity of objects, while places have historical significance associated with cultural characteristics at a specific time. He argued that place names consist of three main components: topographical features, cultural activities, and symbolism [7]. Topographical features include natural environment (soil, hills, lakes, hydrology…) and artificial environment (buildings, streets, monuments…). Cultural activities include events, customs, practices, and community organizations. Symbolism is formed through the experiences and interactions of people in a place. Identity represents the traditional cultural characteristics of each

[1] Edward Relph: geographer with the work "Place and placelessness" published in 1976.

acreage, each region, and each city has its own identity. In Paperport's research,[2] the organization of urban environment construction is based on people's perception of the meaning and functional space of the urban environment. The tangible element brings the initial perception to people, the cultural behavior towards the natural and social environment, creating the foundation for the urban image of a place. This provides urban dwellers with the "meaning" and "function" of an urban space [8]. The context, surrounding natural features, and architectural symbols should all be taken into consideration when choosing solutions to create urban environments and architectural works that will effectively create a sense of place.

Punter and Montgomery's perspective on place, which focuses on exploiting the important factors that shape place in urban public spaces. Punter [9] stated the interweaving of elements: form, perception, and cultural activities in the formation of the place of urban space [9]. Montgomery [10] conducted similar research, which emphasized that the meaning of place is expressed through people's perception of place, form and organization of urban space [10]. Urban identity studies by Relph, Punter, and Montgomery all emphasize the tangible elements and the spatial significance of urban spaces, in which the tangible element of the place is accessible and perceptible. Intangible factors are the cognitive activities and related characteristics perceived behind the tangible element. Depending on our experience in urban spaces, the perception can be different, but basically urban spaces always carry a function associated with a certain visual meaning about that place.

Kaymaz's recent study "Urban Landscape and Identity" argues that urban landmarks are based on the inherent relationship of urban and natural environments. Urban identity is created mostly by human behavior through time, which shapes the urban landscape. As lifestyles, neighborhood organizations, and economies change, so do urban environments. In urban construction planning, it is necessary to consider social factors and people's psychology, to build urban environment, awareness, and community organization. From building the relationship between people and the urban environment, it will contribute to the construction of urban identity [2]. To give a set of description variables to evaluate the identity of urban areas, the case study of Quy Nhon city in Binh Dinh will integrate the findings from the research and the framework for analyzing the cultural landscape.

2.2 *Components of the Cultural Landscape*

In 1925 Sauer presented the definition of "cultural landscape" as "created from the natural landscape by a group of cultures". The continuous interaction between behavioral culture and the environment contributes to the creation of the cultural landscape [11]. Cosgrove (1984) argued that cultural landscape is the social and aesthetic unity

[2] **Amos Rapoport** is an architect and one of the founders of Environmental-Behavior Research (EBS). Author of the book "The Meaning of the Built Environment: A Nonverbal Communication Approach" published in 1990.

between land and human life, accordingly, cultural landscape is an aesthetic view of the environment. He believes that beauty is expressed from the process of cultural construction and natural behavior culture. In Article One of the 1992 UNESCO/ICOMOS convention, the cultural landscape represents the "combination of man-made structures with nature". In 2003, the geographer Jones offered a comprehensive conceptual assessment of the cultural landscape through the fields of study [12]. He emphasized how the term "cultural landscape" had previously been primarily used to describe how humans have altered the natural and social surroundings. In addition to the effects of humans on the environment, he added, it is also important to consider landscape ecology, which highlights the role that people have played in shaping cultural landscapes.

Thus, the cultural landscape illustrates the development of human society, in which culture behaves with the natural environment and the social environment is affected by many factors (natural, social, cultural, and economic). The cultural landscape is a complex system of environmental changes derived from behavioral culture. It is important in the cultural landscape that people's awareness of traditional cultural values helps to shape local identity. Studying cultural landscapes to evaluate previous sustainable land use techniques, considering the characteristics and limitations of natural environment behavior and the relationship of tangible factors with nature (UNESCO 2008). So, to create cultural landscape places in urban public spaces and enhance local identity, an appraisal of traditional cultural values will be necessary (Table 1).

Through analysis (Table 1), all studies focus on considering two main factors: tangible (physical, material) and tangible (not visible). Brown [13] considered major cultural landscape elements as tangible factors (e.g. corridors and intersections, amenities, land cover, sites of cultural importance, industrial trade relationships, etc.) and intangible factors (e.g. cultural practices, local and census boundaries, land use) [13]. O'Donnell [14] classified the main elements of the cultural landscape through their tangible values (identification of the spatial form of the landscape) and intangible values (the meaning of the landscape) [14]. Vogeler (2010) grouped cultural landscape elements into three categories: tangible forms, human behaviors and activities, and cultural meanings and messages [16]. According to Stephenson [15], the qualities of the cultural landscape can be thought of in terms of material forms (represented through descriptions of natural aspects, human impact on nature, tangible construction, etc.), linkages relating to memory, symbolism, ideology, and practice [15]. In recent research by Sonkoly [16], it has been shown that the tangible and intangible characteristics of the typical urban cultural landscape come from the living cultural heritage and the cultural construction process. Living cultural heritage is related to people's perception of life through beliefs, cultural ideologies, and people's living habits [16]. As a result, cultural traits are developed based on the tangible components of urban residents' living spaces. Through the examination of the behavioral culture of the two aspects over time, we analyze the relationship between people and the components of tangible and intangible culture in this study to provide the groundwork for the future growth of the cultural landscape area. As a

Table 1 List the main elements of the cultural landscape

Authors	Key elements of the cultural landscape	Elements
Brown [13]	Tangible elements	Transportation corridors and interchanges, utilities, land cover, sites of cultural importance Major commercial/industrial concerns
	Invisible element	Political and census boundaries, land ownership and use boundaries
O'Donnell [14]	Value of tangible heritage	Natural systems, land use patterns, forms, spatial organization, topographic visual relationships, vegetation, circulation systems, water features, natural landscape structures, artificial landscape, spatial characteristics, form and scale of constructions, network of functional links and objects of use
	Intangible heritage value	Festivals, traditional music, performances, pilgrimages, worship, memorials, traditional practices, indigenous flora gathering place, symbolic community sharing of past and present
Stephenson [15]	Forms	Natural features, vegetation, human intervention in the natural structure
	Relationships	Memory, symbolism, ideology, spirituality, sense of place soul, aesthetics, myth, meaning of place
	Perform	Social organization, ecological system, human activities, historical events, historical processes, traditional activities, national development
Vogeler (2010)	Cultural values and messages	Customs, beliefs
	Human activity/behavior	
	Space form	Topography, vegetation, structure (time, space, visual needs), settlement form
Sonkoly [16]	Cultural heritage	Tangible: products used in religion/culture, household goods (food, costumes, traditional tools), products used in industry
		Intangibility: sense of identity, patterns of spatial use, expression (orally/behavior/customs), belief system, commercial/social/cultural activities
	Built cultural heritage	Tangible: buildings/groups of buildings, parks and gardens, monuments and structures, archaeological sites
		Intangibility: land use pattern, building use pattern, expression (architecture/streetscape/site landscape), perception of the place

Table 2 Main components of cultural landscape

Factors	Key elements	Elements
Tangible	Natural elements	Shape, geography, water surface morphology, vegetation, climate
	Artificial elements	Architectural work, roads, bridges, statues, historical sites, agriculture
Intangible	Beliefs	Religion, thought system, belief, legend
	Legal system (customs)	Social organization, economy, policy
	Customs (behavior)	Places, symbols, lifestyles, festivals
Link (interaction)	Time, progress	History, memory
	Space	Activity, perception, existence, architecture, perception

foundation for use in the process of planning the creation of urban public landscape space, we present some significant components of the cultural landscape in the final section of the research section (Table 2).

2.3 *Framework for Analyzing Cultural Landscape*

In the study of culture, the term "culture" has such a broad definition that it is challenging to comprehend all the components through a single urban function. Consequently, a flexible methodology is required to examine cultural traits and identity-related cultural changes throughout time. The study suggests three aspects in the cultural landscape here since the primary factors have been listed in Sect. 2.2: (V) material culture, (P) intangible culture, and (Q) the interplay of these two factors across time (Table 3). The primary elements are divided into several parts:

V1 (formality): Local identity is shaped by both natural and man-made elements. Local identity is expressed through the urban spatial organization and the blending of natural components in architectural form.

V2 (function): Expressing the space of human activities in the locality (the appearance of rivers in urban areas helps to generate activities of transporting goods, river services, farming, exploiting aquatic products, etc.)

V3 (meaningful): Perceiving the value of physical elements that play the role of creating symbols or carrying local cultural messages.

P1 (formality): The intangible element is important in the organization of physical space. Because the intangible factor is expressed through human perception and perception of the natural environment and the social environment. Thereby, consider how those perceptions affect urban morphology.

P2 (function): The events and activities of the people in the urban space have spiritual meaning, expressing the daily life of the people.

P3 (meaningful): From the phenomenological point of view, urban symbols and spaces are directly related to their immaterial elements. People's social behaviors and beliefs can create a kind of identity for urban public spaces. For example, Ba Dinh Square—Hanoi is considered as a very dignified and converging stage in the country's independence.
Q1 (formality): The process of changing time and space affects the change of physical elements in urban areas. The change in the physical element itself over time represents each period. Architectural style can play the role of identifying the physical element of an urban area in a certain period. Contemporary architectural works (such as Indochina architecture) are typical physical elements of the French colonial period.
Q2 (function): Facilities and achievements over time can be maintained or eliminated over time. Technological, economic, and social development has created many new functional areas for the city. These changes can contribute to creating a new look for the city. For example, the development of means of transport has changed the transport system in the city, enhancing the connection of urban functional areas.
Q3 (meaningful): Symbols and their meanings are subject to change. From a symbolic perspective, metropolitan communities' hidden aspects make up several layers of identity. For instance, certain commemorative occasions may alter how cities look. These event kinds give places a temporal identity. A place-specific identity can be shaped over time by political or economic government activities.

- Material culture factors refer to natural shapes and human-built works, which are expressed through terrain, water surface, buildings, monuments, traffic….
- Intangible cultural elements in the cultural landscape are expressed through different types of beliefs, customs, and behaviors, are controlled by several laws, and appear in the organization of the social environment.
- The link between the two elements Material and Intangible in the cultural landscape is a process in time and space. The study also collects and rearranges the features (discussed in Sect. 2.1) representing the identity of urban space in three forms including: Tangible elements (1); Function (2): operation demonstration; Semantic (3): denotes meaning and symbolism. These elements not only represent the main qualities of the place identity but also encompass many aspects of the metropolis.

3 Current Status of Cultural Landscape in the Downstream Areas of Kon River and Ha Thanh River

Quy Nhon city is one of the important urban areas of the Central region, Vietnam. According to the development orientation of urban space in Quy Nhon city, focus on two main parts: Renovating the old urban space, and strengthening the connection of the urban area, the driving force for the development of Dieu Tri. The selected study

Table 3 Framework for analyzing cultural landscape

Elements	Key elements		Form	Function	Signification
Physical element (V)	Natural elements	Topographic	Natural high	Take a walk, enjoy the scenery, preserve the landscape, practice sports	Typical terrain (mountains, plains, valleys)
		Plants	Vegetation (color, shape)	Agricultural production, forestry, entertainment, environmental activities	Plant symbol in life, native plants
		Climate	Hot, cold, monsoon, Lao wind		
		Hydrology	Pond, lake, river, sea	Fishing, boating, river entertainment	Traditional use of water, symbolic meaning
	Artificial elements	Urban form	Urban morphology (organic, geometric, linear, mixed)	Territorial scope, public space	Reflecting local economic, social, and cultural development
		Traffic	Streets, bridges, accessible spaces	Moving, circulating	Symbols, calculating levels in traffic, forms of movement
		Architecture constructions	Landmarks, fountains, paving	Public works, multi-functional, comfortable, and safe works	Religion, architectural style, symbolism, decoration
Intangible element (P)	Beliefs		Religion, beliefs affect urban spatial layout	Iconic, directional, tourist attraction, devotee	The shape of the building, the meaning of space
	The system of laws (customs)		Events, festivals, boundaries	Area of activities, visiting tourists, pilgrims	Regional specific culture, local code of conduct

(continued)

Table 3 (continued)

Elements	Key elements	Form	Function	Signification
	Customs (behavior)	Folk tales, legends, symbols, signs	Dancing, food, language, crafting traditional items, costumes	Folk songs, proverbs, respect for nature, protection, limits
Interaction (Q)	Time (evolution)	Heritage, period architecture, scientific revolutions	Festival events, labor organizations, traditional music, and dance	Dating, period, change
	Space	Aesthetics, technological changes, material changes, textures	Organization, management, social connections, transportation networks, sales, and marketing	Economic policy, social perception, architectural style change

area is downstream of Ha Thanh river, in Luat Le village, Dieu Tri town, Tuy Phuoc district, which is considered an important part of flood drainage in Quy Nhon city (Fig. 2). In the general planning of Quy Nhon city to 2030 with a vision to 2050, the research area is included in the directional planning for the development of urban green parks [17]. The study combines an integrated analysis of infrastructure data, social, and cultural traits of the research area, both at the micro and macro scales, to develop an acceptable landscape spatial transformation plan for this area. The river systems of the Ha Thanh and Kon cities serve as Quy Nhon's urban infrastructure. To suggest some appropriate functional areas for the area, evaluate the framework for studying the cultural landscape (Table 4).

V1: According to the natural characteristics of the study area, the level of risk assessment in the Kon River and Ha Thanh River areas according to the scenario (2015) emphasizes that this area is mainly affected by floods. The high-risk area is located in the riverside areas of Tuy Phuoc, Tay Son, Phu Cat and An Nhon districts, part of Quy Nhon city, and about one-third of these districts are flooded to a depth of more than 11 meters, even the rain cycle repeats in 10 years [18]. Therefore, in addition to the park function, the study area also plays the role of technical infrastructure to regulate water in the downstream area of Ha Thanh River. Accessible traffic: from the direction of National Highway 1A connecting with Tang Bat Ho street (Dieu Tri town), through Luat Le bridge, into Luat Le village and some embankment routes along Ha Thanh river. Regarding the current status of land use: mainly wet rice cultivation, along both sides of the river are bamboo bushes with mounds surrounding the cultivation area. Regarding population: mainly focus on Luc Le bridge area and along areas reinforced with concrete embankments on both sides of the river.

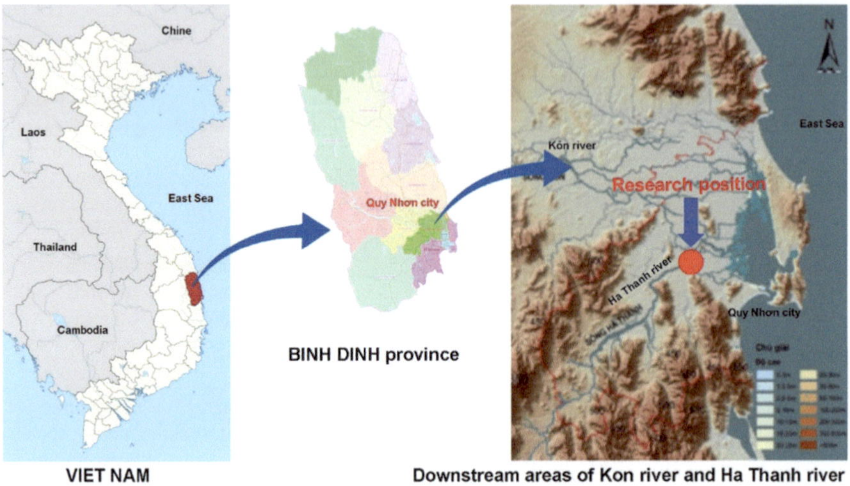

Fig. 2 Location map of public space research area, cultural landscape, Quy Nhon city-Binh Dinh province-Vietnam (author's source)

Table 4 Annotation of relics, cultural history of Quy Nhon city-Binh Dinh

1	Quy Hoa Leprosy Camp	16	Ba Nuoc Man Temple
2	Tomb of Han Mac Tu, Royal Palace, Egg Beach	17	Ong Nui Temple
3	Long Khanh Pagoda	18	Tower of Banh It
4	Nhon Hai Ancient Citadel	19	The Statue of Gautama Buddha
5	Twin towers	20	Church of Luc Le
6	Song Village Church	21	Binh Dinh Citadel
7	Fortune Teller Tower	22	Doing cocain
8	Ba Thuy Temple	23	Go Gang Market
9	Go Thi Market	24	Quy Nhon Citadel Relic
10	Tomb of Dao Tan	25	Do Table Citadel
11	Tomb of Le Dai Cang	26	Canh Tien Tower
12	Go Thi Parish	27	Citadel Relics
13	Binh Lam Tower	28	Thu Thien Tower
14	Go Boi Town	29	Quang Trung Museum
15	Xuan Dieu Memorial Hall	30	Altar to heaven (Tay Son uprising)

V2: When analyzing the function, the research area is located in the downstream area of Kon River and Ha Thanh River, formerly the lifeline of the Champa kingdom and the Nguyen feudal period. The process of urban development has absorbed many cultures: Do Gian Festival in An Thai (martial arts fight each other for sacrifices, wading across rivers, performing martial arts moves to compete in the water), Via Ba festival in Nhon Hanh (it is said that Ba Thuy temple at the river wharf has a whirlpool). The organization of communities along river routes and national roads, particularly with the house space structure "Roof leaves"—Binh Dinh, demonstrates the exceptional feature in the river behavior culture of the people of Binh Dinh and the Central region of Vietnam in the living space. However, the development of science, economy and society is changing the face of the city, changing the perspective in urban construction planning and people's living space. Currently, there are almost no specific activities along the two sides of the Ha Thanh river, except for sand mining and dredging of alluvial deposits along the river. Dike lines are built around the study area. Natural ecosystem associated with wet rice agricultural production. The hydrological regime with the amount of water flowing through the area is unevenly distributed in the dry and rainy seasons. The West borders National Highway 1A and Luat Le Bridge in the North, creating a separation of natural conditions in the area from other areas by the traffic system and surrounding concrete dikes.

V3: Through the analysis of material cultural factors in the study area, the spatial and cultural significance of the area is associated with the element "water", currently only providing water for people in daily life and agricultural production. However, in the minds of Vietnamese people, the element "water" also represents the relationship between humans and nature. Vietnamese people depend on nature, rely on nature rather than dominate. This tendency in the sense of showing respect, the cult of nature; In the act of expressing choices that are adaptive to nature, taking advantage of natural forces rather than conquering nature, using human power to replace nature; In life, activity manifests itself into a lifestyle of harmony, harmony with nature, and attachment to the natural environment [19]. Besides the importance of using water surface, planting trees is also a traditional culture of local people and every Vietnamese. The bamboo tree is like a symbol for Vietnamese people, always reaching high with an upright posture that never gives up. Bamboo has many different types, but all bring great benefits to humans. From the very beginning, under the shade of green bamboo, Vietnamese plowmen-built houses, plowed, and everyone gathered to chat during lunch break. Bamboo can be used for knitting, making many objects in the house. Just like that, bamboo becomes an indispensable close friend of Vietnamese farmers in labor. Bamboo also plays a huge role in protecting the homeland.

P1: Formal analysis of intangible factors demonstrates the significance of river culture to the populace, which is expressed through religion, beliefs, and practices with iconic monuments (such as Saltwater Pagoda, Ba Pagoda, Banh It tower, Lang Song church, etc.), as well as the culture of people's behavior through living spaces (thatched roof house—Binh Dinh) [20]. During the interactions and coexistence of Vietnamese inhabitants with the Cham and Chinese, indigenous traditions are incorporated. Residents of Vietnamese villages produce, exchange, and develop intangible cultural components over time. Particularly, this region is linked to the Tay Son-Nguyen Hue peasant movement and the homes of notable poets from the 20th-century new poetry movement (Han Mac Tu, Yen Lan, Quach Tan, Che Lan Vien, and Xuan Dieu) [21].

P2: The functional space of intangible elements is shown through typical local annual festival events such as: Ngoc Hoi Dong Da victory festival, Cau Ngu festival, Saltwater urban festival, Binh Dinh traditional martial arts festival, along with other forms of performance (Bai Choi folk song, Tuong art). There are also a few traditional spaces typical of the South-Central region being preserved. With the current trend of tourism development, the number of tourists coming to Quy Nhon—Binh Dinh has grown over the years (about 1.9 million visitors),[3] the working age population accounts for 60% of the total population.[4] This will be the object that contributes significantly to the preservation, sightseeing, entertainment, and entertainment activities for the local cultural festival activities.

P3: The consciousness and method of community organization of the Binh Dinh people are consistently linked to the element of "water," even though human society's interest in "water" in the natural world, particularly rivers, is continually changing. New cultural values are adopted in place of the lost ones. Although Quy Nhon city-Binh Dinh has undergone a rapid urbanization process recently with the construction of several architectural works that have altered the city's appearance, the role of the natural water element has never changed. care and defense. In terms of nature, friendliness, love of the weather, and respect for people, research has helped to generalize the kinds of customs, habits, and beliefs of the Binh Dinh people. Typical with the hero's soul hanging in ethnic clothing (Tay Son-Nguyen Hue) [22].

Q1: Stereotypical urban areas are being created because of the process of altering the physical and cultural landscape in the Quy Nhon-Binh Dinh region due to urbanization and globalization. Because the architectural creations are a combination of traditional and modern styles, the regional component is not clearly displayed. In recent years, the system of relics and folk belief establishments of village character tends to be built solidly, expanding the area, or consolidating the boundaries in each village and commune area. Clusters of residents live, creating spiritual fulcrums in the heart of the village and in the new urban area. Cultural

[3] According to the statistical yearbook of Binh Dinh province in 2018.

[4] According to the statistical yearbook of Binh Dinh province in 2018.

activities in Dieu Tri town area are carried out at cultural centers, private establishments, socialization clubs. Most of the communal houses in the Quy Nhon city area only perform religious functions. The role of rivers in the lives of residents along the Ha Thanh and Kon rivers is only in agricultural production and sand mining.

Q2: The process of spatial transformation of the local cultural landscape involves the interweaving of many different types of cultures. In some places, the space of the village's religious establishment was abused, becoming the venue for games with gambling trends such as cockfight betting, gambling, some beliefs such as taking sacred water and burning votive paper still take place. Until now, village wells and some folklore exchanges are only places where people visit when needed, while community exchange activities through these spaces are also gradually reduced. The riverside landscape now acts as a place to exploit real estate, exploit local building materials.

Q3: In addition to ensuring that public spaces serve the requirements of the populace, the process of urban building and development must also take cultural values relating to the natural and social surroundings into account. Additionally, it demonstrates awareness and community organization to transform Quy Nhon city into a popular tourist destination with distinctive cultural values of the south-central coast as well as a hub for meetings, conferences, and exchanges of global academic science (Fig. 3 and Table 5).

4 Solutions to Build Cultural Space in the Study Area

V. This study offers a framework for identifying unique cultural values, assisting design and construction studies of urban parks, through the investigation of the cultural landscape of Quy Nhon city (Table 6). The research area will serve as a link for the growth of the historic urban area of Quy Nhon and the new, dynamic neighborhood Dieu Tri thanks to its location downstream of the Ha Thanh River. The research area will also serve as a link with the network of historical and cultural artifacts in the Quy Nhon-Binh Dinh region. The research also contributes to the strengthening of the local ecosystem by preserving some appropriate local elements (agricultural production areas, cultivation, bamboo banks, native bird habitats, riverside alluviums), designing spaces to showcase traditional culture in conjunction with traditional motifs from Champa architecture, and setting up spaces in accordance with Binh Dinh people's customs. Such representations can offer a visual link between the city's historic and contemporary architecture. The study also set aside indoor and outdoor space for annual exhibitions and traditional festival events, while constructing spillways and re-establishing sand deposits along both sides of the river. The study also allocates indoor and outdoor spaces for exhibitions and traditional festival events every year. In addition, the construction of spillway banks and the re-establishment of sand alluviums along both sides of the river stem from the previous village organization of the people's

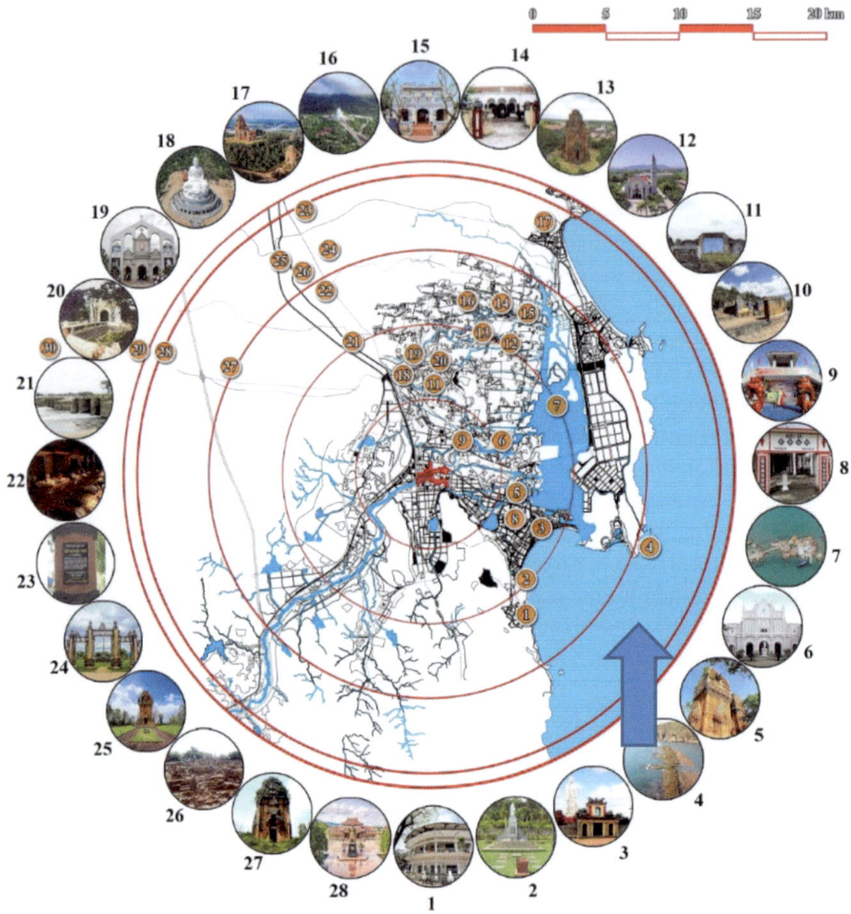

Fig. 3 Map of cultural relics in the downstream areas of Kon and Ha Thanh rivers (author's source)

river life, or from having to cope with, control and conquer restore water by using dikes to prevent saltwater intrusion and flood control. For big floods, it is usually allowed to let the overflowing water pass over the dike to divide the water to flood the whole area to avoid breaking the partial flood dike. To prevent erosion, erosion and limit dam failure, people often build dikes and dams depending on the wind and water flow, running around the terrain around the village, the purpose is to let the dyke not resist the force natural when strong floods flow, so it is not necessary to build straight. For this purpose, the positioning of some picnic spaces in the design of the study provides a traditional cultural space with natural and social features typical of the Central region.

Table 5 Functional spatial orientation of the research area (author's source)

The natural landscape relates to the canal system	Space for linking nature and culture

There is a map of cultural relics and characters in Quy Nhon city area	Use a variety of colorful plants and seasonal plants

Combined natural space and sports landscape	Creating traditional programs and activities in space

Making it possible for birds to live in natural spaces	Connecting spillway to urban drainage networks

(continued)

Table 5 (continued)

Using traditional ecological architectural forms in public spaces	Keeping the natural environment clean

Create a picnic space, entertain people	Define the vision for the landscape space

P: To encourage visitors and locals to take part in regional cultural festivals, the research design process dedicates several open spaces for communal activities. Urban landscape design has value when it combines traditional culture with the design of public areas: In order to give the area a distinctive look, ornamental Champa architecture is used in public spaces; combining traditional cultural spaces with outdoor recreation, physical activity and sports, traditional retail spaces, and the growth of regional flora and fauna (Fig. 4). By performing traditional martial arts, Tay Son war drums, and Dong Da festival, particularly emphasize the historical story of the hero Tay Son-Nguyen Hue.

Q: The river cultural system in the Kon River and Ha Thanh River areas is an indispensable space in the connection of historical, cultural, natural values and the Binh Dinh community. The study tries to connect the Past, Present, and Future values of Binh Dinh's historical culture (Fig. 5). The river in Binh Dinh is still a possibility, nevertheless, and is waiting for the correct development plans, integrating cultural exploitation with the "water" aspect in investments in upgrading and repairing cozy infrastructural floors, directly in existing residential neighborhoods of urban and rural areas (Figs. 6, 7 and 8).

Table 6 Main functions of the park in the study area (source: author)

Park function	Recommended functions
1. Picnic area	Picnic area for people's leisure time Use of land geomorphology along the Ha Thanh River for pedestrian mobility Much of this area is devoted to the restoration of the area's natural vegetation
2. Cultural space	Indoor and outdoor exhibition space for annual exhibitions Traditional festival events Recreating the space of traditional houses with thatched roofs in Binh Dinh, agricultural production of wet rice (crop fields)
3. Sports area	Provide venues for adult/children's sports, exercises, playgrounds Underwater festival space (boat racing, fishing festival, Do Gian festival, martial arts dance underwater, wading across the river, performing martial arts competitions under water)
4. Traditional market space	Using a combination of space for entertainment and shopping, temporary markets follow the model of Truong The market, Go market, and Go Gang market
5. Green space	Bamboo is proposed to be planted along both sides of the river, which has the effect of preventing landslides, keeping soil, draining, and draining water when there is a flood or inundation Encourage tree planting in public spaces
6. Infrastructure services	Design a system of spillways combining open canals and flooded fields, helping to drain water locally in the area. The body of the dam is made up of rocks weighing several kilos, evenly spaced Between the rocks are thorn bushes, an aquatic plant whose roots cling to the rock to hold the wall of the dam

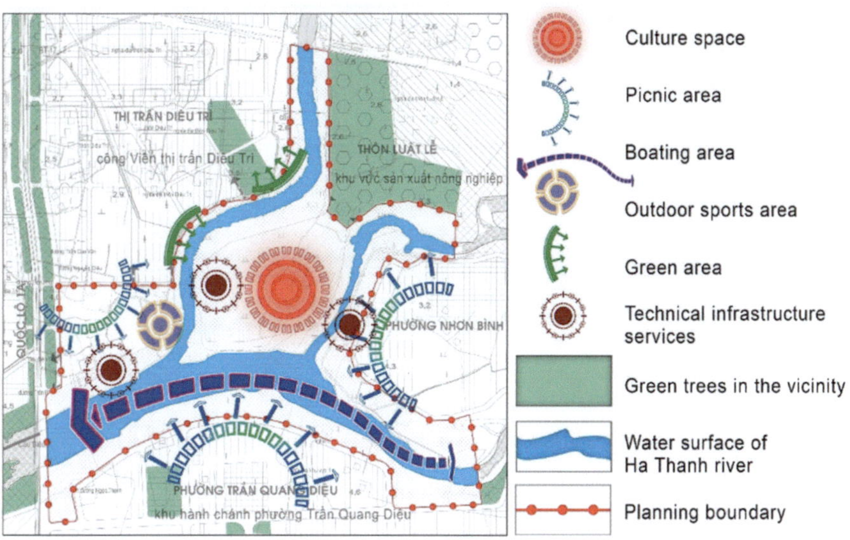

Fig. 4 Orientation map of the cultural landscape of the urban park area (author's source)

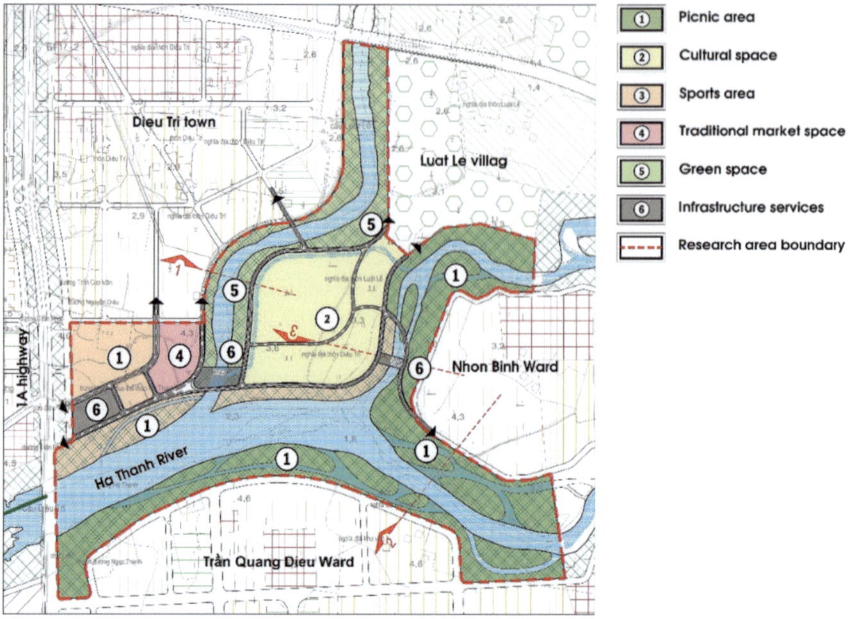

Fig. 5 Structure of land use, cultural landscape, urban park area (author's source)

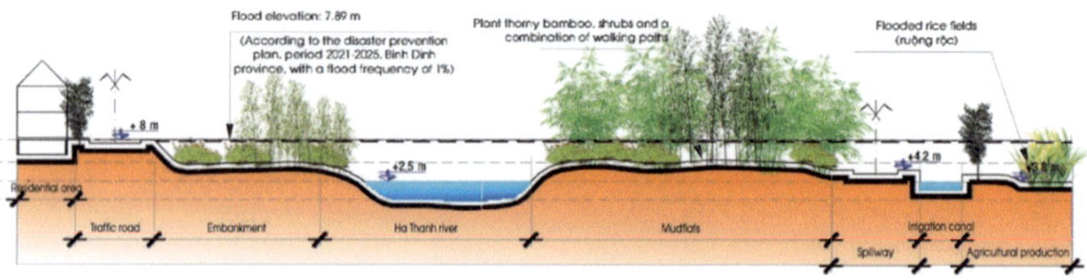

Fig. 6 Cross-section of 1 green space, traditional production space (author's source)

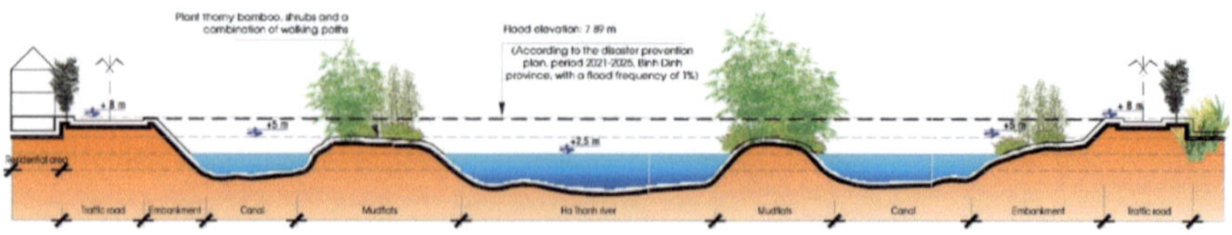

Fig. 7 Cross section of 2 picnic areas (author's source)

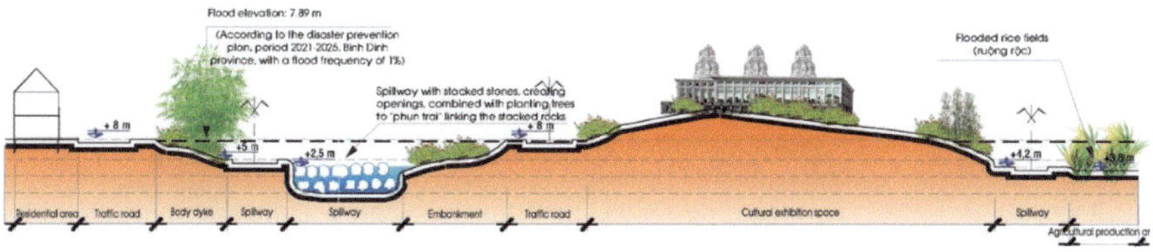

Fig. 8 Cross-section of 3 regional cultural spaces (author's source)

5 Conclusion

Urban planning initiatives are the product of collaboration across a variety of industries, while the cultural component is frequently underemphasized in planning initiatives. Each development period's urban spatial qualities are represented by sociocultural values. Changes in the urban environment of today result in a shift in cultural values. Therefore, maintaining an evolving urban identity is the goal of creating a sustainable culture. The analysis and evaluation of tangible and intangible characteristics, as well as the interaction of these two aspects over place and time, is particularly crucial in improving urban identity to conserve and enhance cultural values in the development of urban public space.

According to the findings of the research framework of the cultural landscape in the case study of the region surrounding Luat Le hamlet, Dieu Tri town, and Tuy Phuoc district, the following stand out as the primary features of intangible cultural factors: Tay Son Dynasty, traditional martial arts, Tuong art, Bai Choi folk song, where famous poets (Han Mac Tu, Yen Lan, Quach Tan, Che Lan Vien, Xuan Dieu); Culinary culture, fishing village, relics of a famous trading port in the 14th century; Elements of material culture (Champa architectural art, space at home "roof leaves"). The study provides a framework for analyzing the cultural landscape by listing the functions, components, and meanings, which are the basis for urban design projects, fully expressing urban cultural aspects. The framework for analyzing the cultural landscape can provide different parameters to the same public space, and at the same time assessing the cultural value in urban planning and design projects.

The study also contributes to the planning orientation until 2035, Quy Nhon-Binh Dinh is one of the marine economic centers of the country, with a development-oriented economy: services-seaports-industry-travel.

References

1. Khoi DM (2018) The attractiveness of public space. Architecture Magazine No. 05-2018
2. Kaymaz I (2013) Urban landscapes and identity. In: Advances in landscape architecture, Croatia, Editor. IntechOpen
3. Carmona M (2021) Public places urban spaces: the dimensions of urban design. Routledge, New York

4. Thuy PTA (2020) Identity of urban landscape architecture—Behavior in the city context. Ho Chi Minh, in PhD Urban and Regional Planning. Ho Chi Minh City University of Architecture
5. Relph E (1976) Place and placelessness, vol 67, Pion London
6. Trancik R (1986) Lost space: theories of urban design. Van Nostrand Reinhold, New York
7. Smith CJ (1978) Geogr Rev 68(1):116–118
8. Rapoport A (1990) The meaning of the built environment: a nonverbal communication approach. University of Arizona Press
9. Punter J (2009) Urban design and the British urban renaissance. Taylor & Francis
10. Montgomery J (1998) Making a city: urbanity, vitality and urban design. J Urban Des 3(1):93–116
11. Carl OS (1925) The morphology of landscape. In: Foundation papers in landscape ecology. University of California Publications in Geography, pp 15–45
12. Jones M (2003) The concept of cultural landscape: discourse and narratives. In: Palang H, Fry G (eds) Landscape interfaces: cultural heritage in changing landscapes. Springer Netherlands, Dordrecht, pp 21–51
13. Brown DG (2001) Characterizing the human imprint on landscapes for ecological assessment. In: Jensen ME, Bourgeron PS (eds) A guidebook for integrated ecological assessments. Springer New York, New York, NY, pp 404–415
14. O'Donnell PM (2008) Urban cultural landscapes and the spirit of place. In: Finding the spirit of place—Between the tangible and the intangible. 16th ICOMOS General assembly and international symposium, Quebec, Canada
15. Stephenson J (2008) The cultural values model: an integrated approach to values in landscapes. Landsc Urban Plan 84(2):127–139
16. Sonkoly G (2017) Vienna and the Vienna Memorandum. In: Sonkoly G (ed) Historical urban landscape. Springer International Publishing, Cham, pp 77–122
17. Binh Dinh Department of Construction (2014) Adjusting the General Planning of Quy Nhon city and surrounding areas to 2035, vision to 2050. People's Committee of Binh Dinh province
18. Central Management Board of Irrigation Projects (2019) In-depth risk assessment for 8 river basins and disaster prevention planning for 10 provinces in the Q.l.t.t.V project area. Nam, Editor
19. Dien CX (2002) Vietnamese cultural foundation. Ho Chi Minh City National University Publishing House
20. Thuy VTT (2013) Culture of dealing with nature through living space of Vietnamese people. Ho Chi Minh City National University, University of Social Sciences and Humanities
21. Quang NT (2020) Hoai Nhon- Quy Nhon- Binh Dinh land and researcher. Culture and Arts Publishing House
22. Tan Q (1967) Non nuoc Binh Dinh. Thanh Quang printing house—34 Phan Thanh Gian—Saigon

The Solutions to Enhance the Demand for Using Sustainable Materials in Interior Design in Ho Chi Minh City

Chau Thi Minh Vo and Bac Thi Minh Le

Abstract This paper aims to find out why sustainable materials have not been widely used in the furniture sector in Vietnam through three main research methods: Desk study of materials literature existing sustainability in HCMC; Survey of the awareness and application of sustainable materials in the interior with two groups of subjects: designers and people with interior design needs; Semi-structural interviews with interior designers in HCMC. From there, find new directions with specific solutions to increase the demand for sustainable materials in interior design.

Keywords Sustainable materials · Sustainable interior design · Green design

1 Introduction

1.1 Environmental Sustainable Interior Design (ESID)

Traditionally, interior design has been concerned with the functionality and aesthetics of a space. Interior design was once described as a relatively conservative profession and focused solely on trends, fashion, and luxury design in a small environment; This approach has bypassed the design to save energy, reduce emissions to the environment, limit the harm to the mental and physical health of users [11].

In recent years, the interior design process has had a positive change. Designers have begun to pay more attention to products designed with environmental and ecological responsibility. The concern for environmental responsibility has sparked the context and demand for sustainable interior design—ESID [5]. Design strategies to promote a healthy and sustainable environment in interior spaces are developed based on selecting environmentally friendly building materials and interior finishes.

C. T. M. Vo (✉) · B. T. M. Le
Nguyen Tat Thanh University, 300A Nguyen Tat Thanh Street, 13 Ward, District 4, Ho Chi Minh City, Vietnam
e-mail: vtmchau@ntt.edu.vn

V. T. Ha et al. (eds.), *Proceedings of the 8th International Conference on Sustainable Urban Development*, Advances in 21st Century Human Settlements,
https://doi.org/10.1007/978-981-99-8003-1_14

Sustainable interior design is defined as "interior design in which all systems and materials are designed with an emphasis on integration into a whole to minimize negative impacts on the environment and occupants and maximize positive impacts on environmental, economic, and social systems over the life cycle of a building" [1]. Sustainable interior design aims to rationalize the use of materials in an innovative way that allows the conservation of natural resources in a way that does not lead to environmental degradation [9]. Currently, the term ESID is gradually becoming a trend in interior design toward a civilized, eco-friendly design environment and people.

1.2 ESID in the Context of Vietnam and Ho Chi Minh City

Vietnam is a country that is open to sustainable design trends, and buildings built according to green building standards are increasingly popular. The interest and practice of green design are strongly realized in big cities such as Hanoi and Ho Chi Minh City. In 2011, the Vietnam Association of Architects published the "Vietnam Green Architecture Declaration with five criteria: 1/ Sustainable location; 2/ Use energy resources efficiently; 3/ Quality of construction environment; 4/ Advanced and identity architecture; 5/ Socialization and humanity. This manifesto shows the commitment of Vietnamese architects to the international community on an inevitable progressive architectural trend; while laying the foundation for the activities toward the sustainable and environmentally friendly design of architects and designers in Vietnam.

Recently, organizations related to architecture and interior design in Ho Chi Minh City have had many activities related to ESID, such as the Webinar "Sustainable Furniture: Greening the Furniture and Furniture Industry" (Ho Chi Minh City Investment and Trade Promotion Center), "Design for life" Design Week (Design Association HCMC Vietnam – VDAS – 2022), Workshop "architecture and construction towards sustainable development" (Saigon Construction and Building Materials Association - SACA - 2022). These ongoing workshops prove the interest of organizations, individuals, architects, and designers who have turned their attention to green design and sustainable interior design.

1.3 Sustainable Materials in Interior Design

The definition of sustainable materials often depends on factors such as 1/ energy consumption and pollutant emissions of the material production process; 2/ the abundance and renewable capacity of resources; 3/ durability and material life cycle; 4/ impacts on the economy; 5/ the impact on the health of the user. Since then, sustainable materials have been defined as "materials that have a relatively positive impact on the community and the environment used to construct, provide services and develop the environment of buildings" [1].

Using sustainable materials is a matter of concern when designing and building interiors for the building. Building furniture with sustainable materials is not only good for the planet and the ecosystem, but it can also save the budget for customers. Moreover, sustainable interior design aims to use materials to protect natural resources in a way that does not harm the environment [1].

1.4 Aims and Objectives

The objective of this research paper is first to develop a comprehensive understanding of what constitutes the choice of sustainable or green materials in interior design and then conduct a study to identify the realistic barriers in Ho Chi Minh City. With the information collected from designers and material users, the research team analyzes and synthesizes information to propose appropriate solutions to increase the demand for sustainable materials in interior design in Ho Chi Minh City.

2 Literature Review

2.1 Materials Selection

In architecture and interior design, materials are used in large quantities. Construction activities worldwide each year consume approximately 3 billion tons of materials, accounting for 40% of the total resource consumption [3]. The choice of materials is an essential issue in the interior design process because materials are the main components that make up the elements in the interior space. Materials can be considered as an independent design element because, through the way they are used, their quantity and distribution in the interior space can give rise to many design ideas [9].

Selecting sustainable materials is a criterion for evaluating green building certifications such as LEED, WELL, Green Mark, etc., but the reality is still far from expectations. Research on the work of interior designers [7] has demonstrated that choosing sustainable materials is not a priority [7]. Sustainable products will cost more than conventional materials, and even the design is somewhat less attractive than cheap popular products. Therefore, sustainable materials are often rejected by investors or customers. Today's designers also spend much time searching for information about sustainable materials because this information source is not widespread and lacks authenticity. The above difficulties are genuine for developed countries and appear in countries gradually adopting sustainable design, like Vietnam. The choice of materials impacts the sustainable results of all interior design projects, especially commercial interior design projects that often have to be changed every 5–7 years, generating large amounts of waste.

2.2 Barriers to Sustainable Materials Selection

A study with interior designers [7] demonstrated that choosing sustainable materials is not a priority [7]. The choice of materials depends on investor requirements, aesthetics, and cost rather than environmental factors. Another critical issue is that the cost of green or environmentally friendly materials is often higher than that of conventional materials. Moreover, investors are only willing to accept higher prices for better quality interior products or criteria consistent with the aesthetic measure.

Moussatche et al. [8] survey shows that interior designers choose materials mainly according to customer preferences, needs, aesthetics, and costs, not considering sustainability as a criterion [8]. Moussatche et al. [8] found that factors such as durability and maintenance were considered necessary by study participants, while the global impact was often listed as secondary [8].

Another study by Kang and Guerin [6] found that designers have significant pressure on schedule and do not have enough time to learn about sustainable materials [6]. Cargo [2] examined the application of a database of sustainable materials (Green Spec., Material ConneXion, …) of interior design students [2]. While these tools help to reduce the time spent learning and selecting materials, they are not yet popular with designers.

The above studies have shown us in a general way the current barriers when choosing sustainable materials. Similar difficulties are encountered in Ho Chi Minh City, so this study identifies the reasons that prevent designers from limiting the use of sustainable materials. From there, propose suitable solutions for sustainable materials to designers and users.

3 Research Methodology

The objective of this study was to understand the factors that govern the selection process of sustainable or green materials in interior design. The authors will then conduct a survey to identify actual barriers in Ho Chi Minh City. The collected results will propose appropriate improvement solutions to increase the demand for sustainable materials in interior design in Ho Chi Minh City. The study uses the following three research methods:

- The desk study for the documents on sustainable materials available in Ho Chi Minh City.
- Surveying the awareness and application of sustainable materials in the interior with two groups of subjects: designers and people with interior design needs.

- Semi-structural interviews with interior designers in Ho Chi Minh City.
- Desk study to determine what materials are considered sustainable and green. It also allows researchers to classify materials according to their application. This form of secondary research involves collecting data from existing sources. It provides a helpful tool to support the primary research of the study, as well as time and cost savings [10].
- A survey on the application of sustainable materials was conducted with interior designers in Ho Chi Minh City (n = 60). They will be asked about the percentages and areas where they typically use sustainable materials. From there, give some reasons why they do not apply them. Another survey will ask people interested in interior design (n = 60) whether they know the concept of sustainable materials and why they do not want to apply this material. The results of this survey will be used to design a set of interview questions for designers in the third research method.
- Semi-structured interviews with a combination of closed and open questions were conducted with the designers (n = 30) to clarify the applicability of solutions to increase demand for durable materials firmly in interior design in Ho Chi Minh City. The questionnaire will be designed based on available studies on barriers to sustainable material selection and the survey result in the second method.

4 Results

4.1 Desk Study

To get an overview of sustainable materials today, the research team collected data from green and sustainable materials sources in Ho Chi Minh City. This format allows researchers to categorize and assess the prevalence of sustainable material supplies. In this study, the authors classified interior space finishing materials: as ceiling finishing, wall finishing, and floor finishing.

4.1.1 Ceiling and Wall Finishing Material

Based on the information gathered as shown in Figs. 1 and 2, with ceiling and wall finishing materials such as metal, plaster, wood, paint, bare concrete, etc., the research team found sustainable alternative materials that meet all forms according to customer needs. For example, in designing products that need the roughness of wood, sustainable materials can be mentioned as green wood, including Ecowood, Greenwood, Conwood, … (over three suppliers). Alternatively, sustainable materials that meet the trend of using bare concrete in interior spaces can be mentioned as green cement and environmentally friendly concrete: EPS ultralight concrete and glass fiber reinforced concrete… (over twelve suppliers). In addition, other alternative materials for bamboo, gypsum, plastic, paint, etc., can find sustainable materials

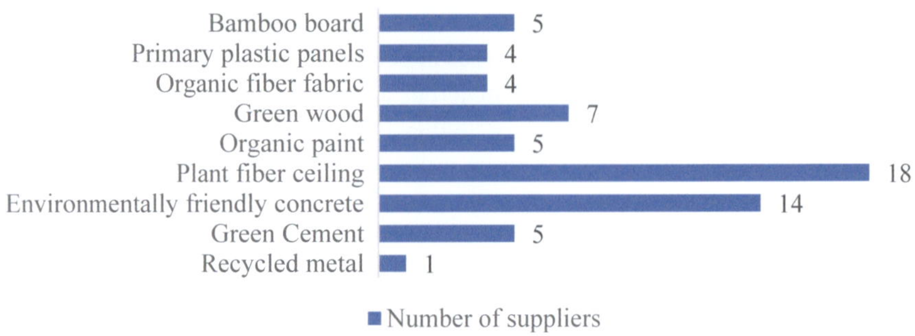

Fig. 1 Number of ceiling finishing material suppliers in Ho Chi Minh City

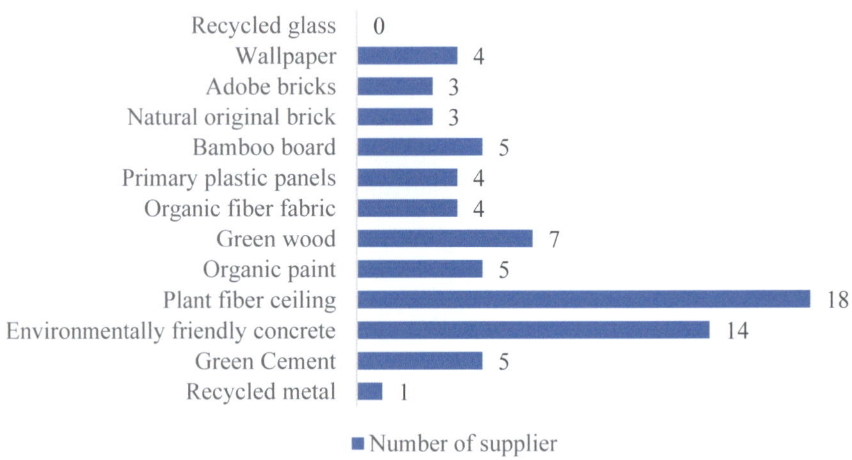

Fig. 2 Number of wall finishing material suppliers in Ho Chi Minh City

such as bamboo boards, mineral fiber panels, 3D plant-based panels, plastic, organic paints, etc. (over three suppliers).

In addition to the ordinary sustainable ceiling and wall finishes shown in the diagram, four groups of materials are mainly used exclusively for walls: wallpaper, tile, and glass. Unburnt bricks, natural bricks, and wallpaper all have more than three suppliers in Ho Chi Minh City. Mainly, recycled glass was once considered a profitable investment from manufacturers in the world, but currently, in Vietnam, there is no supplier for this material.

4.1.2 Floor Finishing Material

Floor finishing materials must ensure many criteria, such as durability, safety, moisture resistance, and bearing capacity. After the desk study, the authors realized that sustainable floor finishing materials still meet the above problems. To replace traditional cement, Ellis et al. state: "By 2060, the number of buildings on Earth is expected to double; this is equivalent to building a New York City every 30 days for the next 40 years. Since each kilo-gram of cement emits nearly 1 kg of CO_2, several gigatons of CO_2 per year will be released from new infrastructure, highlighting the urgency of decarbonizing cement production" [4]. We can use green cement or green label cement such as Insee's "Green Label" certified by the Singapore Green Building Council (SGBC) for eco-friendly cement and concrete products and is recorded in the data system "Green Database" of the Vietnam Green Building Council (VGBC).

To achieve this certification, they have researched and explicitly proven that building a house of 350 m^2 takes 50 tons of cement, with an emission of 50 tons of CO_2. However, 50 tons of cement "Green Label" reduces carbon emissions by at least 24% and saves natural resources for future generations. If all of Vietnam's cement can meet Green Label standards, we can reduce CO_2 emissions by at least 17 million tons annually.

Two groups of sustainable bricks that can replace traditional bricks are unburnt bricks and bricks of organic origin (three suppliers). With the desire to protect natural wood resources, laminated bamboo flooring (five suppliers) and environmentally friendly wood flooring (seven suppliers) are alternatives to natural wood flooring or laminated flooring. Combined with six organic fiber carpet suppliers shows that the popularity can meet the needs of today's floor finishing materials.

Thus, with the desk study results in Ho Chi Minh City, it can be seen in Fig. 3 that sustainable materials substitute for common materials in the interior space is quite diverse. With the majority of statistics on three suppliers, the authors find that the number of suppliers of interior finishing materials is quite common. Therefore, if the barriers are minimized as analyzed, sustainable materials will become the right choice for customer's needs as well as ensure an environmentally friendly trend.

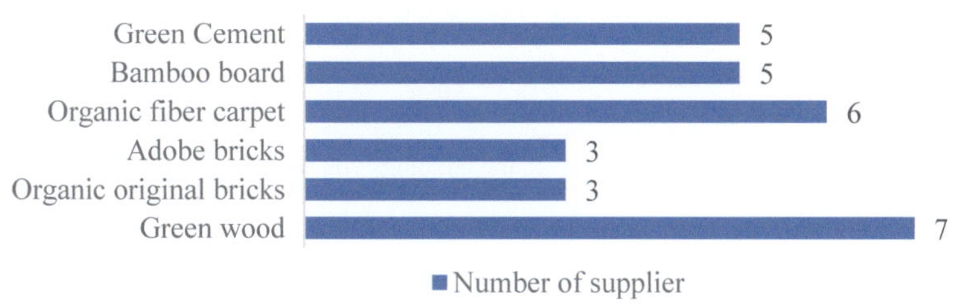

Fig. 3 Number of floor finishing material suppliers in Ho Chi Minh City

4.2 Survey

The research team surveyed two groups of interior designers and people interested in interior design in Ho Chi Minh City to find out the actual causes affecting the demand for sustainable materials in interior design.

The survey is limited to 22–50 years old (n = 60) for designers who have designed various types of works: houses, offices, hotels, resorts, restaurants, cafes, shops, schools, cultural-exhibition centers, shopping malls, etc. The survey question is: What percentage of sustainable materials do you use in your design? As Fig. 4 indicates, The results are: 40% (n = 24) of designers claim to use less than 10%. 33% (n = 20) of designers apply 10–30% green materials in their works (Fig. 4).

Thus, the rate of using sustainable materials in interior design in Ho Chi Minh City is still relatively low.

Convincing owners to use sustainable materials in interior space is a big challenge for designers. Based on survey data result in Fig. 5, the main reasons that sustainable materials are less chosen in the design are:

- Uninterested customers account for 64% (n = 39)
- Not much information, accounting for 46.7% (n = 28)
- No variety of models, accounting for 41.7% (n = 25)
- Expensive accounted for 51.7% (n = 31).

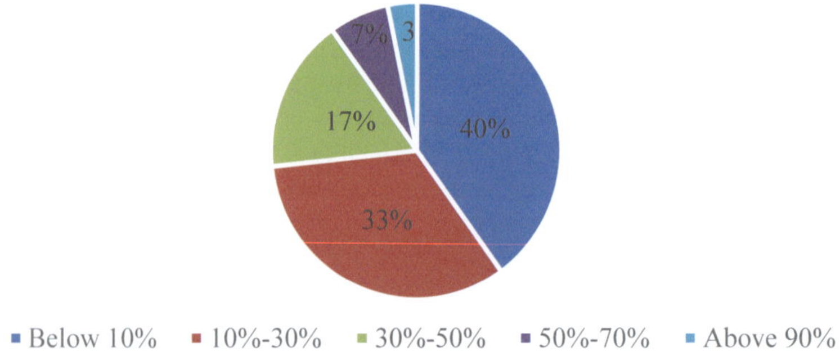

Fig. 4 Percentage of sustainable materials in the buildings you design

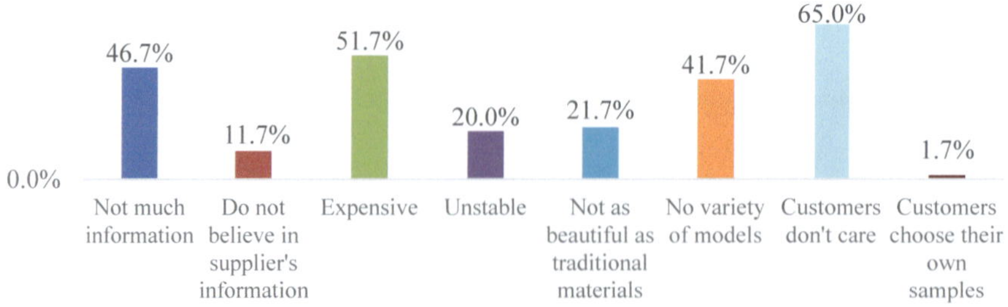

Fig. 5 Why don't you use sustainable materials in your design?

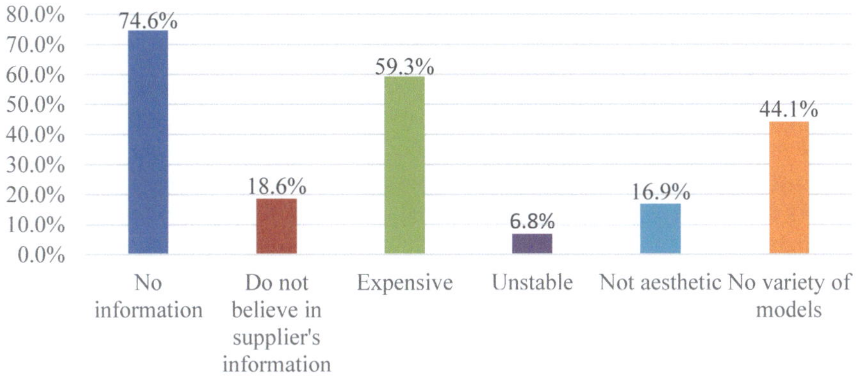

Fig. 6 What do you think is the reason why customers do not choose to use sustainable materials?

The above results have proved that sustainable materials are not yet widely available to both designers and customers; there is not much information needed to consult customers, making it easy for them to use sustainable materials in design.

According to the survey, customers from many different industry groups have interior design needs, ranging from 20–60 years old (n = 60). The results in Fig. 6 showed that 85% (n = 51) of the respondents had heard or known about sustainable materials. However, they are not interested in this material, in line with the survey result from the designer. So, customers generally do not choose sustainable materials because there is little information, accounting for 75% (n = 45). 60% (n = 36) think this material is expensive, and 43.3% (n = 26) consider the design less varied than conventional materials.

Based on the results of two surveys, it is necessary to improve the reasons why these materials are rarely used: providing the necessary amount of information, increasing the variety of models, reducing costs, etc.

4.3 Semi-Structured Interviews

After identifying the barriers when applying sustainable materials in interior works, the research team has orientations to address those causes. However, to increase the practicality of these solutions, the research team continued to conduct semi-structured interviews with 30 random designers in Ho Chi Minh City. The interviews is proceeded with a combination of closed and open questions.

Quick information search is critical, so libraries related to specific fields have become popular. In the design field, 97% (n = 29) of designers said they found no library related to sustainable materials, and the remaining 3% (n = 1) proved a small part exists on the website Archdaily.com. 100% (n = 30) of designers say they would use an online sustainable materials library if it were created.

To apply sustainable materials in interior construction, designers need to understand its features or at least have been introduced to these products. According to the interview, 87% (n = 26) of designers are not trained in green design and sustainable materials knowledge in schools and companies, and 13% (n = 4) receive training during work. 100% (n = 30) of designers confirm that customers have not actively requested or learned about sustainable materials in interior design. Therefore, the authors suggested organizing a workshop to introduce the benefits of sustainable materials: creating a safe space with health and improving the living environment for customers interested in interior design. 100% (n = 30) of designers agreed with this solution and will participate too.

Another view of the government's concern with the community to help the living environment better is: should the state intervene in using sustainable materials in interior design works? The answers prove this is also a viable solution, with 50% (n = 15) of designers agreeing and 50% (n = 15) disagreeing. However, the authors will explain how to intervene appropriately in the discussion.

To find a way to overcome the design barrier of non-diversity of sustainable materials, 16% (n = 5) of designers reported that they regularly interact with manufacturers about the colors and surface details of materials. The remaining 84% (n = 25) of designers received only material catalogs, and manufacturers were challenging to reach. Therefore, 97% (n = 29) of the interviewees think that manufacturers should listen to the feedback of designers to improve the existing problems of sustainable materials and come up with products that match their tastes. The remaining 3% (n = 1) think this is unnecessary because they already have their product research team.

One of the reasons why sustainable materials are less selected is the cost, and we assume the option can reduce the tax on sustainable materials for manufacturers under the government subsidy. 90% (n = 27) of designers agree, and 10% (n = 3) of designers think it is unnecessary and needs more measures to encourage customers to use environmentally friendly materials.

With the numbers obtained from the above interview method and the orientations given to overcome the barriers to using sustainable materials, the research team received many opinions agreeing with the rate of over 70%. These interviews are crucial to create objectivity and practicality with the recommendations that the research team makes in this paper.

5 Discussion

Through the results of three research methods, we have had an overview of the market for supplying sustainable materials, finding objective and subjective causes from a multi-dimensional perspective related to the need for sustainable materials in interior design in Ho Chi Minh City.

From the research results, three main barriers were found:

- Lack of information about materials;
- Models and colors are not diverse;
- High cost.

Therefore, the team focused on analyzing, discussing, and proposing solutions to improve three main barriers to increasing the demand for sustainable materials in Ho Chi Minh City interior design spaces.

Combined with a mix of open and closed questions in the designer interview, we concretize several solutions as follows:

1. Regarding the lack of information related to sustainable materials: the research team proposes four solutions:

- First: create an online sustainable material library. This library provides information about sustainable materials such as origin, cost, composition, color, standard, size, illustration, etc. In this direction, information related to sustainable materials is regularly updated to make searching more convenient. Designers and customers can easily research the benefits of sustainable materials as well as the ability to compare prices with conventional materials
- Second: open training courses on sustainable interior design at professional design schools such as colleges and universities. With a basic knowledge of sustainable materials, designers can quickly advise and encourage customers to use them. In addition, specialized competitions on sustainable interior design can be organized.
- Thirdly: organize a specialized seminar to introduce the benefits of sustainable materials in interior design: safety for health, improvement of the living environment, protection of ecosystems, etc. According to survey results, the places that can attract participating customers are: material fairs, furniture fairs, specialized universities, and specialized companies.
- Fourth: through government organizations, taking advantage of the influence of the media, deploying the application of sustainable materials in public works, public investment buildings, and grade 1 buildings (having a life span of over 100 years). These works are the basis for creating trust and a fulcrum for communication to promote sustainable design.

2. About the diversity of sustainable materials: The research team proposes a two-way interaction between the manufacturer and the interior designer. From the designer's point of view, the choice of changing patterns, layouts, and colors to help sustainable materials be applied more diversely in interior spaces.

3. Regarding the cost of sustainable materials: we have the following solutions:

- First: The government should subsidize or reduce taxes for manufacturers or suppliers of this material (implementing the declaration to bring Vietnam's emissions back to zero by 2050).
- Second: The government focuses on researching and developing environmentally friendly materials through abundant local raw materials combined with modern technology.

- Third: Customers need to see long-term benefits when choosing sustainable materials. Manufacturers can improve incentive packages, and accompanying services, increase warranty period, and reduce maintenance fees.

6 Conclusions

Many sources have proven that CO_2 emissions are constantly increasing year by year. However, mitigating climate change and environmental pollution depends not only on managers but also on the cooperation of the whole community. Through the article, from a professional perspective, the use of sustainable materials in interior design is a practical action. The research team has summarized the current situation of the sustainable materials market in Ho Chi Minh City to provide a relative amount of information on existing common materials. Based on the survey, we found many factors affecting the choice of this material, of which there are three critical reasons. The research team proposed eight related solutions to solve or improve these problems.

The paper's proofs, results, and solutions ultimately aim to increase the demand for sustainable materials in interior design. Many issues still need to be discussed and analyzed in this study. It can be mentioned as the actual effect of using these materials and other necessary solutions for the remaining causes. The research team will implement this in the following projects.

References

1. Alfuraty AB (2020) Sustainable environment in interior design: design by choosing sustainable materials. IOP Conference Series: Materials Science and Engineering 881(1):012035. https://doi.org/10.1088/1757-899x/881/1/012035
2. Cargo A (2013) An evaluation of the use of sustainable material databases within the interior design profession. University of Florida, Senior Capstone Project
3. Eales, A (2012). Sustainability and engineering, an open educational resource for U-Now. University of Nottingham
4. Ellis L, Badel A, Chiang M, Park R, Chiang Y (2019) Toward electrochemical synthesis of cement—An electrolyzer-based process for decarbonating CaCO 3 while producing useful gas streams. Proc Natl Acad Sci 117(23):12584–12591
5. Jones L (2008) Environmentally responsible design: green and sustainable design for interior designers. Wiley
6. Kang M, Guerin DA (2009) The characteristics of interior designers who practice environmentally sustainable interior design. Environ Behav 41(2):170–184
7. Lee E, Allen A, Kim BK (2013) Interior Design practitioner motivations for specifying sustainable materials: applying the theory of planned behavior to residential design. J Inter Des 38(4):1–16. https://doi.org/10.1111/joid.12017
8. Moussatche H, King J, Roger ST (2002) Material selection in interior design practice. In: Interior design educators council international conference abstracts, March 19–14, Santa Fe, NM, pp 26–27
9. Soliman OA (2013) Perception of building materials in architecture. J Eng Appl Sci 60

10. Stewart D, Kamins M (1993) Secondary research: a conceptual approach. Holt, Rinehart and Winston, New York, NY
11. Yang Y, Fenghu W, Xiaodong Z (2011) Contrast study on interior design with low-carbon and traditional design. ICMREE 1(2011):806–809

Mobility Based Transforming Growth Models

Analysis of the Effects of Urban Mobility Concepts on the Socio-Technical Transformation Process in Major Cities: A Qualitative Study of Transforming European Cities

Benjamin Dominic Kraff

Abstract The mobility sector is a major contributor to greenhouse gas emissions, but with appropriate measures, it offers a high potential for emission savings, especially in urban areas. A high level of particulate matter pollution as well as risks for pedestrians and cyclists are additional reasons that already lead certain European cities to implement car-reduced city centres and neighbourhoods. The strategic orientation of the deployed municipal concepts is determined by the socio-technical regime dominating the respective city as well as by overriding trends and local innovations. Under these framework conditions for the concept, planners try to combine different environmental policy instruments to achieve the most desired effects, not only to directly achieve municipal, national and international objectives but also to accelerate the socio-technical transformation process. The aim of this work is to evaluate the effects of environmental policy instruments and concepts on the socio-technical transformation process in the context of a multi-level perspective. The qualitative analysis of this research is based on a literature review and case studies of major European cities with planning papers, municipal resolutions and other documents, as well as expert interviews with planning and management stakeholders. The research highlights and assesses the role of urban mobility strategies, administrative structures and environmental policy instruments in the transformation process. It is shown that the mobility transition requires a continuous implementation of incremental measures as well as a system opening for the diffusion of more radical concepts. To accelerate processes, both key actors and stakeholder networks are crucial to create acceptance and willingness for change. In addition to vertical influences in the multi-level model, horizontal relationships between neighbouring regimes could be identified and elements that weak the path dependencies and make the structures more open to innovation could be shown and integrated into the model.

B. D. Kraff (✉)
Technische Universität Darmstadt, Franziska-Braun-Str. 7, 64287 Darmstadt, Germany
e-mail: kraff@geod.tu-darmstadt.de

V. T. Ha et al. (eds.), *Proceedings of the 8th International Conference on Sustainable Urban Development*, Advances in 21st Century Human Settlements,
https://doi.org/10.1007/978-981-99-8003-1_15

Keywords Urban mobility · Transformation · Socio-technical regime · Multi-level perspective

1 Introduction

The traffic and transportation sector plays a fundamental role in the daily social and economic activities of a society. While on the one hand, the growth in prosperity is attributable to it [30], on the other hand, its growth in the currently existing structures also leads to several challenges. In addition to negative impacts on the environment due to noise and pollutant emissions [19], there are also issues of traffic safety and upcoming debates about land use justice in urban areas [73]. The high emission of greenhouse gases is especially problematic, with serious consequences for the global climate [19].

Although CO_2 emissions in the energy, households and manufacturing sectors have been significantly reduced since 1990, the share of the transport sector has increased by 24% until 2019 [20]. This development is contrary to the EU goals. While the previous EU decisions aimed for a greenhouse gas emission reduction of 80–95% across all sectors in the EU by 2050, the EU is tightening its targets with the European Climate Law [28]. Already by 2030, a reduction of 55% compared to the base year 1990 is to be achieved, and by 2050, complete climate neutrality is to be reached. The law also aims to ensure that all EU Member States make the necessary efforts to contribute to the achievement of the target across all social and economic sectors [61]. To realise the goals in the traffic and transportation sector, a rapid transition to a climate-friendly mobility system is indispensable.

The transformation required for structural change is not only dependent on technological developments but rather on social conventions, state regulations, available infrastructures and local actor networks [23]. With the help of participation formats, citizens are involved in municipal transformation projects [7] to jointly develop viable solutions, increase acceptance of new technologies [5, 72] and thereby break the social inertia of outdated structures [23]. Technological momentum is also due to a network of system builders who stick to existing structures [38]. These structures also need to be fundamentally reorganised from the inside to enable a process of change.

To understand the transformation, it is necessary to look at the framework conditions at different levels. Therefore, the multi-level perspective maps overarching market and global political conditions, current infrastructures and actor networks, as well as innovations under development [23]. The interaction of the levels in the context of sustainable development and the mobility transition as well as global perspectives are subjects of numerous research activities [24, 52, 55]. The interaction of individual components within the socio-technical regime, the meso-level of the multi-level model, still raises numerous questions in the context of mobility transition. This paper, therefore, aims to contribute to the still open research gap with the following research question:

- What structural requirements in the socio-technical regime must be established to enable an accelerated transformation in the urban mobility sector?

To answer this research question, a qualitative research design is developed in the second chapter, considering scientific literature, policy strategies, concepts applied in cities, projects implemented in neighbourhoods and experiences gained in practice. Section 3 then introduces the concept of the multi-level model and focuses on the transformation process of urban mobility. In Sect. 4, the empirical analysis is carried out based on selected European cities. Implications and limitations are pointed out in Sect. 5. The conclusion finally summarises the most important findings and provides an outlook for further research activities.

2 Methodology

At first impression, the EU-wide development of the transport sector seems to be very similar and follows a general growth trend in the emissions sector [20]. However, when looking at international case studies, significant differences in dealing with the problems can be identified. First, the city's urban mobility strategy as a basis for all further measures is essential (see 4.1). Second, the management structures given by the administrative organisation and influenced by political orientation are crucial (see 4.2). Third, the socio-technical transformation is characterised by the application of different concepts using different environmental policy instruments to implement mobility measures (see 4.3). Fourth, the use of different information, participation and cooperation formats that attempt to integrate affected actors and important stakeholders into the processes to varying degrees plays a major role (see 4.4). To highlight these four aspects and their interfaces, a mixed-method approach is applied.

2.1 Research Design and Data Collection

The work aims to provide an overview of current developments in Europe and to identify important mechanisms for accelerating socio-technical transformation processes in urban mobility. For systemic research questions where the multi-level perspective has been applied, qualitative analyses ranging from an individual [23] to extensive comparative case studies [52] have been established in research work. Due to the complexity and uniqueness of mobility and urban structures, quantitative comparability is difficult and not useful for the objective of identifying effects as a result of the interplay of different factors, so qualitative research design is also chosen here. A mixed-method approach is developed, which includes a literature analysis and a scientific contemplation of case studies of European cities. To get data on strategies, measures and also transformation processes, a document analysis is carried out and expert interviews with structure-influencing stakeholders of selected case studies are conducted.

In the first step, literature analysis of primary and secondary literature from studies about socio-technical transformation and transition processes is made. Previous research activities such as those of [55], who also consider the urban mobility transformation in the context of the multi-level perspective, are taken up. The focus of the literature review lies in the practical application of the multi-level model and the identification of current path dependencies of the urban mobility sector. In particular, the structures of the socio-technical regime are taken into the focus of the study. Literature on communal mobility strategies, urban mobility transition concepts and mobility project implementations are examined in the research field's context. In the process of the literature review, stakeholders are gathered who are networked across different levels, who move within existing structures and who also participate to different extents in the opening of existing structures for transition processes.

In the second step, a case study analysis of selected European cities, that are showing off innovative pathways to implement sustainable mobility measures, is carried out. Using the case study method, complex structures and processes are examined [35] and the concept of socio-technical regimes within the multi-level-perspective is applied [23]. With the support of case studies, it is possible to analyse not only internal structures but also the interrelationships with the environment [78] and compare them with each other [35]. The inner structures can thus be revealed [4] and plausibly presented [31]. From the case studies, particularly successful conditions that lead to urban mobility transition processes can be identified.

To be able to compare the different measures more precisely and to better map the organisational structures with their networked interdependencies, the data collection method of document analysis is integrated as part of the case study analysis. It belongs to the interpretative methods [36, 59] and can be used to gain insights into social action and reconstruct social processes and structures [66]. The selection of sources is made in such a way that target-oriented information for the object of investigation and the associated questions can be integrated [62]. From the perspective of social science, official documents and internal documents are particularly relevant document types. Official documents or publications are books, manuscripts of public speeches, magazines, newspapers, brochures or flyers. Internal documents come from public or private organisations and include files, statistics, interview transcripts or other working materials [66, p. 813]. Due to the accessibility, this study mainly works not only with official documents but also partly with published internal documents.

Finally, to be able to classify the different approaches and make a qualitative assessment [67], four 30–45 min long additional expert interviews were conducted with urban mobility stakeholders. With the expert interview, both objective facts and subjective evaluations and attitudes can be obtained (ibid.). For this purpose, interview partners of the case studies were requested who work in the field of urban mobility planning or strategic mobility management. Accordingly, with the objective of this work, a homogeneous group of experts, similar in attitude and positioning

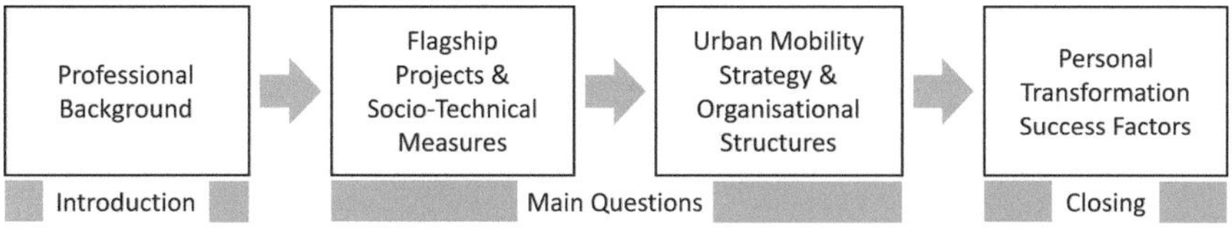

Fig. 1 Structure of the interview guideline

from an empirical point of view, was deliberately chosen to better enable comparability of the case studies [58]. Due to the distance between the European cities, the interviews were conducted as video calls with all experts. A semi-structured interview with a guideline adapted to the municipal circumstances, mobility strategies and special projects was prepared [17]. Open-ended questions were used and made it possible to build up a conversation that was free of predefined answer options [2] and thus provide a more accurate picture of the real situation [58]. The procedure also made it possible to go deeper into statements presented by the interviewee through follow-up questions. The questions were asked in a clear and simple manner, without double negatives, without multidimensionality and judgement [17]. Nevertheless, the interview always contains narrative prompts in order not to lose the focus of the conversation on the research objective [33]. In addition to an introduction, all interviews focused on the implementation of low-car neighbourhoods with environmental policy instruments and participation procedures and looked at the overall urban mobility strategy and the respective organisational and project management structures. Figure 1 shows the interview design.

2.2 *Data Analysis and Evaluation*

The data analysis was started at an early stage of the research to provide a scientific basis for the selection of case studies. In addition, with the help of the knowledge already gained by the literature review, the data collection procedure could be constantly adapted and more effectively aligned with the research focus. The selection of case studies was based on the literature analysis, using cities that have achieved a high profile due to innovative concepts, pioneering strategies and unique measures and have been the focus of numerous scientific research activities. Experts in these cities were contacted with an invitation letter, resulting in four expert interviews with planners and strategic managers from the mobility sector of Brussels, Amsterdam, Hamburg and Freiburg. The expert interviews were conducted by a research associate, who asked the guided questions, and a research assistant, who noted the most important information in a digitally shared document. In addition, an audio-recorded transcript was made to prepare a bullet point transcript afterwards.

The data from the case studies, document analysis and the information from the interviews were collected, digitised and qualitatively evaluated by using the steps of

Table 1 Main coding categories

Upper category	Content examples
(1) Mobility strategy	Mobility goals, mission statement, mobility management, city networks, implementation process
(2) Administrative structures	Hierarchical, functional, interdepartmental interaction, bureaucratic guidelines
(3) Technical measures	Car-reduced/free zones, parking space management, green corridors, priority cycle routes, city tolls
(4) Social measures	Information: information points, flyers Participation: presentations, discussion forums Cooperation: working groups

a deductive category method [49]. In this qualitative procedure, which in contrast to the quantitative evaluation is not aimed at static results, but rather at a representation of the diverse interrelationships, text modules were marked and evaluated with a developed code system. A coding guideline was formulated to specify in which category single text passages had to be placed [50]. For this purpose, well-chosen upper and lower categories were defined from the theory [43]. Coding was carried out with the programme MAXQDA and the upper categories shown in Table 1 were finally defined to map the socio-technical regime of the multi-level perspective.

While working on the data analysis, the upper categories stayed constant, but the lower categories had to be adjusted due to additional coming up. Through the case interpretation, it is possible to make a generalisation or transferability via an empirical type of formation or an internal formation of empirical correlations [4]. In this work, the case studies are compared based on various characteristics and also evaluated in their contexts to do adequate justice to the very different socio-technical regimes and still allow for a comparative interpretation.

3 Theoretical Framework: Urban Mobility Transition

The urban mobility transition encompasses many dynamically changing structures and therefore offers very different research approaches, from social, technical or economic perspectives. The multi-level perspective gives a framework to analyse complex socio-technical systems and was accordingly applied in different studies of several sectors, from power grids [76] to sustainable housing [71] up to mobility. The focus in the field of mobility has so far been on Transitions Theory in particular [44], on transition dynamics with technical changes [8, 25, 56] and on socio-technical structures [55]. This paper also addresses the socio-technical aspects of the transition approach but looks in particular at the interaction and mutual acceleration of different dimensions within the transformation process.

To better understand the urban mobility transition, this chapter explains the innovation process and the meaning of its system, introduces the multi-level model and explores the role of related governance and stakeholder networks.

3.1 Innovation System and Process

In the case studies of the European cities, innovative solutions are applied, which include not only new measures but also new approaches to mobility management. Innovations therefore can be understood as combinations that are newly developed or that are used in a new way, that are functional and that have a commercial benefit or have been successfully implemented in the market [64]. The invention process is also a synthesis of market demand to meet emergent needs [75]. The creative nature of developing new combinations is also intended to ensure competitiveness [29]. In contrast to purely creative processes, innovation is therefore associated with successful commercialisation on the market, which both promotes economic growth and provides a potential for solving social problems. A new combination only acquires a function through the networking of various socio-technical structures [23]. Hughes defines function as an interdependent network of physical artefacts, social structures, available resources, knowledge and legal determinations [39], a functioning configuration [63].

A technology change takes place by changing the socio-technical configurations with incremental or radical innovations. If functions are changed to optimise a product, these are incremental innovations [29]. In radical innovations, on the other hand, existing technologies are substituted by new ones. Here, there is an insistence on the use of existing structures and combinations, as the individual ones show mutual dependencies and are adapted to existing technologies [23].

The innovation process is the linear progression of several individual phases such as idea generation, research, development, production and marketing. The individual stages each represent self-contained units and partly overlap. They are characterised by uncertainties, dependencies, complexity and learning processes [46]. The success of local actors depends on diverse and dynamic learning processes, so one speaks of the socio-economic state of the learning economy (ibid.). The learning processes make it possible to incorporate skills and potentials into routine activities [16] and thus condition innovation and economic growth. Conversely, innovations accelerate learning processes. Both learning and innovation processes depend on local structures [46].

The innovation system comprises a social system on the one hand and an open system on the other. The social system is characterised by innovations that arise through the interaction of economically active actors. The open system stands for constant interaction with the environment (ibid.). The actors acting in an innovation system cooperate in networks, build up mutual trust there and try to achieve common goals in formal as well as informal meetings, taking individual interests into account

[16]. For the analysis of the urban mobility sector, the consideration of the stakeholder structures in the socio-technical regime is of relevance.

3.2 Multi-level Approach

The multi-level approach introduced by Geels and building upon previous research models makes it possible to capture overarching patterns and offers links for the study of transitions. Transition hereby can be defined as the changing process to a new future and requires multiple socio-related transformations [27, 65]. The concept displays complex dynamics in the context of a socio-technical transition [23].

Socio-technical regime. The technological regime comprises the coordination level and is shaped by organisational and mental routines. These routines determine the actions of organisations and the actors working in them. The original concept primarily understood only engineers as a relevant group of actors who shape the technological regime through their almost identical routines. With the same approach and the same thought structure, a technological trajectory emerges when looking at the processes. Technological regimes provide security because their innovation activities lead to incremental improvements along this development line [23].

Rip and Kemp expanded the concept of the technological regime and see it as more strongly anchored in the complex superordinate structures that result from the practical procedures, the technologies and the skills but also how problems are dealt with [63]. They add further social groups such as politics, science and society to the group of actors of engineers.

Geels uses the term socio-technical regime to describe how social groups of actors orient themselves within a stable system with partially coherent rules. Due to the stability, automatic selection and preservation mechanisms take place [23], resulting in only incremental, but not radical innovations [14]. Thus, the unfolding and developing of new combinations are not given in the socio-technical regime [69]. Persistence is the result of the process of variation, selection and conservation [23]. Nevertheless, next to radical innovations, the initiatives and inventions occurring on the regime level are needed as incremental innovation processes to constantly improve the mobility infrastructure [51] and to open it up for system changes.

Socio-technical landscape. The technical line of development can be located in the socio-technical landscape, which has deep structural features and to which actors are subject in their interaction [23].

The socio-technical landscape is characterised not only by the infrastructures present in the space but also by price developments, economic growth, political alliances and social values. It thus relates to factors external to technology, such as cultural changes, demographic changes or changing political orientation. Their change is slower than that of the regime (ibid.).

Niche. Radical innovations may not emerge in the socio-technical regime but in niches that are protected or isolated from the market [68].

Niches are local collaborations between actors such as a contractor or funder and organisations that develop new technologies (ibid.). They provide a space for learning processes, especially since products often perform poorly at the beginning and are uneconomical [23]. In addition to their importance for the development of new technologies and their financing, niches are above all important for the development of new target groups, the formation of new social networks, the expansion of research activities and institutional adaptation to enable market diffusion and further development in the first place [41]. To archive the ambitious but necessary carbon reduction goals, new technologies and smart concepts restructuring the existing patterns in a radical way are essential for the urban mobility transition [42].

Multi-level perspective. The combination of the concepts of macro-level socio-technical landscape, meso-level socio-technical regime and micro-level niche is referred to as a multi-level perspective [23].

New successful products, therefore, emerge in the niche, but under the influence of the evolution of the socio-technical landscape and regime [40]. If the macro-level changes, the new meso-level conditions open opportunities for the uptake of new technologies [23]. Therefore, a closer look at meso- and macro-level processes is necessary, as incremental processes along the structural hierarchy occur at the level of the socio-technical regime [14]. Geels has identified seven dimensions that are of particular importance at the regime level [23]:

- technology
- user experience and areas of application
- (symbolic) value of the technology
- infrastructure
- corporate structure
- policy
- technical-scientific knowledge.

Due to the relationships between the different elements, the stability of the existing socio-technical elements is brought about. Interpersonal activities build on these elements and reproduce other interrelated elements. Through the different structures, groups are formed whose activities adapt to the other groups. Even though the different characteristics are interrelated, they have a momentum of their own. This creates tensions in the socio-technical regime that softens the interdependencies and consequently allows for change (ibid).

The process of technology change is subject to various learning processes and has a cascading course. It is based on niche cumulation (ibid.), which means that there is no sudden change from an old to a new regime but rather a gradual process of reconfiguration [18].

The breakthrough of radical innovations at the level of the regime can be understood as a technological add-on or hybridisation since the new product connects to already established technologies and covers the open needs there [23]. It can be

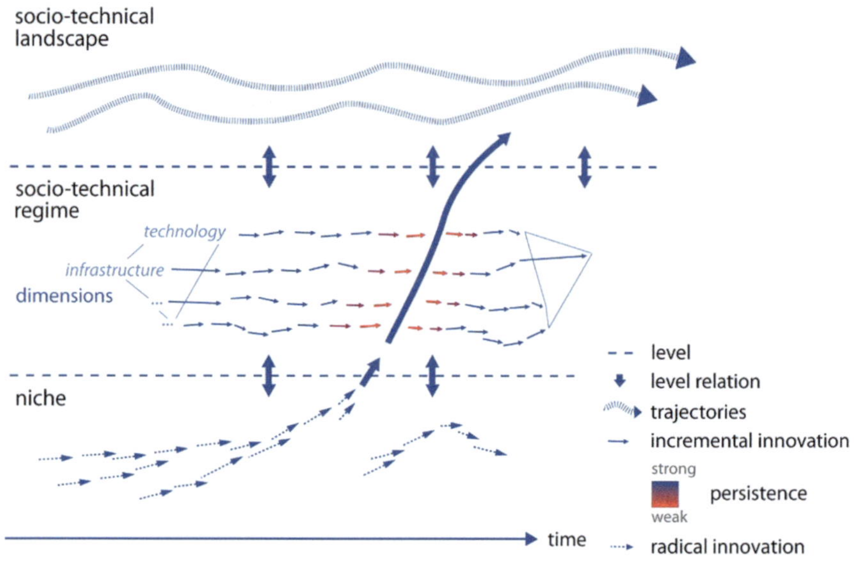

Fig. 2 Dynamic multi-level perspective (based on [23])

concluded from this that there does not always have to be competition between old and new technology but that symbiosis can also arise [57]. The dynamic multi-level perspective is illustrated in Fig. 2 and shows how processes at all levels are influenced and shaped by dependencies and changes in each other's levels.

3.3 *Governance and Stakeholder Network*

Infrastructures consist of various subsystems and stakeholders active within them, which are interrelated in complex ways and whose interaction can influence accelerated change [48]. Cameron and Geels [12] examined the conditions for politically accelerated change and distinguished between a top-down influence from the landscape to the regime and a bottom-up interaction from the niche to the regime. Thus, in addition to the interests of economic actors, mass phenomena in society and new technologies, internal policy developments could be identified as factors in political action that have a facilitating or inhibiting effect on change [12]. The multi-level approach makes it clear that a transformation process always takes place through cross-level actor interaction [60]. Technological transformation is therefore not only the result of technological changes but also the result of changed practices, regulations, networks, infrastructures and social values [23]. For this reason, the political-administrative system is often the starting point of a systematic transformation at the local level, so it is often seen as the initiator of the transformation process. At the meso-level, political-institutional framework conditions are created to facilitate innovation-promoting regional and economic development organisations to support regional policy that implements measures that promote cooperation [46]. Thus, the policy has the task of driving the learning processes in the learning economy by providing network-promoting structures [15].

Successful governance can only take place in a balance between the overarching interests on the one hand and the interests of the local level on the other [74]. To achieve this balance, a joint and coordinated approach by private and public actors is required [22]. Regional governance networks are formed between companies, the private sector, the state and civil society actors. Cooperative and network-like approaches predominate [74]. If an innovation system is successfully established with a self-reliant administrative dimension, policies can be implemented at the meso-level and public funds can be invested in the development of innovative structures [46]. A high level of willingness to perform, the exchange of information, and the cooperation and coordination of participants are elementary, to build trust through personal interaction and thereby release knowledge as well as potential [45]. The task of mediating between the different activities, often assumed by companies [37], is increasingly being replaced by informal and individual relationships in a steadily growing network of actors [48].

On the one hand, there should be a certain degree of consistency in the general orientation and activities. On the other hand, if there is too much fixation on one technology, new development can hardly be integrated. This so-called lock-in [21] reduces the likelihood of radical innovations breaking out of the niche and rearranging the socio-technical regime [26, 76].

4 Analysis of European City Strategies

The focus of the analysis lies on the creation of local structures in the socio-technical regime that meet overarching objectives and open possibilities for change in the urban mobility sector. For this purpose, municipalities, in cooperation with different stakeholders, realise projects and city strategies that directly or indirectly influence the predominant traffic and transport infrastructure, promote sustainably oriented trajectories in the landscape and implement innovative technologies and concepts coming out of the niche [23]. It is assumed that even small structural changes and certain driving elements in the city can accelerate the transformation process beyond the scope of the actual measure. Therefore, in the following, the mobility strategies of selected European cities are presented with the structural set-up of the authorities, the respective implemented technical measures and the stakeholder management. Table 2 shows an excerpt of the urban analysis and highlights the strategies, diverse structures and different environmental policy instruments.

Table 2 Urban mobiliy transition in major European cities (study excerpt)

City	Amsterdam	Barcelona	Brussels (region)	Copenhagen	Freiburg	Hamburg	Helsinki	Oslo	Paris
Population (2022)	0.94 mio	1.63 mio	1.07 mio	0.6 mio	0.23 mio	1.86 mio	1.33 mio	1.07 mio	0.63 mio
Cities' modal split	(2019)	(2016)	(2018)	(2017)	(2016)	(2017)	(2020)	(2015)	(2019)
– Car	29%	26%	49%	31%	21%	36%	25%	34%	25%
– Public transport	25%	40%	45%	21%	16%	22%	31%	32%	17%
– Bicycle	29%	2%	2%	28%	34%	15%	10%	7%	2%
– Foot	16%	32%	4%	20%	29%	27%	34%	28%	46%
Urban mobility strategies and concepts	Amsterdam Attractively Accessible	Superblock	Good move	Green mobility	Climate mobility plan	Green city plan	Mobility as a service	Car-free Livability	Embellish your neighbourhood
Administrative structures	Integral project management model: project		Combined management on a regional level		Technical department and mobility unit, hierarchical	Networked mobility authority			

(continued)

Table 2 (continued)

City	Amsterdam	Barcelona	Brussels (region)	Copenhagen	Freiburg	Hamburg	Helsinki	Oslo	Paris
Environmental policy									
– Regulatory instruments	Temporary car-free days; car-free district; low emission zones	Speed reduction; driving barriers	Temporary car-free days; speed limit; innovation-open legislation	Cycling prioritisation;	Car-reduced district; reduced parking space key; bicycle key	Car-reduced districts; pedestrian zones	Innovation-open legislation	Banning inner-city parking	Temporary car-free days; car-free streets
– Financial instruments	High parking fee			High parking fee				High parking fee	
– Planning instruments	Metro line; cycle path network	Block planning; pedestrian zones; green axes: car-free in long-term	Digitalisation; cycle path network; circulation pentagon	Cycle path network; pedestrian zones, car-free planning	Cycle path network	Metro line; cycle path network	Digitalisation; on-demand services; metro line; pedestrian zones	Renew walking and cycling paths; parklets; connect pedestrian axis	Cycling infrastructure; pedestrian zones
Communication and participation	Co-development of a district	Information and permanent participation	Strategy participation	Intense public participation	District planning participation	Alliance for walking and cycling	Networks; initiatives; lobbying		

4.1 Urban Mobility Strategies

In the first step, the study focuses on the development of a city-wide concept and the related clear communication of the mobility strategy. The strategies follow a mobility shift that is intended to improve the transport system but also the urban quality of life. The problems of air pollutants, safety, land consumption and the need for climate protection are addressed. All concepts are aimed at promoting alternative mobility options and, to various degrees, at reducing car traffic. Alternative mobility offers included in the strategic concepts are characterised above all by the expansion of cycling and walking networks, the improvement of public transportation and the opening of structures for new mobility service providers. The spectrum of car reduction can be distinguished from car-lite measures, temporary car-free measures, permanent local measures and permanent large-scale measures aiming at car-reduced or car-free streets, neighbourhoods or cities [77]. Car-lite measures like reduced speed limits and shared spaces are easy to implement and can be found in all major cities. The so-called superblock concept in Barcelona combines speed-reduction areas with driving barriers that make it necessary to drive out of the blocks to get from one place to another and thus cause car-reduced spaces. Amsterdam, Brussels and Paris use temporary car-free days as an instrument to show how public places and streets can be used without car traffic. Freiburg owns a permanent car-reduced quarter and two new ones are in the planning process. Amsterdam is also going to restructure an existing quarter in its first car-free district. Oslo aims to become a car-free city by banning all parking spaces out of the inner-city part. Numerous cases show that different concepts are being pursued to reduce the share of cars in the modal split.

New mobility concepts are more likely to establish themselves as momentum in the regime and replace existing structures if they are supported by a broad network of actors, funded more extensively and have regulatory openness [24]. The importance of stakeholder management can be seen in Brussels with the Good Move Strategy, which has brought together over 100 stakeholders, or the Hamburg Alliance for Walking and Cycling, in which 28 alliance partners exchange ideas via various platforms. For the concept formulation, stakeholders are involved here to define common goals and standards. Nevertheless, there needs to be a changing culture that enables the implementation of new technologies and innovative concepts. Next to acceptance, political willingness and networked administrative structures must be given.

4.2 Administrative Enabling Structure

The mobility transition mostly requires a change of the current organisational structures of the traffic and transport sector, especially of the administrative organisation. Many interfaces on the one hand and a lack of exchange between technical departments on the other hand are challenges that the cities studied were partly able to overcome with new constellations. Municipality structures are often hierarchical top-down organised and subdivided according to specialised technical departments. As urban planning is in one administrative department, an exchange process with other topics of urban development located in different departments must be established. In fact, in these top-down structures, most information transfer occurs through personal contacts or a long-lasting process of implementing the topic on other departments' agendas. Moreover, it is cost inefficient, it has a high demand for personal resources and some departments also work against each other for various reasons. Having these structures, a strategic informal bottom-up approach must be established by promoting persons. Better coordination must be built up and a change to cross-over project teams needs to be realised in the long term. A political transform-oriented agenda must be established, the administrative structures must be adjusted and public communication should be united considering the formulated mobility strategy.

Brussels Mobility is one institution that is set up to combine the coordination of the city's Good Move Strategy, the management of the stakeholders and the mobility planning processes. With the creation of the concept faced in Brussels 19 communes with a high level of independence, Brussels Mobility managed to formulate less narrowed regulations to open its structures for mobility services and future developments by using expert platforms and participation processes. Amsterdam uses the integral project management model, a project management team with a technical manager responsible for design and designers' coordination, a project manager and a project controller as well as an environmental manager. The environmental manager takes over the social interaction, represents the municipality and is in contact with the neighbourhood's population. Hamburg finally has built up an Authority for Transport and Mobility Change including a senator playing a key role in the concept formulation, measurement realisation and stakeholder participation. It is important to note that influential actors who promote a mobility transition must also act in a system-open way. In a study, [55] found that key actors often only pursued certain strategies and other measures and concepts were denied the opportunity to experiment in the regime.

All viewed administrative setups include cross-functional structures as well as networked key actors and promote both, the system's improvement and the system change through implementing well-placed and communicated measures.

4.3 Environmental Policy Instruments

Environmental policy instruments are used to bring about material and immaterial changes to existing structures of the socio-technical regime that directly or indirectly influence social and economic activity, in other words, the movement within these structures, to achieve political objectives [10]. According to the degree of state intervention and the mechanisms, they can be divided into regulatory instruments with a very strong intervention, economic instruments with a financial incentive element, planning instruments with concepts and construction measures, and informal instruments with an information and cooperation function [11]. In the literature and especially in behavioural economics, environmental policy instruments are also classified slightly differently based on other criteria [6], but a concrete classification of measures is not always possible, as mixed forms can also occur. Since the classification according to Böcher and Töller [11] is well suited for infrastructure measures, it is also used to classify and evaluate the measures implemented in the case studies. For the evaluation, the degree of ecological effectivity, cost efficiency, dynamic incentive effect as well as acceptance and political enforceability can be considered (ibid.). Since the comparison at the European Cities level focuses on the structural counterparts and the overarching actions, this study does not evaluate individual measures but creates a picture of the overall transformation and especially the acceleration process.

Individual instruments or a combination of instruments can enable the integration of new technology or service into the existing system and, rather, also the intermodal integration [55]. Helsinki has set itself the goal of making its cars obsolete and implementing a sustainable post-car system by using mobility as a service. A single app provides access to all types of transport, which on the one hand simplifies sharing and on the other hand also significantly reduces the inhibition threshold. The challenge is that innovation often fails when existing infrastructures do not provide the necessary framework conditions [24]. Therefore, information and cooperation formats were used and even more important according to the study of [3], a vision for the city was formulated and innovation-open legislation by using regulatory instruments was created. The Good Move Strategy from Brussels is as well characterised by the fact that based on comprehensive participation processes the very regulated mobility infrastructure is being opened for further services with the help of corresponding legislation. The development project Kleineschholz in Freiburg is also confronted with barriers of the building law regarding the implementation of sharing concepts. But Freiburg also uses other regulatory instruments as the reduced parking space key, while qualitative bicycle parking spaces are requested as a substitute under building law. Regulatory instruments can be used to create new structures in new projects and to prevent an adaption of path dependencies from neighbourhoods. These examples show that intervention in existing neighbourhoods requires much stronger political support. The city of Vienna is a pioneer in this respect, relying on regulatory instruments and even banning car traffic in the first district. In return, public transport and the expansion of the cycling infrastructure are strongly promoted. Also, the inner-city of Oslo and the new quarter Frans Halsbuurt of Amsterdam become areas where car parking is strictly prohibited.

Many cities meanwhile use other instruments as the state intervention of regulatory instruments is high and societal and political acceptance causes challenges. When not rejecting parking spots, the increase in parking fees is a financial instrument that is used to reduce the willingness for car parking in inner-city areas. The price leading cities in Europe, after London, are Oslo and Amsterdam. Nevertheless, financial instruments must be used wisely as the usage of a car should not become a question of the individual socio-economic situation.

Therefore, planning instruments as concepts and infrastructure projects have a high potential to realise measures adjusting to specific structures. Copenhagen, the city with one of the highest cycling shares in Europe, aims with its Bicycle Strategy the structural adaptation, widening and improvement of cycle paths, the development of new neighbourhood connections and the construction of new cycling tunnels and bridges. Even in cities where cycling has played a minor role in the past, but a major role throughout Germany, such as Hamburg and Freiburg, the focus has increasingly been placed on the expansion of bicycle priority routes. Public transport is also being expanded with the extension of the metro lines in Helsinki, Amsterdam and Hamburg but also in many cities with on-demand services that are much more flexible and thus more customer-friendly. Not only the functional advantage of the final state but also the rebuilding phase must be designed in such a way that projects are socially supported. The impact on residents and road users as well as other affected stakeholders must be kept to a minimum. Hamburg, for example, operates a construction site management system that prevents as far as possible traffic flow disruption.

This analysis and previous studies [1] show that, next to an overall project strategy, the implementation of measures should include participation processes with local stakeholders as well as key players. But, moreover, the alignment to the unique environment is important as this enables a mobility change, strengthens network dynamics, reduces path dependencies and thus accelerates the mobility transition.

4.4 Change Communication and Participation

The described projects do not only meet with favour from residents, companies and other organisations. The highly complex and heterogeneous technical systems [54] are widely interconnected and have immaterial and social system components in addition to technical ones [13]. Structural changes, therefore, affect mobility and communication behaviour, the labour market and thus also the economic situation of the region [53]. To uncover these resistances at an early stage, to break down barriers to innovative technologies and to avoid project delays, participation formats—some of them already mentioned in Sect. 4.3—must be implemented [70]. In this way, participation fulfils a conflict avoidance and acceleration function [47]. In addition to the public or private project promoter, the municipality, specialised administrations, legitimised committees, organised interest groups, economic players and citizens can also take part in a participation process [5].

The communication and participation formats can be divided according to the direction of the information flow and assigned to process steps ranging from pure information transfer to cooperation [7, 47]. For the more established formats, a distinction is therefore made between the categories of information, participation and cooperation. E-participation can also be cited as a hybrid form [32]. Figure 3 provides an overview of the information flows of participation categories.

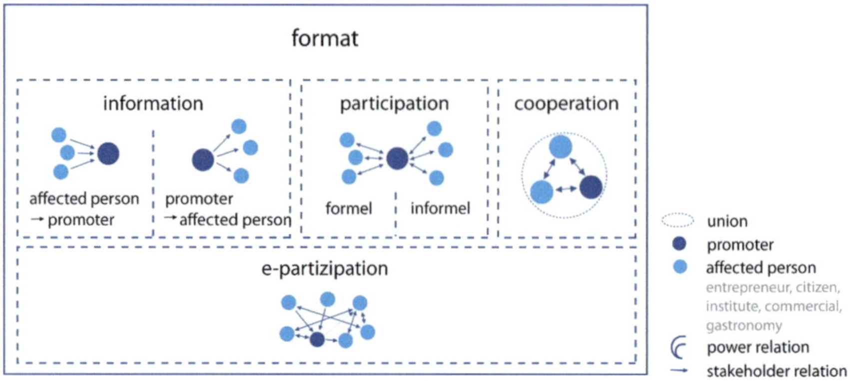

Fig. 3 Categories of participation formats

Information is unidirectional, so there is no two-way exchange, which clearly distinguishes it from communication. The flow of information can be from citizens, companies, research institutions or other organisations to the project promoter (interviews, questionnaires) as well as in the opposite direction (flyer, local press, town hall meeting, presentation, discussion, excursion, site visit). If, on the other hand, there is a mutual exchange of information through dialogue or communication, we speak of participation (public inspection, committee, workshop) [7]. If, for example, local projects are implemented that affect certain groups of people, such as users of public transportation, bi- or multi-directional communication is necessary. Beckmann [5] argues that this is the only way to create understanding and acceptance of new structures or technologies. A permanent communication process that aims at solving concrete problems is cooperation (working group, round table). While participation is subordinated to a central decision-making process in which third parties can participate, cooperation is a negotiation and decision-making process between defined actors [7].

The expert platform and the participatory process as part of the Brussels strategy, as well as the alliance partners in Hamburg, play an important role in the transformation process. With the help of a permanent exchange, concerns can be identified, evaluated and integrated into the planning of individual measures at an early stage. Amsterdam's integral project management model enables a continuous exchange between technical experts, project management and local stakeholders, but also with governors and politicians. In the city's project Frans Halsbuurt, information formats were used to raise awareness and participation formats allowed citizens and local business operators to make their demands on the design of public spaces, which also fulfilled an acceptance function. Moreover, cooperation formats with designers were used to include inhabitants in the creation process. Nevertheless, it is important to emphasise that the implementation of such measures in other neighbourhoods may require significantly different environmental policy instruments and thus also participation formats, as each neighbourhood is characterised by a unique social and urban structure. In the new Kleineschholz neighbourhood in Freiburg, informal participation processes were also carried out in addition to the formal ones, but these were comparatively easy due to the positive attitude towards the project. These cultural attitudes and values have long trajectories at the level of the landscape. They are brought about in Freiburg by continuous, incremental sustainable mobility measures as well as more radical projects such as Freiburg's low-car neighbourhood Vauban, which has achieved international prominence. Already realised projects must be highlighted, making mobility change visible, tangible and experienceable. With such pilot projects, municipalities can include excursions as an information format in follow-up projects and demonstrate opportunities for the mobility transition for other city districts. For all actors, the experience gained there is beneficial serves as an impulse generator and promotes the learning and transfer process. In addition to an activating and motivating function, mutual understanding is built up through the social interaction of actors with different positions [7].

The administrative actors must also be involved. Late objections from technical departments due to a lack of information flow led to a significantly greater effort for changes in the planning and thus also to a delay. Stakeholder networks over all groups play a key role in the change process. They are the social backbone that, together with other framework conditions, makes the diffusion of new technologies and services into the socio-technical regime possible. In Helsinki, the regulatory conditions would not have been sufficient to successfully implement mobility as a service strategy. In fact, networks were developed, initiatives created and informal lobbying carried out for the willingness of the transport operators to disclose data and thus enable a product that can be brought to market.

Participation processes are carried out by the analysed cities mainly when it comes to larger projects with different alternatives and design options. If it is a small adaptation or renewal or if there are no other alternatives, direct implementation is preferred, in order not to slow down the processes. When managing projects, social structures must be identified and incremental steps must be taken to open existing path dependencies. Finally, it is important to identify good alternatives and take opportunities to initiate a change in attitude and eventually also in behaviour. The advantages of new concepts must be highlighted, the disadvantages of the current structure must be communicated and networks with key players must be established.

5 Discussion

The transformation process is driven by political willingness and encouragement, individual key actors and decision-makers, structural conditions, cultural attitudes and values, governmental structures and sophisticated stakeholder management. In addition to municipal departments, the mobility sector consists of a complex network with service providers (public transport providers, taxi companies, car-sharing providers, …), institutions and industries (retail, housing industry, automotive industry) as well as civil society and civically organised interest groups

(Fridays4Future). The socio-technical system is accordingly characterised by many actor relationships that favour a permeability of the boundary between the meso- and micro-levels. But interdependencies between the actor groups and the dimensions at the level of the socio-technical regime shape the overall picture.

5.1 *Multi-level Model*

The multi-level model mainly considers vertical relationships between landscape, regime and niche. However, the mutual influence of neighbouring regimes due to increasing social mobility and networking is not considered. The interviews pointed out that city networks as well as regional exchanges on the one hand and technical system connections to neighbouring regimes on the other hand are essential for the transformation process. Figure 4 shows the horizontal influence.

The stability of the dimensions' structures at the regime level changes over time. It can be assumed that continuous changes and mutual influences take place, which can lead, for example, to a weakening of the existing infrastructure's reliance on technology. However, this influence does not necessarily have to be strong enough to replace the existing technology. If, for example, new forms of mobility are offered in a municipality, the existing technology may fade into the background due to the initial public attention. To promote the change, acceptance must be established through open discussions about the mobility transition, participation and cooperation with stakeholders at an early stage by showing limitations and using environmental policy instruments that are tailored for the target groups [9]. If there is no system penetration as the new mobility offers only become established for a short time, the previously used technology stabilises again. Figure 5 visualises the dynamic influence on the stability of persistence.

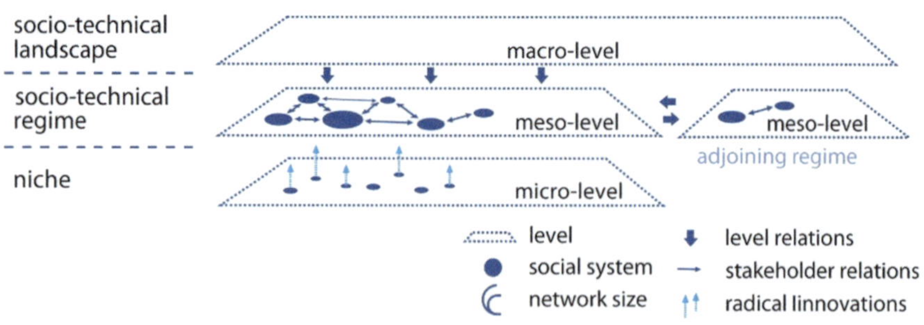

Fig. 4 Adjoined multi-level model (added model of [23])

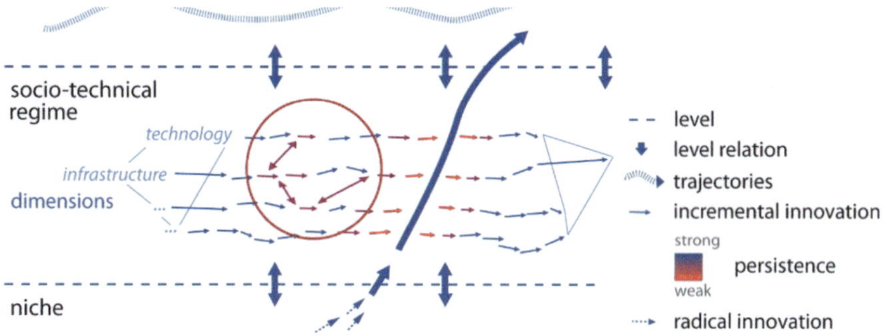

Fig. 5 Persistence influenced multi-level perspective (added model of [23])

5.2 Practical Implications

The case studies make clear that socio-technical regimes can develop very dynamically but political requirements are also needed to create an open system for change. In this context, planning measures need to be linked with regulatory frameworks and economic incentives, as well as communicative work.

The case studies show that the initiative of politics is not only important but also vital for implementing new structures and realising innovative concepts. Therefore, political majorities must be in place and there must be acceptance among the population. The difficulty of establishing innovations in a car-oriented regime is confronted with the chicken or egg causality dilemma. Without general social acceptance of change, the broad political mood will tend to insist on established structures. Without political majorities and thus missing promoters, it will be difficult to implement radical measures that foster acceptance. However, if there is acceptance and an awareness of societal challenges that should be met by the public, the scope for political actors to take measures is clearly in a better position. Nevertheless, it is necessary to build up an administrative structure that fosters fast information exchange and fruitful cooperation between different technical departments and parties in the process.

Once the preconditions for the development of innovations have been created, forums and other platforms must be offered for the integration of new technologies into municipal structures to achieve networking with social groups of actors. Knowledge can be communicated, concerns about new developments can be reduced and acceptance of change can be created through various participation formats. It is not without reason that the German legislator recommends in § 25 Part 3 VwVfG to involve the public in infrastructure projects at an early stage, which is particularly necessary if a system transformation is to be expected because of the measures. Although personal interaction is the most important aspect of participation, e-participation should always be included in participation processes to also consider the interests of population groups with weak articulation. Such structures can then also enable measures and actor networks to act as drivers for the further realisation of mobility projects and thus accelerate the change process.

5.3 Limitations

The limits of technological transformation in a socio-technical system can be technological, for example, if the developed product does not have the desired technical function and therefore cannot penetrate the market [64]. As shown in the analysis, function encompasses more than just its technical components, as it is also defined by social structures, societal values and legal frameworks at the regime and landscape level [39].

The challenges that arise can hardly be met with incremental innovations, so more radical solutions must be developed. The radical innovations required for this often fail due to society's persistence with existing technologies [23]. Therefore, the successful implementation of transformation processes requires a change in social attitudes and an openness to change. As the case studies show, this change is met with resistance from certain groups of people.

Similarly, legal framework conditions often prevent developed technologies from breaking out of their niche, for example, if competing products receive government subsidies and thus the economic viability and the market penetration of the new technology are not given.

To open the structures, knowledge and information exchange between the actors from science, business, society and politics are needed. This is not only a matter of adapting product development to local demands but also of creating acceptance with participation formats and enabling access to the use of new offers. Thus, an environment must be built up [45] that strikes a balance between the overarching and regional interests and steers the development process within politics. Since the traffic and transport infrastructure is a complex and highly networked system, every change has a cascading effect on other system elements. This has the consequence, those dependencies can indeed be identified to a certain extent, but the estimation of the dynamic development of the infrastructure in case of intervention is hardly possible [34]. This uncertainty significantly limits the political will to commit to change. It requires courage to try new approached.

6 Conclusion

The multi-level perspective has been established in infrastructure research to look at complex problems on an abstract scale and especially to work out the meaning of element relations in systems. Using a literature review and case study research with the help of documents and expert interviews enabled to identify the main factors that are drivers or inhibitors of a transformation process in the traffic and transportation sector. The analysis has shown that in addition to a united political community with the willingness for change, individual key actors in particular drive the process due to their commitment, networks and influence. Administrative structures must be changed and internal exchange processes must be made significantly easier. The technical environment requires a continuous process of change not only by improving and adapting current mobility structures but also by opening and allowing new and more radical concepts. For this, on the one hand, regulations that are based on path dependencies need to be disentangled, and on the other hand, the inertia in society and the economy, in addition to the above-mentioned politics, needs to be broken down. A social majority must be positively disposed towards the change process, which means that with participation processes, stakeholders must be informed or even involved at an early stage. Involvement does not only mean considering the concerns of the different interests but also cooperating and shaping the processes together.

Even though many factors such as city structures, existing mobility offers, regional climate, local topography, social values and cultural attitudes make it impossible to transfer urban mobility strategies, parallel driving elements in the process and stakeholder management can be identified in all case studies. As social attitudes are crucial in all socio-technical regimes, a much deeper look at social value settings, persistence, individual meaning of participation and empowerment must be taken in the context of mobility transition. According to these social factors, the promoting relationship between administrative structures and stakeholder networks must be conducted for accelerating the mobility transition with a well-adapted urban mobility strategy.

References

1. Arrto K, Kujala J, Dietrich P, Martinsuo M (2008) What is project strategy? Int J Project Manag 26:4–12
2. Atteslander P (2003) Methoden der empirischen Sozialforschung. In: Methods of empirical social research. 10th edn. Walter de Gruyter, Berlin, New York
3. Audouin M, Finger M (2018) The development of mobility-as-a-service in the Helsinki metropolitan area: a multi-level governance analysis. Res Transp Bus Manag 27:24–35
4. Baur N, Lamnek S (2005) Einzelfallanalyse. Individual case analysis. In: Mikos L, Wegener C (eds) Qualitative Medienforschung: Ein Handbuch. Qualitative Media Research: A Handbook. UVK Verlagsgesellschaft, Konstanz, pp 241–253
5. Beckmann KJ (2012) Das Kreuz mit den Ungleichgewichten—Information und Kommunikation in der Beteiligung. The cross with imbalances—Information and communication in participation. In: Beckmann KJ (ed) Bürgerbeteiligung in Kommunen: Anmerkungen aus der Stadtforschung zu einer aktuellen Herausforderung. Citizen participation in municipalities: notes from urban research on a current challenge, vol 3. German Institute of Urban Affairs, Berlin, pp 7–12
6. Beer DL (2016) Ökonomische Bildung für eine nachhaltige Entwicklung: Eine phänomenographische Untersuchung in der Lehrerinnenbildung. Economic education for

sustainable development: a phenomenographic investigation in teacher education. In: Commission on education for sustainable development. Schriftenreihe Ökologie und Erziehungswissenschaften. Ecology and Education Series. Verlag Barbara Budrich
7. Bischoff A, Selle K, Sinning H (2005) Informieren, Beteiligen, Kooperieren: Kommunikation in Planungsprozessen: Eine Übersicht zu Formen, Verfahren, Methoden und Techniken. Inform, participate, cooperate: communication in planning processes: an overview of forms, procedures, methods and techniques. Dortmunder Vertrieb für Bau- und Planungsliteratur, Dortmund
8. Van Bree B, Verbong GPJ, Kramer GJ (2010) A multi-level perspective on the introduction of hydrogen and battery-electric vehicles. Technol Forecast Soc Chang 77(4):529–540
9. Burkhard, H (2021) Strategiepapier "Klimaschutz und Mobilität". Strategy paper "Climate protection and mobility". Freiburg
10. Böcher M, Töller AE (2007) Instrumentenwahl und Instrumentenwandel in der Umweltpolitik. Instrument choice and instrument change in environmental policy. In: Jacob K, Biermann F, Busch P-O, Feindt PH (eds) Politik und Umwelt. Politics and the environment, PVS-Special Issue 39. VS Verlag für Sozialwissenschaften, Wiesbaden, pp 299–322
11. Böcher M, Töller AE (2012) Umweltpolitik in Deutschland. Environmental policy in Germany, vol 50. Springer VS, Wiesbaden
12. Cameron R, Geels FW (2019) Conditions for politically accelerated transitions: historical institutionalism, the multi-level perspective, and two historical case studies in transport and agriculture. Technol Forecast Soc Chance 140:221–240
13. Chatzis K (1999) Designing and operating storm water drain systems: empirical findings and conceptual developments. In: Coutard O (ed) The governance of large technical systems. Routledge, London, New York, pp 73–90
14. Clark KB (1985) The interaction of design hierarchies and market concepts in technological evolution. Res Policy 14:235–251
15. Cooke P, Boekholt P, Schall N, Schienstock G (1996) Regional innovation systems: concepts, analysis and typology. In: EU_RESTPOR conference: global comparison of regional RTD and innovation strategies for development and cohesion, vol 96. Restpor, Brussels
16. Cooke P (2001) Regional innovation systems, clusters, and the knowledge economy. Ind Corp Change 10(4):945–974
17. Diekmann A (2006) Empirische Sozialforschung: Grundlagen, Methoden, Anwendungen. Empirical social research: foundations, methods, applications. 15th edn. Rohwohlts Taschenbuch Verlag, Reinbek
18. Van den Ende J, Kemp R (1999) Technological transformations in history: how the computer regime grew out of existing computing regimes. Res Policy 28:833–851
19. European Environmental Agency Homepage (2022). https://www.eea.europa.eu/de/themes/transport/intro
20. European Environmental Agency Homepage (2022). https://www.eea.europa.eu/data-and-maps/data/data-viewers/greenhouse-gases-viewer
21. Fritsch M (2013) Das regionale Innovationssystem. The regional innovation system. In: Institute for Higher Education Research, Martin-Luther-University Halle-Wittenberg (eds) HoF-Handreichungen "die hochschule". HoF handouts "die university", vol 2. Halle-Wittenberg, pp 15–18
22. Fürst D, Benz A (2003) Erfolgsbedingungen für 'Regional Governance'—Resümee. Conditions for Success for 'Regional Governance'—Summary. In: Adamaschek B, Pröhl M (eds) Regionen erfolgreich steuern: Regional Governance - von der kommunalen zur regionalen Strategie. Successfully steering regions: Regional governance—from municipal to regional strategy. Verlag für Sozialwissenschaften, Gütersloh, pp 189–212
23. Geels FW (2002) Technological transitions as evolutionary reconfiguration processes: a multi-level perspective and a case-study. Res Policy 31:1257–1274
24. Geels FW (2012) A socio-technical analysis of low-carbon transitions: introducing the multi-level perspective into the transportation studies. J Transp Geogr 24:471–482
25. Geels FW, Hekkert MP, Jacobsson S (2008) The dynamics of sustainable innovation journeys. Technol Anal Strategic Manag 20(5):521–536

26. Geels FW, Schot JW (2007) Typology of sociotechnical transition pathways. Res Policy 36(3):399–417
27. Geels FW, Schot JW (2010) The dynamic of transitions: a socio-technical perspective. In: Grin J, Rotmans J, Schot J (eds) Transtitons to sustainable development: new directions in the study of long Termin transformative change. Routledge, New York, London, pp 11–103
28. German Environment Agency HOMEPAGE, Sep 20 (2022). https://www.umweltbundesamt.de/daten/klima/europaeische-energie-klimaziele
29. Glanz S, Pechmann A, Tampe-Mai K (2011) Innovationen im Netz: Die Rolle von Beziehungen zwischen Wissenschaft und Wirtschaft für den Wissens- und Technologietransfer. Innovations on the Net: The role of science-business relations for knowledge and technology transfer. In: University of Stuttgart, Institute for Social Science, Department of Technology and Environmental Sociology (eds) Stuttgarter Beiträger zur Risiko- und Nachhaltigkeitsforschung. Stuttgart Contributors to Risk and Sustainability Research, vol 2, no 21. Stuttgart
30. Gudmundsson H, Hall RP, Marsden G, Zietsman J (2016) Sustainable transportation. springer texts in business and economics. Springer, Berlin, Heidelberg
31. Hammersley M, Forster P, Gomm R (2000) Case study and theory. In: Gomm R, Hammersley M, Foster P (eds) Case study method. Sage, London, Thousand Oaks, New Delhi, pp 234–258
32. Hekmati B (2012) Echtzeitpartizipation: Potenziale von sozialen Anwendungen der Kommunikationstechnologien für die Motivation Jugendlicher und junger Erwachsener zur Beteiligung an Partizipationsprozessen in der Stadtentwicklung in Deutschland. Real-time participation: Potentials of social applications of communication technologies for motivating youth and young adults to participate in participatory processes in urban development in Germany. TU Darmstadt, Darmstadt
33. Helfferich C (2019) Leitfaden- und Experteninterviews. Guided and expert interviews. In: Baur N, Blasius J (eds) Handbuch Methoden der empirischen Sozialforschung. Handbook of Methods of Empirical Social Research. 2nd edn. Springer Vieweg, Berlin, Bonn, pp 669–686
34. Hempel L, Kraff BD, Pelzer R (2018) Dynamic interdependencies: problematising criticality assessment in the light of cascading effects. Int J Disaster Risk Reduct 30:257–268
35. Hering L, Jungmann R (2019) Einzelfallanalyse. Single case analysis. In: Baur N, Blasius J (eds) Handbuch Methoden der empirischen Sozialforschung. Handbook of Methods of Empirical Social Research. 2nd edn. Springer Vieweg, Berlin, Bonn, pp 619–632
36. Hitzler R, Honer A (1997) Sozialwissenschaftliche Hermeneutik. Social science hermeneutics. Leske + Budrich UTB, Opladen
37. Hodson M, Marvin S, Bulkeley H (2013) The intermediary organization of low carbon cities: a comparative analysis of transitions in Greater London and Greater Manchester. Urban Stud 50(7):1403–1422
38. Hughes TP (1983) Networks of power: electrification in western society, 1880–1930. John Hopkins University Press, Baltimore
39. Hughes TP (1987) The evolution of large technological systems. In: Bijker WE, Hughes TP, Pinch T (eds) The social construction of technological systems: new directions in the sociology and history of technology. MIT Press, Cambridge, pp 51–82
40. Kemp R, Rip A, Schot JW (2001) Constructing transition paths through the management of niches. In: Garud R, Karnoe P (eds) Path dependence and creation. Lawrence Erlbaum, Mahwah, pp 269–299
41. Kemp R, Schot JW, Hoogman R (1998) Regime shift to sustainability through processes of niche formation: the approach of strategic niche management. Technol Anal Strategic Manag 10:175–196
42. Kern F (2012) Using the multi-level perspective on socio-technical transitions to assess innovation policy. Technol Forecast Soc Change 79(2):298–310
43. Klein H (2014) Zeitungsartikel. Newspaper article. In: Baur N, Blasius J (eds) Handbuch Methoden der empirischen Sozialforschung. Handbook of Methods in Empirical Social Research, pp 841–846. Springer Vieweg, Berlin, Bonn
44. Kohler J, Whitmarsh L, Nykvist B, Schilperoord M, Bergman N, Haxeltine A (2009) A transition model for sustainable mobility. Ecol Econ 68(12):2985–2995

45. Kohrsmeyer M (2007) Wissen in Verbindung. Steuerung regionaler Innovationssysteme am Beispiel der Öresundregion. Knowledge in Connection. Steering regional innovation systems using the example of the Öresund region. In: Faculty of Social Science, Ruhr University Bochum (eds) Diploma thesis. Leibniz Institute fort he Social Science, Dortmund
46. Koschatzky K (2001) Räumliche Aspekte im Innovationsprozess: Ein Beitrag zur neuen Wirtschaftsgeographie aus Sicht der regionalen Innovationsforschung. Spatial aspects in the innovation process: a contribution to the new economic geography from the perspective of regional innovation research. In: Gaebe W, Grotz R, Nuhn H, Schamp EW, Schätzl L (eds) Wirtschaftsgeographie. Economic geography, vol 19. LIT Verlag, Münster
47. Langhagen-Rohrbach C (2005) Raumordnung und Raumplanung. Spatial planning and development. WBG, Darmstadt
48. Mattes J, Huber A, Koehrsen J (2015) Energy transition in small-scale regions—What we can learn from a regional innovation systems perspective. Energy Policy 78:255–264
49. Mayring P (2014) Qualitative content analysis: theoretical foundation, basic procedure and software solution. Klagenfurt
50. Mayring P, Fenzl T (2019) Qualitative Inhaltsanalyse. Qualitative content analysis. In: Baur N, Blasius J (eds) Handbuch Methoden der empirischen Sozialforschung. Handbook of Methods of Empirical Social Research. 2nd edn. Springer Vieweg, Berlin, Bonn, pp 633–648
51. McLean A, Bulkeley H, Crang M (2015) Negotiating the urban smart grid: sociotechnical experimentation in the city of Austin. Urban Stud 53(15):3246–3263
52. Medina-Molina C, de la Sierra Rey-Tienda M, Suárez-Redono EM (2022) The transition of cities towards innovations in mobility: searching for a global perspective. Int J Environ Res Public Health 19(12):7197
53. Monstadt J (2007) Großtechnische Systeme der Infrastrukturversorgung: Übergreifende Merkmale und räumlicher Wandel. Large-scale technical systems of infrastructure supply: overarching characteristics and spatial change. In: Gust D (ed) Power supply change and spatial policy. Verlag der ARL, Hannover, pp 7–34
54. Monstadt J, Naumann M (2004) Neue Räume technischer Infrastruktursysteme: Forschungsstand und -perspektiven zu räumlichen Aspekten des Wandels der Strom- und Wasserversorgung in Deutschland. New spaces of technical infrastructure systems: state of research and perspectives on spatial aspects of the transformation of electricity and water supply in Germany. netWORKS-Papers 10:1–64
55. Moradi A, Vagnoni E (2018) A multi-level perspective analysis of urban mobility system dynamics: what are the future transition pathways? Technol Forecast Soc Chang 126:231–243
56. Nykvist B, Whitmarsh L (2008) A multi-level analysis of sustainable mobility transitions: niche development in the UK and Sweden. Technol Forecast Soc Chang 75(9):1373–1387
57. Pistorius CWI, Utterback JM (1997) Multi-mode interaction among technologies. Res Policy 26:67–84
58. Pohl J (1998) Qualitative Verfahren: Methoden und Instrumente räumlicher Planung. Qualitative techniques: methods and instruments of spatial planning. In: Academy for territorial development in the Leibniz Association. Hannover
59. Prior L (2003) Using documents in social research. Sage, London
60. Raven R, Schot J, Berkhout F (2012) Space and scale in socio-technical transitions. Environ Innov Soc Trans 4:63–78
61. Regulation (EU) 2021/1119 of the European Parliament and of the Council of 30 June 2021 establishing the framework for achieving climate neutrality and amending Regulations (EC) No 401/2009 and (EU) 2018/1999 ('European Climate Law')
62. Reh W (1995) Quellen- und Dokumentenanalyse in der Politikforschung: Wer steuert die Verkehrspolitik? Source and document analysis in policy research: who controls transport policy? In: von Alemann U (ed) Politikwissenschaftliche Methoden. Methods in political science. Wiesbaden, pp 201–259
63. Rip A, Kemp R (1998) Technological change. In: Rayner S, Malone EL (eds) Human choice and climate change, vol 2. Battelle Press, Columbus, pp 327–399
64. Roberts EB (1987) Generating technological innovation. Oxford University Press, Cambridge

65. Rotmans J (2003) Transitiemanagement. sleutel naar een duurzame samenleving. Transition management. Key to a sustainable society. Van Gorcum
66. Salheiser A (2019) Natürliche Daten: Dokumente [Natural data: documents]. In: Baur N, Blasius J (eds) Handbuch Methoden der empirischen Sozialforschung [Handbook of methods of empirical social research], 3nd edn. Springer Vieweg, Berlin, Bonn, pp 1507–1522
67. Schnell R, Hill PB, Esser E (2008) Methoden der empirischen Sozialforschung. Methods of empirical social research. 8th edn. Oldenbourg Verlag, München
68. Schot JW (1998) The usefulness of evolutionary models for explaining innovation: the case of the Netherlands in the nineteenth century. History Technol 14:173–200
69. Schumpeter JA (1934) The theory of economic development: an inquiry into profits, capital, credit, interest, and the business cycle. Harvard Business Press, Cambridge
70. Selle K (2000) Was? Wer? Wie? Warum? Voraussetzungen und Möglichkeiten einer nachhaltigen Kommunikation. What? Who? How? Why? Prerequisites and possibilities for sustainable communication. Dortmunder Vertrieb für Bau- und Planungsliteratur, Dortmund
71. Smith A (2007) Translating sustainabilities between green niches and socio-technical regimes. Techol Anal Strategic Manag 19(4):427–450
72. Strauss W-C (2012) Partizipation und Neue Medien. In: Beckmann KJ (ed) Bürgerbeteiligung in Kommune: Anmerkungen aus der Stadtforschung zu einer aktuellen Herausforderung. Deutsches Institut für Urbanistik, vol 3. Berlin. pp 27–36
73. Strößenreuther H (2014) Wem gehört die Stadt? Der Flächen-Gerechtigkeits-Report. Mobilität und Flächengerechtigkeit. Eine Vermessung Berliner Straßen. Who owns the city? The Land Justice Report. Mobility and land justice. A survey of Berlin streets. Berlin
74. Trienes M. (2013) Innovation und Governance über Grenzen? Perspektiven eines grenzüberschreitenden Regionalen Innovationssystems. Das Beispiel der roten Biotechnologie in der Euregio Maas-Rhein. Innovation and Governance across Borders? Perspectives of a cross-border regional innovation system. The example of red biotechnology in the Euregio Meuse-Rhine. Aachen
75. Tushman ML, Moore WL (1982) Managing innovation over the product life cycle. In: Tushman ML, Moore WL (eds) Readings in the management of innovation. Pitman, Boston, pp 131–150
76. Verbong GPJ, Geels FW (2010) Exploring sustainability transitions in the electricity sector with socio-technical pathways. Technol Forecast Soc Change 77(8):1214–1221
77. Wright L (2005) Car-free development. sustainable transport: a sourcebook for policy-makers in developing cities. In: Deutsche Gesellschaft für Technische Zusammenarbeit (GTZ) GmbH (ed) Division 44. Environmental and Infrastructure Sectore projects "Transport Policy Advise", Module 3e. Eschborn
78. Yin RL (2009) Case study research. Sage, Thousand Oaks

Barriers for E-Scooter Adoption for Last-Mile Urban Delivery in Hanoi, Vietnam

Thanh Thi My Truong

Abstract Vietnam's e-commerce market is quickly becoming one of the largest in Southeast Asia. To serve the last-mile delivery for e-commerce, urban freight transport is a necessary activity to enable the movement and deployment of goods and services. However, it also accounts for a significant portion of carbon dioxide (CO_2) emissions. E-scooters are a solution for sustainable transport. E-scooters can contribute to improve air quality while reducing the use of fossil fuels, greenhouse gas emissions, and adverse effects of climate change and global warming. However, the extent to which they perform is still based on how many crowdsourced shippers are ready to adopt them. The purpose of this study is to identify the main barriers that affect people's behavioral intentions about e-scooters. The results show that public infrastructure and vehicle performance in terms of charge range and battery life are the top three concerns of crowdsourcing shippers about the adoption of e-scooters for last-mile delivery. Early adopters of e-scooters may include crowdsourced shippers between the ages of 26 and 35 who have more than 5 years of professional experience. Such findings facilitate policymakers and logistics providers to develop strategies that will enhance interest in adopting e-scooters for last-mile delivery in urban freight transport.

Keywords E-Scooters · Green transport · Urban logistics · Last-Mile delivery

1 Introduction

An on-demand economy is being created by growing online trends, giving consumers immediate satisfaction. Logistics is not an exemption in the e-commerce era. Most customers prefer on-demand delivery to their door [1]. E-commerce retailers and logistics providers pay consideration to last-mile delivery to achieve success in the

T. T. M. Truong (✉)
University of Transport Technology, 54 Trieu Khuc Street, Thanh Xuan District, Hanoi, Vietnam
e-mail: thanhttm@utt.edu.vn

V. T. Ha et al. (eds.), *Proceedings of the 8th International Conference on Sustainable Urban Development*, Advances in 21st Century Human Settlements,
https://doi.org/10.1007/978-981-99-8003-1_16

online market. Nevertheless, the last-mile logistics needed to provide on-demand deliveries become more difficult as consumer expectations in e-commerce rise.

In Vietnam, the high-speed development of e-commerce has been observed in recent years. Based on the E-Commerce Index Report (EBI) 2022 by Vietnam E-commerce Association (VECOM), Vietnam's e-commerce will reach a growth rate of over 20% and reach a scale of over 16 billion USD. This rate will be much higher in 2021 thanks to the good control of the COVID-19 pandemic by the Vietnamese Government. In parallel with the development of e-commerce, last-mile delivery services are also increasing rapidly [2, 3].

The delivery services in the large area in Hanoi and Ho Chi Minh city are significantly based on the crowd-sourced shippers to adapt to the huge demand of online shopping. It has been estimated that there are 62.000 shippers working in Ho Chi Minh city area and 14.500 in Hanoi capital. Motorcycles are widely used by shippers for last-mile delivery services [4, 5]. Therefore, the shipping traffic is contributing significantly to the urban transport situation and environmental status in large cities in Vietnam [5].

Many studies revealed that the transport sector is a significant source of air pollution, especially of particulate matter and nitrogen dioxide, and is the main source of environmental noise in many cities [6, 7]. Therefore, the production of electric vehicles has expanded in the mobility market. Previous studies have revealed that an e-scooter uses 72% less energy than a typical internal combustion engine (ICE) scooter and produces 6.16 tons less CO_2 equivalent than greenhouse gas (or 45% less) over its 8-year lifespan [8]. E-scooters can gradually replace gasoline-powered motorcycles in many Asian cities and can be considered a potential solution for air quality improvement and noise mitigation [9, 10].

Although e-scooters are environmentally friendly, they pose several problems during their purchasing and operating, including the high purchasing cost, the charging system, the battery life, and the limitation in driving range.

The Vietnamese government has issued some incentive policies to encourage electric vehicle production and adoption. However, such policies have primarily focused on four-wheel vehicles via import tax reductions and the production of automotive spare parts. In addition, there is no subsidy for operating costs such as road-use tax, electricity rate, and annual vehicle tax. Additionally, the local government has not actively expanded the charging infrastructure, and there is a lack of incentives for Hanoi residents to purchase e-scooters. Therefore, e-scooter adoption is still achieving a very low rate among transport users. Especially, e-scooters are seldom used by shippers for last-mile delivery. Therefore, the study aims to investigate barriers to e-scooter adoption over ICE scooters, as it is crucial to help recommend innovative policy incentives and the wider adoption of e-scooter for last-mile delivery in large cities in Vietnam. For this purpose, crowd-sourced shippers were selected for the survey to assess the barriers to e-scooter adoption for cargo delivery. The next section examines previous studies on barriers for e-scooter adoption. The experimental design and data collection are described in Sect. 3. Section 4 explains the data analysis and discussion, and the conclusions are presented in Sect. 5.

2 Literature Review

Many researchers investigated the advantages of e-scooters for sustainable transportation. Numerous studies have examined market penetration by examining the perspectives of manufacturers, investors, and governmental organizations. To identify possible barriers to the utilization of e-scooters by crowd-sourced shippers for last-mile delivery, this study focuses on the customer side. Many other impediments to the adoption of e-scooters have been found in numerous research published in the literature. These topics are addressed in this section, and Table 1 provides a summary.

2.1 Cost-Related Challenges

Cost-related barriers to e-scooter adoption include high purchase costs, battery costs, awareness of fuel and maintenance costs, and resale value. Compared with ICE scooters, e-scooters have limited functions [12, 26, 27]. Due to a lack of economies of scale and the fact that they are a relatively new technology, e-scooters are sometimes

Table 1 E-scooter barriers selected from the literature for analysis

Barrier Type	Code	Actual Barrier	Explanation	Sources
Cost-related challenges	CC1	E-scooter purchasing price	E-scooter purchasing price in the market without subsidy	[11]
	CC2	E-scooter battery cost	Cost to replace a battery when it reaches the end of life	[12–14]
	CC3	E-scooter fuel cost	Electricity cost of driving e-scooters	[15]
	CC4	E-scooter maintenance cost	Routine maintenance cost of e-scooters	[16, 17]
	CC5	E-scooter resale value	The resale value of e-scooters once sold as used motorcycles	[18]
Vehicle-related challenges	VC1	Driving distance on a charge	Longest range of driving per one full charge	[19]
	VC2	Battery life	Battery lifespan caused by degradation	[20–22]
	VC3	Charging time	Time usage to fully charge an e-scooter	[18, 23]
	VC4	Engine power	Maximal speed and acceleration of e-scooters	[24, 25]
Infrastructure-related challenges	IC1	Public infrastructure availability	Public charging stations	[25]
	IC2	Infrastructure availability at home	Charging stations at residence areas or condominiums	[26]

more expensive for customers to acquire than ICE scooters. As a result, the high cost of e-scooters is frequently cited as a key deterrent in consumer surveys [17, 28]. Another obstacle is the price of the batteries, which makes up a sizeable amount of the e-scooter's price [29]. Size, price, and range all increase with battery capacity [30, 31].

E-scooters, however, have benefits in terms of fuel and maintenance expenses. Electricity, which is less expensive and has fewer direct emissions than gasoline, serves as the e-scooters' fuel source [32]. Due to the e-scooter motors' simplified design, maintenance costs are also lower than for ICE scooters [33]. Nevertheless, in addition to technology and usefulness, various elements influence customer purchasing decisions [18].

2.2 Vehicle-Related Challenges

Performance limitations for e-scooters include range, engine power, reliability, battery life, charging speed, safety, size, and style. The performance and range of e-scooters are important obstacles to their acceptance, according to numerous studies [31, 34–38]. The amount of battery left over cannot be used by drivers to determine how far they could travel or to extend their journey [19, 25].

As a result, battery loss happens when driving. Drivers worry about the restricted range of e-scooters, which causes range anxiety on long journeys [39]. Especially, functions relating to the engine power, driving distance, and ability to transport large and heavy parcels are all significant for cargo shippers. Charge times, safety, and reliability are three more unsatisfactory aspects of e-scooter performance that respondents who test-drive e-scooters mention [40]. The restricted selection of sizes and types of e-scooters on the market also worries consumers [39].

2.3 Infrastructure-Related Challenges

The availability of charging infrastructure is necessary to promote the widespread adoption of e-scooters because the driving range of these vehicles is a significant barrier to adoption, like how filling stations for ICE scooters enable ICE scooter adoption [41]. For the demand and competitiveness of e-scooters, public charging stations are crucial [42]. To improve customer convenience and the security and safety of automobiles, overnight home charging is crucial [43].

Numerous studies in the literature have examined various obstacles to the widespread adoption of e-scooters. However, the research hasn't discussed how all these obstacles affect the adoption of e-scooters in Vietnam. To fill this research gap, this paper examines all these obstacles in the context of Vietnam.

3 Methods and Data Collection

3.1 *Survey and Questionnaire Design*

To determine obstacles to the adoption of e-scooters in Vietnam, a poll was conducted online utilizing a questionnaire. Three components made up the questionnaire. Section 1 of the report focused on demographic data, such as gender, age, educational background, household income, occupation, family size, car ownership, working experience, awareness of different types of e-scooters, and expectations regarding e-scooter prices. A five-point Likert scale was employed in Sect. 2 to identify adoption hurdles for e-scooters. For the e-scooter adoption survey, 11 hurdles (Table 1) were selected from the literature review. On a scale of 1 (not at all important), 2 (somewhat important), 3 (important), 4 (very important), and 5 (extremely important), respondents were asked to rate the importance of the impediments to the adoption of e-scooters. In Sect. 3, the public acceptance of e-scooters was examined. Respondents were questioned about their readiness to purchase, recommend, and own additional e-scooters. The following scores were requested from respondents: 1, very unwilling; 2, a little unwilling; 3, willing; 4, partially willing; and 5, willing. To allow respondents to offer suggestions and thoughts regarding the adoption of e-scooters, an open question was included at the end of the survey.

The questionnaires were conducted by face-to-face interviews in November 2019 for 15 days to ensure the respondents fully understand the meaning of each question. By applying the following criteria, invalid questionnaires were eliminated: respondents had to be shippers and respond to questions without repeating or receiving the same score on the variable group of Likert scales. 94.2% of the total responses, or 565 out of 600 questions, met the criterion.

Table 2 illustrates the characteristics of the respondents. It was not surprising that 87% of the respondents were males and only 13% were females. Working as motorcycle—shippers, people must travel all day with large parcels behind. Therefore, it is observed that more males are working in this area than females. Furthermore, some parcels are expensive and require significant efforts for protection.

Relating to the age of respondents, the young age group is dominant since 68% of them are under 35 years old. It also reveals that crow-sourced shipping is an emerging job in Vietnam. The shippers must be familiar with the smartphone, installing the shipping apps and having well-understood functions of the apps to follow task requirements. The young age group has more advantages to access and adopt this new generation of working technology.

The working experience of respondents was mostly early since 22% of them worked for less than 1 year and 51% worked for 1–3 years. This phenomenon is consistent with the early-development stage of crowd-sourced shipping companies in Hanoi. Many of them are start-up businesses and adopt last-mile delivery for such young e-commerce platforms as Shopee, Lazada, Tiki, and Grab-Food. Interestingly, the shipping job requires low experience of the candidate, therefore, many students at

Table 2 Respondent demographics (N = 565)

Demographic	Category	Percentage (%)
Gender	Male	87
	Female	13
Age	Under 25 years	42
	26–35 years	26
	36–45 years	13
	46–55 years	11
	Over 56 years	8
Working experience in delivery job	Less than 1 year	22
	1–3 years	51
	3–5 years	19
	More than 5 years	8
Monthly income	Less than 3mil	2
(VND*)	3mil–5mil	11
	5mil–10mil	47
	10mil–15mil	23
	Over 15mil	17

* *Note* 1USD = 24.100VND (The exchange rate in October 2022)

university are working part-time jobs as shippers. It also means that the labor market has a high percentage of new entrance and exit people.

The monthly income of respondents significantly depended on the number of parcels they delivered and the distance for each parcel. Based on the interview results, many respondents (47%) have a monthly income of VND 5–10 million. Compared with other jobs where candidates have minimum working experience, shippers have an acceptable monthly salary. For respondents who have experience in organizing the parcels in the cargo box and well-organized and combined delivery points, the monthly income even reaches VND 10–15 million (23%) or over VND 15 million (17%).

With a low requirement of working experience and monthly income mostly based on the "hard-working" of people, crowd-sourced shippers in big cities in Vietnam are becoming more and more popular jobs. This trend is also consistent with the rapid development of e-commerce platforms in big cities such as Hanoi, Da Nang, and Ho Chi Minh.

The delivery characteristics of respondents are also unique concerning the travel distance and the number of delivery points to fulfill their job requirements (Table 3). The interview results showed that many respondents (45%) must travel from 20 to 40 km in a day, 26% of them travel 41–60 km, and 17% travel over 60 km to deliver their daily parcels. Each shipper has shifts for two parcel picking-up (morning and afternoon). To deliver the parcels, their routes are recommended by the shipping apps. However, many of them even better utilize the delivery routes than the app

Table 3 Delivery characteristics of the respondent (N = 565)

Delivery characteristics	Category	Percentage (%)
Daily delivery distance	Less than 20 km	12
	20–40 km	45
	41–60 km	26
	Over 60 km	17
Number of delivery points	Less than 5 points	5
	5–10 points	66
	More than 15 points	29

reference based on their travel experience on the small alleys in Hanoi. Traveling to those alleys facilitates them to avoid congested roads at peak hours.

Survey results (Table 3) also revealed that the delivery points largely depended on the working time of respondents. Since crowd-sourced shippers are allowed to have flexible working time, many of them utilize their free time to work this additional job, for instance, the students, workers, and office-men. Therefore, 66% of respondents who work as a shipper in a haft-day delivery 5–10 points. For those who work a full-time job, the delivery points reach over 15 points (29%). 5% of respondents who just started this job delivered less than 5 points. Most of them were young students and utilized their free school time to work and have additional income.

3.2 Chi-Squared Test

A statistical test is known as the chi-squared test and is used to establish statistical relationships between variables under the null hypothesis that there is no relationship between a set of groups and the responses they generate. The null hypothesis was rejected if the p-value was less than 0.05 because the significance level was set at 0.05. The test statistic was

$$\chi^2 = \sum_{i=1}^{c}\sum_{j=1}^{r}\frac{(O_{i,j} - E_{i,j})}{E_{i,j}}$$

and

$$E_{i,j} = \frac{n_i n_j}{N}$$

χ^2 is Pearson's cumulative test statistic, which approaches a non-normal distribution with (r − 1) (c − 1). Row variables had a degree of freedom, but column variables were independent. $O_{i,j}$ denotes how many observations are present in row i and column j. $E_{i,j}$ stands for the predicted frequency value in columns j and i. The number of cells in the table equals n, where n is the total number of observations. In this study, the relationship between e-scooter adoption hurdles and personal and trip characteristics was investigated using the chi-squared test.

3.3 *Questionnaire Reliability*

As indicated in Table 3, the reliability or internal consistency of the barriers was assessed using Cronbach's alpha. It is advised that infrastructure, performance, and financial obstacles have Cronbach's alpha values above 0.70, which denote high reliability [44]. In terms of theory, Cronbach's alpha is a reliability coefficient that may be determined using the following equation:

$$\alpha = \frac{k \times \overline{c}}{v + (k - 1)\overline{c}}$$

where α refers to Cronbach's alpha; k is the number of items; c is the average of all covariance between items; and v is the average variance of each item.

4 Results and Discussion

Using descriptive statistics, samples were examined to learn 565 respondents' perspectives on e-scooter barriers.

4.1 *Descriptive Analysis*

Table 4 presents a summary of the findings from the descriptive analysis. The highest average score was for driving distance on a charge (VC1), with engine power (VC4), charging time (VC3), and public infrastructure availability (IC1) coming in second, third, and fourth place, respectively. This shows that most respondents are worried about the capabilities of e-scooters and the availability of public infrastructure for delivery employment. Even though some businesses offer a 5- to 8-year battery warranty and fast charging technology is available, the requirements for driving distance on a charge and e-scooter range were not met. Other factors like current battery life, safety concerns, and charging times also pose potential barriers to e-scooter adoption for last-mile delivery.

Vehicle-related challenges were the most concerning among respondents for e-scooter adoption for cargo delivery in Hanoi. This suggested that the respondents' confidence in the performance of e-scooters remained poor. Respondents were concerned about driving distance on a charge since shippers must travel mostly all

Table 4 Survey reliability among barriers variables

Barrier Type	Category	Cronbach's Alpha
Cost-related challenges	CC1, CC2, CC3, CC4, CC5	0.766
Vehicle-related challenges	VC1, VC2, VC3, VC4	0.823
Infrastructure-related challenges	IC1, IC2	0.845

Table 5 Descriptive analysis of barriers to e-scooter adoption

Barrier ID	Barrier factors	Min	Max	Mean ± SD	Rank
VC1	Driving distance on a charge	1	5	4.22 ± 1.025	1
VC4	Engine power	1	5	4.18 ± 1.008	2
VC3	Charging time	1	5	4.16 ± 0.985	3
IC1	Public charging infrastructure availability	1	5	4.16 ± 1.046	4
CC1	E-scooter purchasing price	1	5	4.14 ± 1.087	5
VC2	Battery life	1	5	4.06 ± 1.078	6
CC2	E-scooter battery cost	1	5	4.02 ± 1.016	7
IC2	Infrastructure availability at home	1	5	4.02 ± 1.127	8
CC4	E-scooter maintenance cost	1	5	3.97 ± 1.060	9
CC3	E-scooter fuel cost	1	5	3.94 ± 1.064	10
CC5	E-scooter resale value	1	5	3.90 ± 1.155	11

day to fulfill the delivery requirements. Furthermore, the engine power is one of the significant concerns because shippers must travel with very large and heavy parcels. This heavy good highly impacts the speed of vehicles, therefore, impacts the productivity of the shipper's job performance. The charging time is also a big consideration since it impacts the continuous working time of respondents. Waiting for too long time for vehicle charging will delay the delivery schedule.

The lack of public charging infrastructure, particularly on roads and in homes, was the fourth barrier to the use of e-scooters. The difficulty in having specific charging stations is largely due to the distance between cities as well as the high cost of property in major cities and metropolitan areas. As a result, more respondents selected a public charging infrastructure than a household or highway infrastructure. Charging stations should be easy to access along the delivery corridor of the shippers, they can charge their vehicles even during the short-break time.

Despite battery cost being a factor that Meisel found to be a barrier to e-scooter adoption, pricing, and battery cost were not given high rankings [30]. Nevertheless, the battery replacement cost of VND 17,200,000 (two batteries for an e-scooter VinFast Klara S) was too high and unacceptable to respondents. This cost is equivalent to a new ICE scooter.

The least significant barrier to responders was the resale value of e-scooters. The respondents' perception of a scooter as a tool for working conditions could be the cause. Therefore, they seldom think of selling their vehicles unless they are out of function. The fuel cost is also low ranking, this could be explained that electric fuel is still cheap in Vietnam, and it is much cheaper than gasoline fuel.

Given that e-scooters have fewer parts and a simpler engine than ICE scooters, the maintenance cost of e-scooters is ranked higher than its fuel cost, which may indicate that respondents currently have limited knowledge of maintenance costs.

4.2 *Reveals from Respondent Characteristics and E-scooter Adoption*

Gender, age, employment history, monthly income, daily delivery distance, and the number of dropping locations were used to categorize respondents. Each of these factors has a big impact on whether people use e-scooters. Table A1 in the Appendix lists the barriers that were found and compares them into several categories.

4.2.1 Gender and Age

Most challenges were viewed differently by men and women. There were statistically significant opinions that were at odds with a study by Peters et al. which discovered that males and females held similar opinions regarding most barriers and purchase intention. On the other hand, Plotz et al. discovered that men tended to adopt e-scooters more than women [17].

The age differences between groups CC1 (purchasing price), CC4 (maintenance cost), and VC5 (charging time) were statistically significant. The adoption of e-scooters was found to be limited among respondents between the ages of 26 and 35. These respondents have recently begun a family; thus, they are more responsible with their financial assets. The respondents who were above 45 had the lowest score. According to the study, people in the 36–45 age range are more likely to adopt e-scooters for last-mile delivery. This result contrasts with one from a study conducted in Germany by Peter and Dütschke who discovered that who found that early adopters were middle-aged men who were living with their families [17].

4.2.2 Work Experience and Monthly Income

Respondents were separated into three categories based on their work experience: those with less than a year, those with between one and three years, and those with more than five years of experience as shippers. Battery cost (CC2) and charge time (VC5) both had statistically significant results. The expense of batteries, according to experienced respondents, is a barrier to e-scooter adoption. This suggested that experienced drivers were more concerned about maintenance costs than they were about purchasing costs. Experienced drivers rated the charging time a high score;

this may be because it takes longer to charge an e-scooter than it does to fill up with gas at a gas station. Long-term, experienced shippers adopted e-scooters strongly, at an average rate of 52.1%.

Two barriers—VC2 (engine power) and VC4 (battery life)—had statistically significant effects on monthly income. The difference in purchase intention between groups was statistically significant; respondents with higher incomes adopted e-scooters more frequently (55%), which is consistent with research from Indonesia, Thailand, and China [13, 34, 45].

4.2.3 Daily Delivery Distance and Dropping Points

Regarding daily delivery distance, six barriers were statistically significant, including CC1 (E-scooter purchasing price), CC2 (E-scooter battery cost), CC3 (E-scooter fuel cost), VC1 (Driving distance on a charge), VC2 (Battery life), VC3 (Charging time). Respondents who must travel a long distance to deliver cargo thought that VC1 (Driving distance on a charge), VC2 (Battery life), and VC3 (Charging time) significantly impacted the ability to complete their daily workload, hence, were impeditive to e-scooter adoption. However, the advancement of quick charging technology, longer battery life, and overnight home charging may all help to overcome this issue. This result is comparable to that in [39], where the authors discovered that consumers valued battery life and charging speed higher than the availability of public charging stations.

Table A shows a statistically significant intergroup difference, i.e., CC3 (E-scooter fuel cost). Respondents who had short travel distances to deliver goods scored more than those with longer delivery distances. Respondents with shorter travel distances revealed a greater purchase intention, 39.6%, compared to 13.2% for longer delivery distances. This implies that respondents with longer delivery distances were more concerned about fuel costs.

For dropping points, six barriers were statistically significant: CC1 (E-scooter purchasing price), CC2 (E-scooter battery cost), CC3 (E-scooter fuel cost), CC5 (E-scooter resale value), VC2 (Battery life), and VC3 (Charging time). Regarding the purchase intention of respondents who had less than 5 and more than 15 dropping points, only 6.1% showed interest in adopting an e-scooter for their last-mile delivery. Of respondents who had 5–10 dropping points, 23.6% were willing to adopt an e-scooter. This implied that junior shippers and shippers working full-time jobs were more hesitant to adopt an e-scooter since its technical barrier (charging time, battery life) highly impacted the working performance. Improvement in technology such as engine power or driving distance might overcome those barriers.

4.2.4 Expectations on Electric Vehicles

Figure 1 illustrates the responses to an open question about expectations or recommendations for the adoption of e-scooters from 521 out of the 565 respondents.

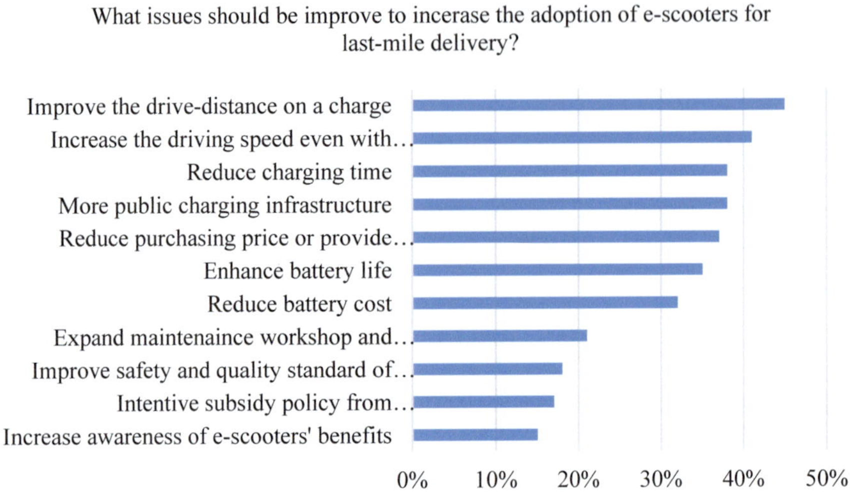

Fig. 1 Respondent expectations on e-scooters

The top three expectations of the respondents were "improve the drive distance on a charge," "increase the driving speed," and "reduce charging time," stated by 45%, 41%, and 38% of respondents, respectively. This suggests that the improvement of technical-related issues of the e-scooters is vital to motivate shippers to adopt e-scooters. With the unique working environment, for instance, long driving distance, heavy parcels to carry, stopping in many locations and dropping points, the crowd-sources shippers have a special expectation for the e-scooter to use them as long-term and daily vehicles. Previous research, which considers the willingness of normal passengers to adopt the e-scooter in daily commuting trips, reveals that purchasing price and maintenance cost are major concerns of customers. This research finds special expectations of last-mile delivery shippers to adapt to their working performance.

Moreover, respondents expected more public charging infrastructure, enhance battery life, and reduce battery cost, mentioned by 38%, 35%, and 32% of respondents respectively. This implied that people are unsatisfied with the current e-scooter performance. This implied that the battery-related barriers were a big concern of respondents. E-scooters are adopted widely in Vietnam. However, the adoption of e-scooter for last-mile delivery is still emerging in many cities. Shippers who are working with large parcels and travel very long distances still hesitate to buy an e-scooter to replace their current ICE scooter.

Policy support was also expected by many respondents: reduce purchasing price or provide subsidy (37%), incentive subsidy policy from the government (17%), and increase awareness of e-scooters' benefits (15%). About one-third of respondents expressed concern about cost-related challenges. This suggests that early in the adoption process, e-scooter subsidies are necessary. In the long run, nevertheless, the total cost of ownership can be more alluring than the beginning cost.

Numerous academics have looked into the adoption of e-scooters. For instance, Hard-man discovered that the number of charging stations is a significant determinant in the uptake of electric vehicles in California, USA [28]. It was found in Sweden that more public charging stations are linked to a higher uptake of electric vehicles [36]. Norway also reported the finding of similar phenomena [46]. In Canada, the adoption of electric vehicles was greatly influenced by age [35]. According to a study by [16, 18], the constraints in this study that prevent the adoption of electric vehicles are also typical in India. Studies conducted in the USA and Korea found that policy incentives for the adoption of electric vehicles are essential [38, 47]. Some barriers identified in this study were also common in other studies, such as [42], where the authors investigated many different barriers and drivers to international acceptance of electric vehicles. However, those studies focus mostly on passenger transport. The adoption of e-scooter in last-mile delivery is seldom. Alberto analyses the performance of electric bicycles and electric scooters for the delivery of goods in urban areas and tests the use of these vehicles in seven European countries with twenty-five companies and measures the effects in terms of reduction of CO_2 emissions and energy savings in urban transport [48]. A study conducted in Greater Manchester revealed that the connotations of risk, comfort and speed, and levels of awareness of e-cargo bikes remain low are major barriers of e-cargo adoption [49]. It was recommended in the study of Comi that the impact of weather conditions on the type of urban delivery mode, as well as the use of electric vehicles (including autonomous ones), the estimation of delivery demand according to study area features (e.g., number of shops, residents, and employees), and its integration within a decision support system for supporting city logistics scenario assessments by city administrations and planners, should be considered [50].

5 Conclusions and Recommendations

Crowd-sourced shipping is an emerging mobility that growing with the high-speed development of e-commerce in Vietnam. The adoption of e-scooter in last-mile delivery promises a positive feature of emission reduction and urban sustainability. A survey of 565 respondents was conducted to investigate the barriers to e-scooter adoption for last-mile delivery in Hanoi. Respondents were aware of e-scooters and their benefits as a sustainable choice for transport. However, shippers have unique working performance with long travel distances and heavy parcels to transport. E-scooters are a new technology that brings a new car market; thus, respondents have a wait-and-watch attitude when reacting to adoption. The biggest adoption barriers are financial, performance, and infrastructure, which challenge governments, policymakers, stakeholders, and the business sector to overcome them. Driving range on a charge and engine power, which have a significant impact on the performance of transporters, are the two most concerned e-scooter hurdles. Additionally, they have a lot of concerns about public charging infrastructure, which may mean that the public is not satisfied with the infrastructure that is now in place. The early adoption of electric scooters for last-mile delivery necessitates an extensive network of charging stations. In large cities including Hanoi, Ho Chi Minh City, and Da Nang, Vietnam has pushed public charging stations as a pilot project. The adoption of e-scooters depends on a convenient infrastructure and a universal charging method. As a result, policymakers should concentrate on a variety of charging stations, locations, infrastructure networks, distribution, and private sector cooperation with a new technology business model. Finally, because they have a great interest in adopting e-scooters, experienced drivers between the ages of 26 and 35 could be targeted as early adopters.

To make e-scooters economically competitive with ICE scooters in the early stages of adoption, financial incentives should also be offered. Consumer impression of e-scooter technology should improve adoption rates. For environmental awareness and sustainable transportation, a uniform approach to battery deterioration should be enacted into law.

Considering that e-scooters are a relatively new technology in Vietnam, this study's limitation is the fact that while the respondents may have known of them, they did not have any experience utilizing them. After taking the vehicle for a test drive or having some experience, respondents might change their decision [16]. To conveniently facilitate the adoption of e-scooters, a survey of large cities is required for planning future e-scooter adoption as a pilot region.

Acknowledgements This research was supported, in part, by Dong-A Corporation Joint Stock Company. Any opinions, findings, conclusions, or recommendations expressed in this material are those of the authors and do not necessarily reflect the views of Dong-A Corporation Joint Stock Company.

Appendix

See Table 6

Note χ^2 = Pearson's chi squared. * p value < 0.05; ** p value < 0.01

Table 6 Pearson chi-squared test of respondent's characteristics

	Gender (df = 4)		Age (df = 16)		Working Experience (df = 12)		Monthly Income (df = 20)		Daily delivery distance (df = 4)		Delivery points (df = 4)	
	χ^2	p-Value	χ^2	p-Value	χ^2	p-Value	χ^2	p-Value	χ^2	p-Value	χ^2	p-Value
CC1	10.888	0.028 *	29.139	0.023 *	15.491	0.216	7.700	0.994	20.785	0.008 **	29.285	0.012 **
CC2	18.207	0.001 *	14.797	0.303	28.425	0.005 **	24.361	0.227	18.132	0.020 *	19.212	0.034 *
CC3	34.928	0.000 **	25.714	0.282	15.677	0.206	18.981	0.523	27.466	0.001 **	22.134	0.028 **
CC4	15.323	0.004 *	40.780	0.001 **	11.085	0.522	20.341	0.437	9.450	0.306	10.231	0.401
CC5	26.046	0.000 **	21.310	0.446	9.432	0.666	20.561	0.423	10.254	0.248	27.923	0.016 **
VC1	19.001	0.001 **	17.127	0.034 *	23.023	0.011 *	29.887	0.072	26.410	0.001 **	11.254	0.298
VC2	17.026	0.002 **	45.722	0.211	13.646	0.004 **	33.211	0.032 *	19.510	0.012 *	18.665	0.025 *
VC3	4.190	0.381	21.780	0.942	16.298	0.888	21.687	0.358	18.213	0.020 *	19.120	0.036 *
VC4	12.764	0.012 *	22.039	0.314	25.990	0.804	36.131	0.015 *	12.493	0.131	11.872	0.202
IC1	3.588	0.032 *	13.102	0.303	7.272	0.839	25.895	0.169	10.923	0.206	9.781	0.311
IC2	5.364	0.465	17.823	0.282	7.752	0.210	17.640	0.944	4.719	0.787	5.291	0.872
	(df = 2)		(df = 8)		(df = 6)		(df = 10)		(df = 2)		(df = 2)	
Purchase intention	28.660	0.000 **	15.435	0.051	37.200	0.000 **	2.287	0.683	35.045	0.000 **	18.419	0.001 **

References

1. Dablanc L, Morganti E, Arvidsson N, Woxenius J, Browne M and Saidi N 2017 The rise of on-demand 'Instant Deliveries' in European cities *Supply Chain Forum: An International Journal* **18** 203–17
2. Thanh Truong TM, Ngoc AM (2020) Parking behavior and the possible impacts on travel alternatives in motorcycle-dominated cities Transportation Research Procedia 48:3469–3485
3. Thanh Truong T M, Ngoc A M (2020) The city-airport connection, implications for urban transport planning in motorcycle-dominated cities *Transportation Research Procedia* **48** 3232–44
4. Truong T M T (2021) Impacts of digitalization on travel behaviour in Hanoi city center *AIP Conf Proc* **2428**
5. Truong T M T (2021) A proposal for electronic ticketing based on travel behavior, towards the integrated public transport for smart city in Hanoi, Vietnam. *AIP Conf Proc* **2428**
6. Hernandez M, Kockelman K M, Lentz J O, Lee J (2019) Emissions and noise mitigation through use of electric motorcycles. Transportation Safety and Environment
7. Jaller M, Pineda L, Ambrose H, Kendall A (2021) Empirical analysis of the role of incentives in zero-emission last-mile deliveries in California J Clean. Prod 317:128353
8. Cruz C, Montenon A (2016) Implementation and Impacts of Low Emission Zones on Freight Activities in Europe: Local Schemes Versus National Schemes Transportation Research Procedia 12:544–556
9. Assessment A I, Conditions C, July C (2007) Air Emissions Issues Related to Two- and Three-Wheeled Motor Vehicles. *Control*
10. de Assis Brasil Weber N, da Rocha B P, Smith Schneider P, Daemme L C, de Arruda Penteado Neto R (2019) Energy and emission impacts of liquid fueled engines compared to electric motors for small size motorcycles based on the Brazilian scenario. *Energy*
11. Yan Q, Qin G, Zhang M, Xiao B (2019) Research on real purchasing behavior analysis of electric cars in beijing based on structural equation modeling and multinomial logit model Sustainability (Switzerland) 11:1–15
12. Bjerkan KY, Nørbech T E, Nordtømme ME (2016) Incentives for promoting Battery Electric Vehicle (BEV) adoption in Norway. Transp Res D Transp Environ 43 169–80
13. Li L, Wang Z, Chen L, Wang Z (2020) Consumer preferences for battery electric vehicles: A choice experimental survey in China Transp Res D Transp Environ 78:102185
14. Hardman S, Chandan A, Tal G, Turrentine T (2017) The effectiveness of financial purchase incentives for battery electric vehicles – A review of the evidence. Renewable and Sustainable Energy Reviews 80 1100–11
15. Nocerino R, Colorni A, Lia F, Luè A (2016) E-bikes and E-scooters for Smart Logistics: Environmental and Economic Sustainability in Pro-E-bike Italian Pilots *Transportation Research Procedia* vol 14 (Elsevier B.V.) pp 2362–71
16. Quak H, Nesterova N, van Rooijen T (2016) Possibilities and Barriers for Using Electric-powered Vehicles in City Logistics Practice Transportation Research Procedia vol 12 (Elsevier B.V.) pp 157–69
17. Plötz P, Schneider U, Globisch J, Dütschke E (2014) Who will buy electric vehicles? Identifying early adopters in Germany. Transp Res Part A Policy Pract
18. Kumar RR, Alok K (2020) Adoption of electric vehicle: A literature review and prospects for sustainability J Clean. Prod 253:119911
19. Franke T, Neumann I, Bühler F, Cocron P, Krems JF (2012) Experiencing Range in an Electric Vehicle: Understanding Psychological Barriers. Appl Psychol 61:368–391
20. Gass V, Schmidt J, Schmid E (2014) Analysis of alternative policy instruments to promote electric vehicles in Austria. Renew Energy
21. Mersky AC, Sprei F, Samaras C, Qian ZS (2016) Effectiveness of incentives on electric vehicle adoption in Norway. Transp Res D Transp Environ
22. Ajanovic A, Haas R (2016) Dissemination of electric vehicles in urban areas: Major factors for success. Energy 115:1451–1458

23. Napoli G, Andaloro L, Sergi F, Randazzo N, Antonucci V (2013) Electric vehicles for urban logistics improvement 2013 World Electric Vehicle Symposium and Exhibition (EVS27) (IEEE) pp 1–4
24. Poullikkas A (2015) Sustainable options for electric vehicle technologies. Renewable and Sustainable Energy Reviews
25. Tu J C, Yang C (2019) Key factors influencing consumers' purchase of electric vehicles. Sustainability (Switzerland) 11
26. Lai IKW, Liu Y, Sun X, Zhang H, Xu W (2015) Factors influencing the behavioural intention towards full electric vehicles: An empirical study in Macau Sustainability (Switzerland) 7:12564–12585
27. Asghar R, Rehman F, Ullah Z, Qamar A, Ullah K, Iqbal K, Aman A, Nawaz AA (2021) Electric vehicles and key adaptation challenges and prospects in Pakistan: A comprehensive review. J Clean Prod 278
28. Hardman S, Berliner R, Tal G (2019) Who will be the early adopters of automated vehicles? Insights from a survey of electric vehicle owners in the United States Transp Res D Transp Environ 71:248–264
29. Hidrue M K, Parsons G R, Kempton W, Gardner M P (2011) Willingness to pay for electric vehicles and their attributes. Resour Energy Econ
30. Meisel S, Merfeld T (2018) Economic incentives for the adoption of electric vehicles: A classification and review of e-vehicle services Transp Res D Transp Environ 65:264–287
31. Haddadian G, Khodayar M, Shahidehpour M (2015) Accelerating the Global Adoption of Electric Vehicles: Barriers and Drivers. Electricity Journal 28:53–68
32. Yong JY, Ramachandaramurthy V K, Tan K M, Mithulananthan N (2015) A review on the state-of-the-art technologies of electric vehicle, its impacts, and prospects. Renewable and Sustainable Energy Reviews
33. Steinhilber S, Wells P, Thankappan S (2013) Socio-technical inertia: Understanding the barriers to electric vehicles. Energy Policy 60:531–539
34. Guerra E (2019) Electric vehicles, air pollution, and the motorcycle city: A stated preference survey of consumers' willingness to adopt electric motorcycles in Solo, Indonesia. Transp Res D Transp Environ
35. Azarafshar R, Vermeulen WN (2020) Electric vehicle incentive policies in Canadian provinces. Energy Econ 91:104902
36. Vassileva I, Campillo J (2017) Adoption barriers for electric vehicles: Experiences from early adopters in Sweden. Energy 120:632–641
37. Nie Y, Wang E, Guo Q, Shen J (2018) Examining Shanghai consumer preferences for electric vehicles and their attributes Sustainability (Switzerland) 10:1–16
38. Kwon Y, Son S, Jang K (2018) Evaluation of incentive policies for electric vehicles: An experimental study on Jeju Island Transp Res Part A Policy Pract 116:404–412
39. de Mello BRA, Goes GV, Schmitz Gonçalves DN (2019) D'Agosto M de A and Oliveira C M de. Electric vehicles in the last mile of urban freight transportation: A sustainability assessment of postal deliveries in Rio de Janeiro-Brazil Transp Res D Transp Environ 67:491–502
40. Jensen AF, Cherchi E, de Dios Ortúzar J (2014) A long panel survey to elicit variation in preferences and attitudes in the choice of electric vehicles. Transportation (Amst)
41. Ehrler V C, Schöder D, Seidel S 2021 Challenges and perspectives for the use of electric vehicles for last mile logistics of grocery e-commerce – Findings from case studies in Germany. Res Transport Econom 87 100757
42. Delang CO, Cheng WT (2013) Hong Kong people's attitudes towards electric cars International Journal of Electric and Hybrid Vehicles 5:15–27
43. Tran DD, Vafaeipour M, el Baghdadi M, Barrero R, van Mierlo J and Hegazy O (2020) Thorough state-of-the-art analysis of electric and hybrid vehicle powertrains: Topologies and integrated energy management strategies. Renewable and Sustainable Energy Reviews
44. Ben-Akiva M, Morikawa T, Shiroishi F (1992) Analysis of the reliability of preference ranking data. J Bus Res 24:149–164

45. Kongklaew C, Phoungthong K, Prabpayak C, Chowdhury M S, Khan I, Yuangyai N, Yuangyai C, Techato K (2021) Barriers to electric vehicle adoption in Thailand Sustainability (Switzerland) **13**
46. Mersky AC, Sprei F, Samaras C, Qian ZS (2016) Effectiveness of incentives on electric vehicle adoption in Norway Transp Res D Transp Environ 46:56–68
47. Jin L, Searle S, Lutsey N (2014) Evaluation of State-Level U.S. Electric Vehicle Incentives
48. Lia F, Nocerino R, Bresciani C, Colorni A, Luè A (2014) Promotion of E-bikes for delivery of goods in European urban areas: an Italian case study
49. Blazejewski A, Publisher N (2020) Delivering the last mile : scoping the potential for E-cargo bikes
50. Comi A, Savchenko L (2021) Last-mile delivering: Analysis of environment-friendly transport. Sustain Cities Soc 74

Energy Demand Model Towards Sustainable On-Road Transportation in Ho Chi Minh City

Hiep Thi Do

Abstract The transport sector is the largest emitter in Ho Chi Minh City (HCMC), accounting for 45% of the total greenhouse gas emissions (GHG) in 2021. Therefore, the energy demand transition in transportation from high carbon emission to low or zero emission sources plays a central role in the city's green development. This paper studies the energy demand model for on-road vehicles—the primary mode of transport in the city, with significant emissions, diverse vehicle types, and many actors. Four alternative demand scenarios are prepared and analyzed using long-range energy alternatives planning (LEAP): business as usual (BAU), energy efficiency (ENE), alternative fuel (ALF), and sustainable transition (SUT). The findings indicate that the ALF and SUT scenarios consume significantly less energy than the BAU and ENE. The SUT relative to the BAU contributes to a 56.25 billion kWh (41.6%) decrease in the total energy demand and 5,424 million USD (58.56%) fuel cost saving in 2050. Moreover, the GHG emissions under the SUT are only half of those under the BAU, thanks to the combination of energy efficiency, alternative fuel measures, and lower transportation demand growth.

Keywords Energy transition · Transportation · Greenhouse gas · Energy efficiency · Alternative fuel

1 Introduction

Fighting climate change is increasingly urgent globally. Without solid and sustainable efforts, the constraint of global warming to levels well below 2 °C compared with pre-industrial levels will not be satisfied [1].

Urban areas, including the city and the surrounding area, contribute significantly to greenhouse gas (GHG) emissions. These areas consume around 80% of the world's

H. T. Do (✉)
Faculty of Industrial and Energy Management, Electric Power University, 235 Hoang Quoc Viet, Bac Tu Liem, Ha Noi, Vietnam
e-mail: hiepdt@epu.edu.vn

V. T. Ha et al. (eds.), *Proceedings of the 8th International Conference on Sustainable Urban Development*, Advances in 21st Century Human Settlements,
https://doi.org/10.1007/978-981-99-8003-1_17

energy and account for 71–76% of CO_2 emissions despite covering only 2% of the world's surface [2].

Along with stationary energy uses (residential, commercial, and industrial buildings), transportation dominates city GHG emissions. In 2020, the transport sector accounted for 29% of global total final energy consumption and contributed 25% to the total GHG emissions. Urban transport was responsible for around 40% of the energy used in the transport sector and 37% of the transport-related GHG emissions [3]. In one-third of 167 cities surveyed by [4], more than 30% of the total GHG emissions were from on-road vehicles.

Energy demand transition in transportation, mainly on-road transport, has recently gotten global attention. However, despite the high potential reduction of fossil fuel consumption, the implementation of the energy demand transition and the associated measures for the transport sector has still been low compared to the energy transition in the power sector. Consequently, it does not keep pace with the needs of GHG emissions [5].

At the recent United Nations Climate Change Conference (COP26), Vietnam committed to taking action to contribute to the mitigation of climate change [6]. Concretely, Vietnam aims to reach its net-zero carbon emission target by 2050, phase out coal-fueled power generation by 2040, and reduce greenhouse gas emissions (GHG) by 9–27% compared to the business as usual scenario by 2030 [7]. To fulfill those commitments, the government of Vietnam specifies six groups of emissions based on related processes, sources, and sinks, including (1) energy, (2) agriculture, (3) waste, (4) land use, land-use change, and forestry (LULUCF), and (5) the industrial processes (IP). The energy group covers fuel combustion-based sectors, including transportation [8].

Recently, numerous energy demand transition models for Vietnam covering household electricity consumption, industry, power sector supply, and transportation have been studied [9, 10]. These studies design low-carbon pathways at the national scale using the Energy Forecasting Framework and Emissions Consensus Tool (EFFECT), with the support of stakeholders and experts. Study results have become good references for the authority in identifying policy and technical measures to help realize GHG emission potential. The study [11] became a big project focusing on the low-carbon pathways for the transportation sector in Vietnam. However, the use of 2014 as the baseline data year is considered too old compared to the rapid economic development in Vietnam.

On the city scale, few attempts have been made to understand the city's GHG inventory and analyze the future energy demand. The GHG inventory of Ho Chi Minh was studied for the first time in 2017 [12], resulting in 45% of GHG emissions from transportation. Focusing on transportation, using the International Vehicle Emission model, [16] indicated that in 2013 HCMC had the highest annual emission compared to several other cities in Asia. Using a bottom-up approach, [13] evaluated that private vehicles contribute the most significant to the total carbon emissions in on-road passenger transportation in cities in Vietnam. All in all, the previous studies have only focused on GHG inventory but have not assessed future energy needs, solutions, and potential for GHG emission reduction. Moreover, it is a fact that the emission sources from transportation and transport technologies change over time. Continuous development of emission inventory, low-carbon emission scenarios, and city mitigation opportunities are essential for the city to take action to reduce GHG emissions.

HCMC made up over 20% of the total GHG emissions in Vietnam in 2013. In 2020, the transport sector overtook the stationary energy users to become the largest emitter in HCMC [14]. On-road transport contributed 84% to the city's transportation-related GHG emissions [12]. So far, a completed study on common pathways of HCMC is unavailable. Therefore, developing an energy demand transition model for sustainable on-road transportation in HCMC is crucial and urgent. This model aims to (1) estimate the future energy demand, (2) forecast the future number of vehicles, (3) evaluate the impact of energy demand scenarios on GHG emissions, and (4) estimate the fuel cost savings for on-road transport in HCMC.

The remainder of this paper is organized as follows. Section 2 introduces the modeling tool, model framework, data sources, and scenario designs. In Sect. 3, the results of the HCMC on-road transportation model, including future energy demand, fuel demand, vehicle number, GHG emissions, and fuel cost savings, are presented. Section 4 is devoted to the conclusions and policy implications.

2 Proposed Model

2.1 Modeling Tool and Framework

This study uses Long-range Energy Alternatives Planning (LEAP) to develop alternative energy demand models for on-road transportation in HCMC between 2019 and 2050. Developed by the Stockholm Environment Institute and used by over 5,000 users in 179 countries, for the demand side, this tool supports the methodologies from bottom-up, end-use accounting techniques to top-down macroeconomic modeling [15]. Integrated scenario managers in this tool can describe individual policy measures and then have combined them in different combinations to create alternative scenarios.

As shown in Fig. 1, the model framework uses two input modules: key assumptions for population and average income, the demand module for activity level, final energy intensity, demand cost, and average environmental loading. Four alternative scenarios are prepared for long-range energy planning in HCMC, including business as usual (BAU), energy efficiency (ENE), alternative fuel (ALF), and sustainable transition (SUT). The expected output is energy demand forecasting, fuel demand forecasting, vehicle number forecasting, GHG emissions forecasting, and fuel cost forecasting.

For the demand analysis, the activity level, energy intensity, and fuel cost of vehicles are calculated using the following equations:

$$\text{Vehicle kilometers: } VKM = TIV * ADT \quad (1)$$

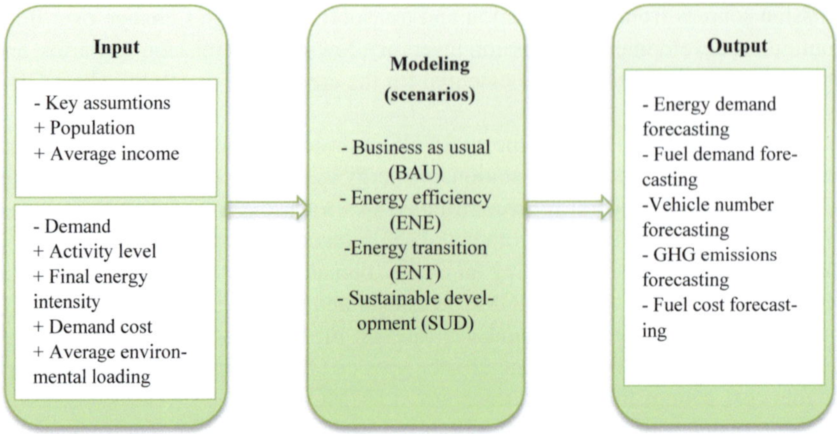

Fig. 1 Energy demand model framework for on-road transportation in HCMC

$$\text{Passenger kilometers: } PKM = VKM * OR \tag{2}$$

$$\text{Ton kilometers: } TKM = VKM * LF \tag{3}$$

$$\text{Passenger energy intensity:} PEE = \frac{AFE}{OP} \tag{4}$$

$$\text{Freight energy intensity:} FEE = \frac{AFE}{LF} \tag{5}$$

$$\text{Passenger fuel cost:} PFC = PEE * FP \tag{6}$$

$$\text{Freight fuel cost:} FFC = FEE * FP \tag{7}$$

where: VKM: vehicle kilometers (billion km), TIV: total in-use vehicles (billion vehicles), ADT: average distance traveled per vehicle (km), PKM: passenger kilometers (billion pass-km), OR: occupancy rate of passenger vehicles, TKM: ton kilometers (billion ton-km), LF: load factor of freight vehicles, AFE: average fuel efficiency (l/km, kg/km, or kWh/km), PEE: passenger energy intensity (l/pass-km, kg/pass-km, or kWh/pass-km), FEE: freight energy intensity (l/ton-km, kg/ton-km, or kWh/ton-km), PFC: passenger fuel cost (USD/pass-km), FFC: freight fuel cost (USD/ton-km), FP: fuel price (USD/GJ, USD/MMBTU, or USD/kWh).

Concerning the output, the following equations are used:

$$\text{Energy consumption:} EC = VKM * AFE \tag{8}$$

$$\text{GHG emissions:} EP = EC * EF \quad (9)$$

$$\text{Number of vehicles:} TIV = \frac{VKM}{ADT} \quad (10)$$

$$\text{Total passenger fuel cost:} TPFC = PFC * PKM \quad (11)$$

$$\text{Total freight fuel cost:} TFFC = FFC * TKM \quad (12)$$

where: EC: energy consumption (billion-liter, km, or kWh), EP: emissions (ton CO_2), EF: emission factor of fuel (ton/TJ, kg/TJ, ton/MWh), $TPFC$: total passenger fuel cost (million USD), $TFFC$: total freight fuel cost (million USD).

2.2 Data Sources

The HCMC on-road transportation model input data includes population, GDP per capita, transportation demand (pass-km, ton-km), transportation structures by vehicle and fuel, final energy intensity, fuel cost, and average environmental loading.

Situated in the southeast region of Vietnam, HCMC is the largest urban area in Vietnam. In 2019, the city had a population of 8,371,428 million people, living in nineteen urban and five suburban districts [16] (Fig. 2). With a GDP of 61.7 billion USD, the city's GDP per capita reached 6,862 USD in 2019 [17].

The data related to activity level of transportation, final energy intensity in the base year 2019 is extracted or calculated from (1) official statistics [20], (2) legal documents [21], (3) published research literature [14, 22–25], (4) research reports

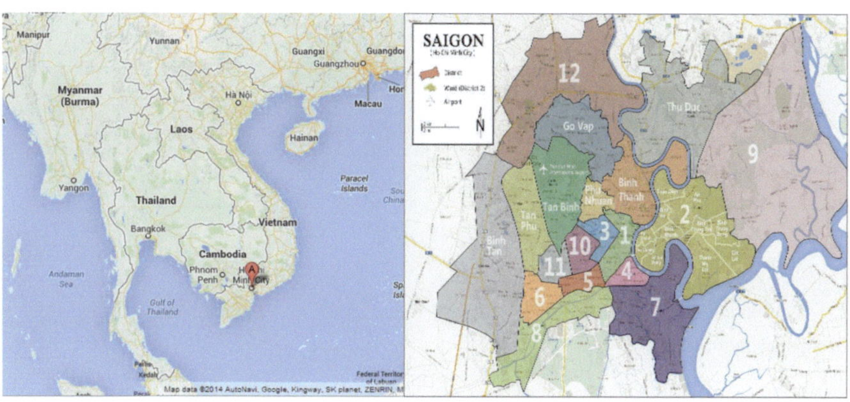

Fig. 2 Ho Chi Minh City [18, 19]

published by relevant institution, such as ADB [10], GIZ [26], IEA [27, 28], and (5) websites such as Statista [29], Drive [30], Ministry of Transportation [31], Sai gon Giai Phong [32], BNEWS [33], Vietnam Plus [34], and Sustainable-bus [35]. Table 1 shows the base year data for the HCMC on-road transportation model.

Along with the rapid economic and urbanization growth, the number of on-road vehicles has increased dramatically. By 2019, 9,186,855 on-road vehicles, including passenger and freight vehicles, were registered. Four passenger vehicles consist of

Table 1 Vehicle classes and fuel consumption in on-road transportation in HCMC in 2019

Vehicle types		Passenger				Cargo	
		Two-wheelers	Passenger cars	Light-duty passenger cars	Heavy-duty buses	Light-duty trucks	Heavy-duty trucks
Total registered vehicles (9,186,855)		8,120,000	722,384	20,475	2,261	204,752	21,000
Total in-used vehicles (5,971,456)		5,278,000	469,550	13,309	1,470	133,089	76,039
Share of vehicles by fuel (%)							
Fuel type	Gasoline	73.56%	66.13%	66.13%	0%	0%	0%
	Diesel	0	8.47%	8.47%	79%	100%	100%
	Gasohol	21.54%	19.36%	19.36%	0%	0%	0%
	CNG	0	0.00%	0.00%	21.10%	0%	0%
	Electric	4.90%	5.40%	5.40%	0%	0%	0%
	Hybrid	0	0.64%	0.64%	0%	0%	0%
Average daily vehicle kilometers traveled (km/day)		10	14	14	80	60	60
Average fuel consumption (kg/100 km for CNG, kWh/100 km for electricity, l/100 km for others)							
Fuel type	Gasoline	1.92	7.1	7.6	–	–	–
	Diesel	–	6	6.5	22.8	8.5	14.5
	Gasohol	1.8	7	7.5			
	CNG	–	–	–	30.18	–	–
	Electricity	2.8	15	20	135	25	–
	Hybrid (use gasoline as a fuel source)	–	4.26	4.56	–	–	–
Occupancy rate (passengers/vehicle), load factor (tons/vehicle)		1.2	1.5	2	38	2	6.7

two-wheelers (mopeds, motorcycles), passenger cars (personal cars and taxis), and light-duty passenger cars (sport utility vehicles), heavy-duty buses. For freight transport, there are light-duty trucks (pickup trucks, vans, and minivans, with a gross vehicle weight of no more than 3,500 kg) and heavy-duty trucks (trucks with a mass of more than 3,500 kg). Two-wheelers dominated the on-road vehicle structures with 88.39%, followed by passenger cars at 7.86%, light-duty trucks at 2.23%, heavy-duty trucks at 1.27%, and light-duty passenger cars and heavy-duty buses with only 0.22% and 0.02%, respectively.

Five types of fuel consumed by vehicles in HCMC include gasoline, gasohol (E5—fuel mixture of 5% ethanol and 95% gasoline), diesel, compressed natural gas (CNG), and electricity. Gasoline was the most demanded, accounting for more than 47.23% of the total energy consumption, followed by diesel with 38.84%, and gasohol with 12.90%. CNG and electricity represented around 1%. In other words, the energy demand transition potential to greener fuels, such as gasohol, CNG, and electricity, is still high.

Regarding the fuel cost, the fuel price for gasoline, diesel, and natural gas between 2019 and 2050 is extracted from [36]. According to [37], the E5 price equals 93.06% of the gasoline price. The average electricity price in Vietnam is referred from [38].

In terms of GHG emissions, the three major greenhouse gases are CO_2, CH4, and N_2O. The GHG values measure each emission potential relative to a ton of CO_2. The emission factors from gasoline, diesel, and CNG are based on the Technology and Environmental Data (TED) in the LEAP, which includes IPCC Tier 1 Default Emission Factors [39]. The emission factors from gasohol motorcycles are referred from [40] for CO_2, [41] for CH4 and N_2O. For electricity vehicles that have zero tailpipe emissions, emissions may be produced by power plants that depend on electricity source structures. The higher penetration of renewables is the lower the emission factor from electricity is. The average emission factor of the Vietnamese electricity grid is referred from [42–44].

2.3 Scenario Design

2.3.1 Business as Usual (BAU) Scenario

This scenario assumes no new policy interventions for reducing fuel use and emissions will be approved. The historical transportation demand growth, vehicle structure, and energy intensity remain in the upcoming years. Specifically, the average transportation demand growth is expected to remain at 8.43% for passengers and 10.67% for freight [20]. However, thanks to the higher future income [45], more people will shift to using private cars instead of two-wheelers [10]. This study sets the target for passenger cars at 40% and two wheelers at 30% of passenger transportation in 2050. According to [46], HCMC expects the bus to meet 15% of travel demand by 2025 and 25% by 2030. However, this plan may be too optimistic considering the current bus share. This study's target number for buses is around 27% of city travel demand in 2050.

Regarding socioeconomic parameters, the population is forecasted to grow at the same rate as the history at 2.71%/year. According to [45], HCMC expects to reach a GDP per capita of 13,000 USD in 2030 and 37,000 USD in 2045. With the same GDP per capita growth rate, the GDP per capita in HCMC in 2050 may reach 45,000 USD.

2.3.2 Energy Efficiency (ENE) Scenario

In this scenario, energy efficiency measures of transport modes and vehicle technology are applied to reduce energy intensity. Historical data on passenger and freight transport energy efficiency in eight EU countries indicate no improvement during the past three decades [47]. Similarly, the energy intensity of light-duty vehicles and motorcycles in the USA between 2007 and 2016 shows no improvement [48]. These facts reveal that increasing energy efficiency is not easy to achieve.

In Vietnam, statistics or research related to the time series data on the energy efficiency of vehicles are unavailable. However, the Vietnam government has improved the circular on solutions for energy efficiency in transportation since 2011 [49]. The initiative of energy labels for vehicles is expected to orient customers towards the use of high-energy performance, gradually eliminating those with outdated technology and reducing energy consumption [50, 51]. With the above affords, the author assumes that Vietnam may achieve an annual energy intensity reduction of 0.2% for gasoline, gasohol, diesel, and CNG engines and 0.5% for electric, and 0.3% for hybrid vehicles between 2019 and 2050.

2.3.3 Alternative Fuel (ALF) Scenario

This scenario is designed with the application of alternative fuel measures. Regulating the emission standards for on-road vehicles is an initiative toward alternative fuels [52]. HCMC set up eight voluntary motorcycle emission testing points accessible for people to detect motorcycles with emission concentrations higher than the standard to record the cause and solution [53]. Also, the government of Vietnam has approved the decisions on promoting and up-scaling the use of E5 and hybrid cars [54] and electricity in on-road transportation [55].

According to [55], three phases of electric upscaling in on-road transportation in HCMC have been proposed. The start-up phase is from 2022 to 2030, with the expected percentage of electric vehicles sold by 2030 will reach 20% with motorcycles/motorcycles/cars, 10% with taxis, and 50% with buses. For the rapid growth phase, from 2030 to 2040, electric vehicles will account for 50% of motorcycles/ motorcycles, 60% of cars, 20% of taxis, and 100% of buses in 2040. With the stable growth stage, from 2040 to 2050, electric technology accounts for 90% of motorcycles/motorcycles/cars, 60% of taxis, and 100% of buses.

Considering the current transport structure by fuel, the above plan may be too ambitious. This study assumes that the 2050 vehicle structure would be 80%/20% for electric and gasohol two-wheelers and 60%/20%/10%/10% for electric, gasohol, hybrid, and diesel passenger cars. For buses, 30% electricity, CNG 40%, and diesel 30% are expected.

No electric trucks are currently used in Vietnam. Also, policies to promote the use of this truck technology are unavailable. However, along with the world trend, the author expects that light-duty electric trucks will account for around 5% of the total light-duty trucks in 2050. For heavy-duty trucks, long-distance freight has been the trucking segment least amenable to electrification since the increased weight of batteries, relative to fuel, detracts from payload capacity, and the alternative, more frequent recharging, detracts from the delivery time. Therefore, no heavy-duty trucks will be adopted by 2050.

2.3.4 Sustainable Transition (SUT) Scenario

For the design of this scenario, all the above energy efficiency measures and alternative fuel plans are integrated. In addition, a slight decrease (10%) in future travel and freight demand growth rate is expected compared to the historical growth rate. Specifically, this scenario considers a growth demand rate of 7.59% for passengers and 9.60% for freight.

Recently, the action plan for green energy transition and reducing carbon and methane emissions in the transport sector has been approved [56]. Although this plan is uncertain, it signals stakeholders to act toward greener and more efficient vehicles.

3 Results and Discussion

To facilitate analysis and ensure comparability after data collection, data expressed in the various units of different fuels are converted to the same units. For example, because electricity is expected to be the most demanded energy in the future, energy demand and fuel demand are expressed in billion kWh.

3.1 Energy Demand Forecasting

The computation results reveal the different energy demand growth rates among the scenarios, with a clear difference from 2030. Between 2019 and 2050, the energy demand will increase 18.63 times for the BAU, 17.50 times for the ENE, 15.79 times

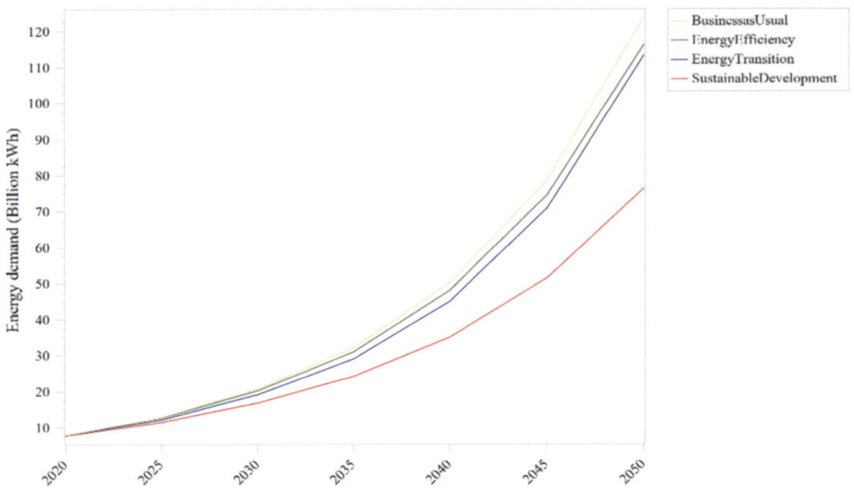

Fig. 3 Forecasted total energy demand for on-road transportation in HCMC from 2020 to 2050

for the ALF, and 10.88 times for the SUT. In 2050, compared to the BAU, in contrast to a slight decrease (8.20 billion kWh) in energy demand for the ENE, the energy demand decreases significantly (20.85 billion kWh) if alternative fuel measures are fostered. The combination of energy efficiency, alternative fuel, and transportation demand reduction measures can decrease 56.25 billion kWh in 2050. The average energy demand in 2050 will be 113.8 billion kWh (Fig. 3).

3.2 *Fuel Demand Forecasting*

For the BAU and ENE, the fuel demand structure is similar between 2020 and 2050. However, diesel will take over the highest energy consumption proportion by gasoline for the first time in 2028 for the BAU and 2029 for the ENE and will remain in the same order in the following years. In 2050, the fuel demand structure would be 55% diesel, 34% gasoline, 9% gasohol, 1.4% CNG, and only 0.6% electricity (Fig. 4).

Thanks to the upscaling of electric vehicles, electricity demand will accelerate dramatically to reach 28.36% for the ALF and 26.53% for the SUT in 2050. At the same time, the CNG and gasohol demand will increase to more than 3.25% in 2050. In contrast, gasoline consumption would peak in 2040 and decrease to only around 2.5% in 2050 (Fig. 4).

The energy demand transition for trucks is much more difficult. However, freight transport demand increases significantly annually; diesel demand will reach more than 55% of the total fuel demand under all scenarios in 2050 (Fig. 4).

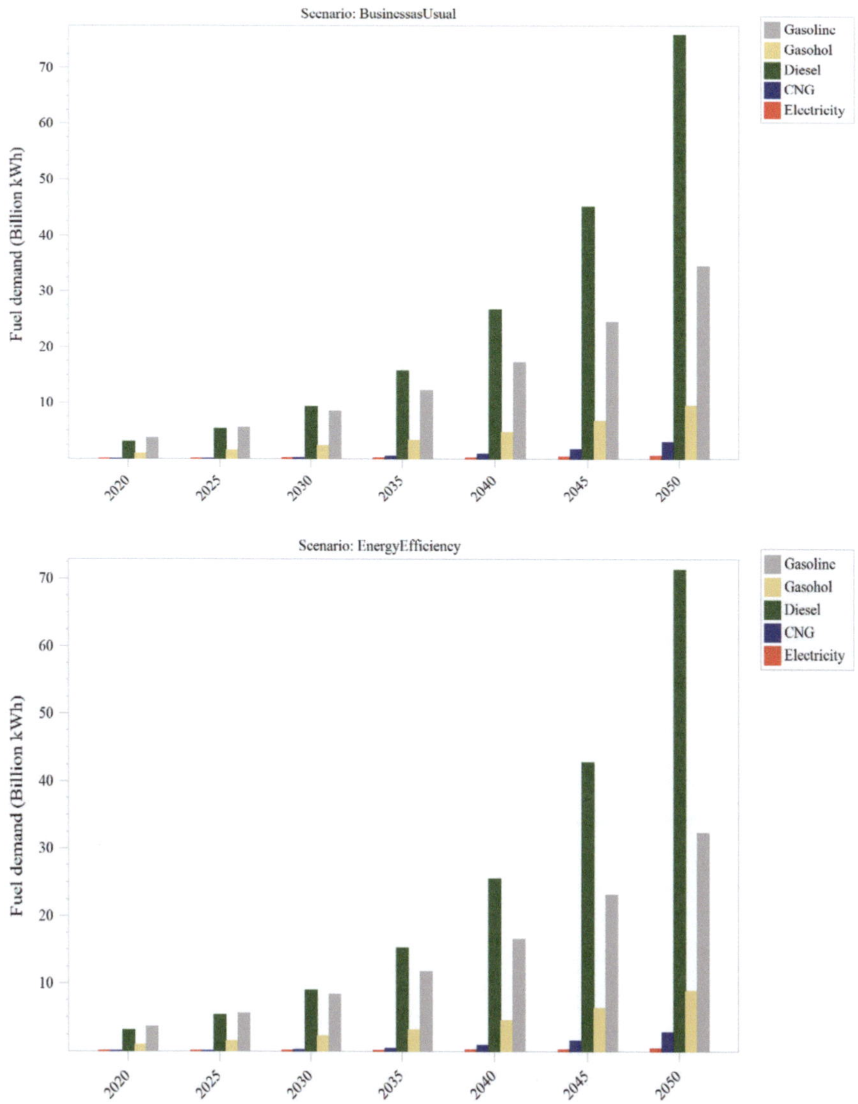

Fig. 4 Forecasted fuel demand for on-road transportation in HCMC from 2020 to 2050

3.3 Vehicle Number Forecasting

With the same transportation demand growth rate, the total vehicles in the BAU, ENE, and ALF are the same. The SUT's total vehicles are growing less between 2019 and 2050 due to the lower transportation demand growth rate. The total in-use

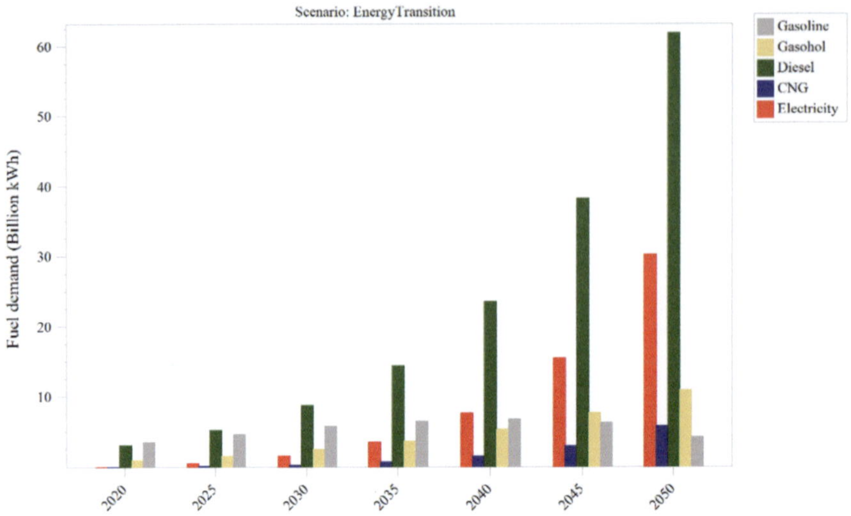

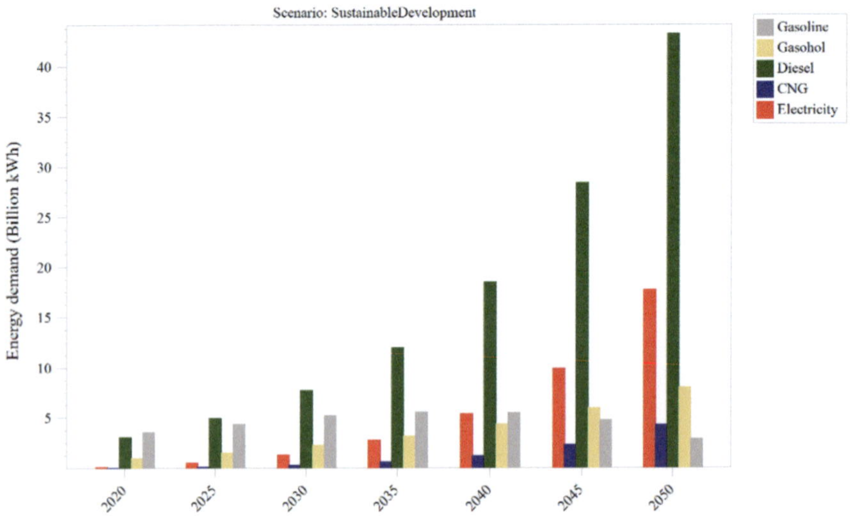

Fig. 4 (continued)

vehicles will increase 8.7 times from 5.97 million in 2019 to 52 million in 2050 for the BAU, ENE, and ALF and 6.8 times to 40 million for the SUT (Fig. 5).

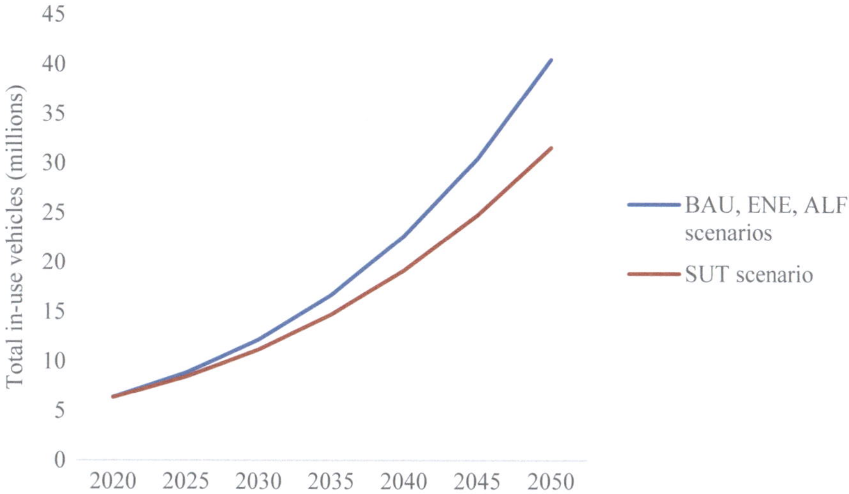

Fig. 5 Forecasted total in-use vehicles for on-road transportation in HCMC from 2020 to 2050

3.4 GHG Emission Forecasting

The GHG emissions will increase significantly between 2020 and 2050, reaching 21.9 $MtCO_2e$ for the BAU and 20.6 $MtCO_2$ for the ENE in 2050. The increase in GHG emissions is much lower for the ALF and SUT compared to the BAU and ENE. By 2050, the GHG emissions will reach around 18.6 for the ALF and 13.1 for the SUT, thanks to alternative fuel measures in the ALF, or combined with energy efficiency, lower transportation demand growth in the SUT (Fig. 6).

3.5 Fuel Cost Forecasting

Figure 7 shows the fuel cost estimation for each scenario's fuel consumption until 2050. Between 2019 and 2050, the fuel cost will increase significantly (16–17 times) for the ENE and BAU. Relative to the BAU, in 2050, HCMC will save 3,882 million USD for the ALF and 5,425 million USD for the SUT. These results indicate that energy efficiency and alternative fuel measures benefit fuel cost savings.

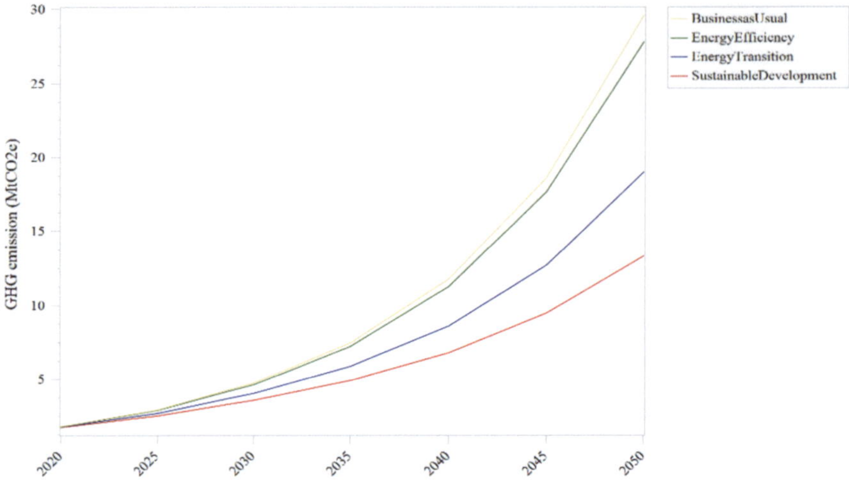

Fig. 6 Forecasted GHG emissions for on-road transportation in HCMC from 2020 to 2050

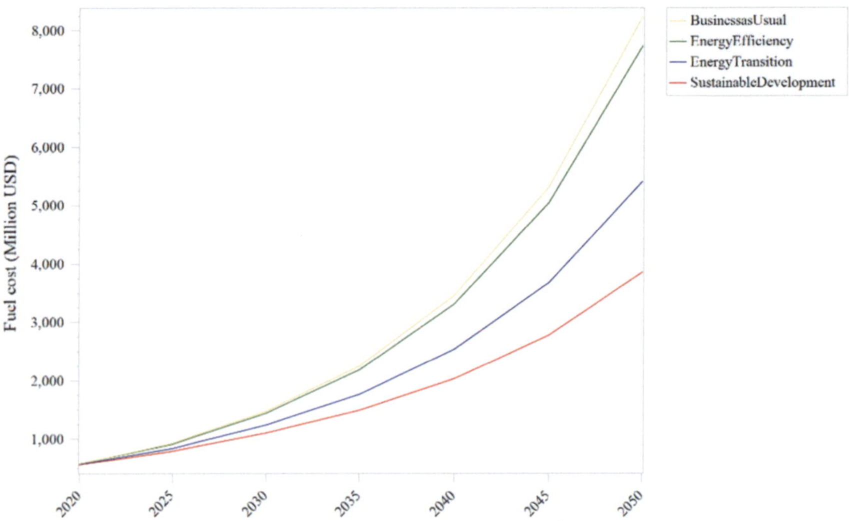

Fig. 7 Forecasted total cost of fuel for on-road transportation in HCMC from 2020 to 2050

4 Conclusions and Policy Implications

This paper has indicated the high energy demand for on-road transportation in HCMC over the next three decades. To meet that demand growth, four long-term energy alternative demand scenarios have been prepared with the integration of different energy policies. The main findings and policy implications can be summarized as follows:

Firstly, energy efficiency measures can help to decrease energy intensity in on-road vehicles. Therefore, the authorities should support the research and development of energy-efficient vehicle technologies while providing strict regulation on energy benchmarking for newly registered vehicles.

Secondly, alternative fuel measures, mainly up-scaling electric vehicles (electric mopeds, electric motorcycles, and electric cars), play a vital role in the pathway toward sustainable on-road development in HCMC. Between 2019 and 2050, the energy transition to cleaner fuels can decrease 47 billion kWh of energy demand compared to the BAU. The energy demand will decrease by 118 billion kWh if combined with energy efficiency measures. Accordingly, compared to the BAU, GHG emissions will be reduced by 33–45%, and fuel costs will be saved by 34–48%. To succeed in up-scaling electric vehicles, city authorities need to provide favorable policies and charging station infrastructure.

Thirdly, due to the difficulty in applying alternative fuel measures in freight transport vehicles, the diesel demand is forecasted to continue increasing significantly to meet the high freight transport demand growth. This fact limits the energy transition speed in on-road transportation.

This study has identified specific opportunities and challenges for HCMC's sustainable on-road transportation development. To take advantage of opportunities, the city needs to have action plans for implementing energy efficiency and alternative fuel measures in both the medium term and long term. In addition, the city can create mechanisms and policies to overcome difficulties and challenges in implementing energy efficiency and alternative fuel measures in the transport sector.

Although this paper is only concerned with the energy demand for on-road transportation in HCMC, the methodology used in this study can be applied to analyzing the energy demand scenarios and GHG emissions strategies in other provinces. In addition, this paper has only studied the fuel cost savings, not the marginal abatement cost. Therefore, a comparison of marginal net cost per unit of abatement of energy efficiency measures and alternative fuel measures can be studied in future research.

References

1. United Nations Environment Programme (UNEP) (2020) Emissions Gap Report 2020
2. UNHabitat (2022) Climate Change. https://unhabitat.org/topic/climate-change. Accessed: 04-Aug-2022
3. REN21 (2021) Renewables in Cities 2019 Status Report
4. Wei T, Wu J, Chen S (2021) Keeping Track of Greenhouse Gas Emission Reduction Progress and Targets in 167 Cities Worldwide. Front. Sustain. Cities 3(July):1–13
5. Renewable Energy Policy Network for the 21st century (REN21) (2021) Renewables 2021 Global Status Report
6. UN, "COP26: Together for our planet," Climate Action, 2021. https://www.un.org/en/climatechange/cop26. Accessed: 23-Feb-2022
7. Shira D (2021) COP26 and Climate Change: Vietnam's Commitment to Reducing Emissions," Vietnam Briefing. https://www.vietnam-briefing.com/news/cop26-climate-change-vietnams-commitment-reducing-emissions.html/. Accessed: 21-Apr-2022

8. The government of Vietnam, "Updated Nationally Determined Contribution (NDC)," no. July, 2020
9. P. Audinet, B. Singh, D. T. Kexel, S. Suphachalasai, P. Makumbe, and M. Kristy (2016) Exploring a Low-Carbon Development Path for Vietnam Environment and Sustainable Development
10. ADB (2017) Pathways to Low-Carbon Development for Viet Nam
11. Oh JE et al (2019) Addressing Climate Change in Transport
12. JICA (2017) GHG Inventory of Ho Chi Minh City
13. A. Leung, "Carbon Emissions Assessment of Urban Transport A New Approach for 10 Major Vietnamese cities," no. March, pp. 9–10, 2018
14. T. Thi Quynh Nguyen, W. Takeuchi, P. Misra, and H. Sachiko, "Technical note: Emission mapping of key sectors in Ho Chi Minh City, Vietnam, using satellite-derived urban land use data," Atmos. Chem. Phys., vol. 21, no. 4, pp. 2795–2818, 2021
15. EnergyPLAN, "LEAP." https://www.energyplan.eu/leap/. Accessed: 05-Aug-2022
16. T. H. Duong, T. T. Thao, and S. T. Tung, Housing and Transportation in Vietnam's Ho Chi Minh City Case studies in urban social development. 2017, p. 36
17. Ho Chi Minh City People's Committee, "Report on Ho Chi Minh city's economical - cultural - social situation in 2019, key tasks and solutions in 2020," 2019
18. P. Hoang, "HCMC District Map In 2021 – Guide to Live, Travel in Ho Chi Minh." https://rentapartment.vn/hcmc-district-map/. Accessed: 05-Aug-2022
19. World Easy Guides, "Ho Chi Minh City on Map of Vietnam." http://www.worldeasyguides.com/asia/vietnam/ho-chi-minh-city/ho-chi-minh-city-on-map-of-vietnam/. Accessed: 05-Aug-2022.
20. General Statistics Office of Vietnam, "Volume of freight carried by the road by province." https://www.gso.gov.vn/en/px-web/?pxid=E0917&theme=Transport%2C Postal Services and Telecommunications. Accessed: 17-Aug-2022
21. Ministry of Transportation of Vietnam, Circular on Promulgating economic-technical framework norms for public passenger transport by bus. 2014, pp. 1–6
22. Y. Y. Lam, K. Sriram, and N. Khera, "Strengthening Vietnam's Trucking Sector: Towards Lower Logistics Costs and Greenhouse Gas Emissions," World Bank, vol. 135753, 2019
23. Ngoc AM, Nishiuchi H, Van Truong N, Huyen LT (2022) A comparative study on travel mode share, emission, and safety in five Vietnamese Cities. Int J Intell Transp Syst Res 20(1):157–169
24. N. T. K. Oanh and H. H. Van, "Comparative Assessment of Traffic Fleets in Asian Cities for Emission Inventory and Analysis of Co-Benefit From Faster," Int. Emiss. Invent. Conf., pp. 1–10, 2015
25. D.-S. Tran, H. Le, and Z. Yang, "Two-wheelers in Vietnam: A baseline analysis of fleet characteristics and fuel consumption in 2019 and 2020," 2022
26. K. D. Contreras, Al. Mejia, P. Q. Nhu, and M. Tacderas, "Tracking Sustainable Transport in Vietnam: Data and Policy Review for Energy Efficiency and Climate Change 2015," 2015
27. IEA, "Vehicle Fuel Economy in Major Markets 2005–2019," 2021
28. G. Pistoia, "Electric and Hybrid Vehicles," Electr. Hybrid Veh., no. June, pp. 1–7, 2010
29. Statista, "Vietnam: passenger car share by drive-type 2019," 2021. https://www.statista.com/forecasts/1278117/passenger-cars-in-vietnam-by-fuel-or-drive-type. Accessed: 24-Aug-2022
30. T. Fraser, "What is a good energy consumption figure for electric vehicles?," drive.com.au, 2022. https://www.drive.com.au/caradvice/what-is-a-good-energy-consumption-figure-for-electric-vehicles/. Accessed: 13-Sep-2022
31. Ministry of Transportation of Vietnam, "Determination of fuel consumption in building economic and technical norms for CNG buses for Hanoi city," 2021. https://mt.gov.vn/khcn/tin-tuc/71264/xac-dinh-luong-nhien-lieu-tieu-hao-trong-xay-dung-dinh-muc-kinh-te-ky-thuat-cho-loai-hinh-buyt-cng-cho-dia-ban-thanh-pho-ha-noi.aspx. Accessed: 25-Aug-2022
32. L. Phong, "Consumption of E5 bio-fuel falls dramatically," SSGP, 2021. https://www.sggpnews.org.vn/business/consumption-of-e5-biofuel-falls-dramatically-91083.html. Accessed: 14-Sep-2022

33. Q. Toan, "E5 fuel consumption equivalent to gasoline Ron92," BNEWS, 2018. https://bnews.vn/xang-e5-tieu-hao-nhien-lieu-tuong-duong-voi-xang-ron-92/73237.html. Accessed: 14-Sep-2022
34. VietnamPlus HCM City's eco-friendly buses face fuel shortage," 2019. https://en.vietnamplus.vn/hcm-citys-ecofriendly-buses-face-fuel-shortage/155469.vnp. Accessed: 24-Aug-2022
35. sustainable-bus, "Vietnam, first e-bus service launched in Ho Chi Minh City," 2022. https://www.sustainable-bus.com/electric-bus/vietnam-saigon-ho-chi-minh-e-buses/. Accessed: 24-Aug-2022
36. Fuel price projections for Vietnam. Background to the Vietnam Energy Outlook Report 2019, 2019
37. Anh N (2018) Compare the price of biofuel E5 RON 92 and mineral gasoline RON 95," BNEWS, 2018. https://baodautu.vn/so-sanh-gia-xang-sinh-hoc-e5-ron-92-va-xang-khoang-ron-95-d75326.html. Accessed: 23-Sep-2022
38. Hiep DT, Hoffmann C (2019) Estimation of the Future Electricity Price Surcharge for the Integration of Wind and Solar Power into the Vietnamese Electricity System," 2019 IEEE Asia Power Energy Eng. Conf., pp. 263–268
39. Eduardo CB, Guendehou S, Limmeechokchai B, Pipatti R (2006) 2019 Refinement to the. IPCC guidelines for national greenhouse gas inventories. 2019:1–15
40. Tuyen PH, Tuan PM, Yamamoto K, Karin P (2021) Updating emission factors for in-use motorcycles fueled by gasoline, e5 and e10 in Vietnam. ASEAN Eng. J. 11(3):199–206
41. Clairotte M et al (2020) Exhaust emission factors of greenhouse gases (GHGs) from European road vehicles," Environ. Sci. Eur., vol. 32, no. 1
42. Quang Anh H (2020) Final Report on the Study and Development of Emission Factor (Ef) for Vietnamese Electrical Grid in 2018
43. The Asia Investor Group in Climate Change (AIGCC), "Power of ASEAN Accelerating clean energy in Vietnam and Indonesia (2021)
44. Vietnam's climate transparency report (2021) Climate Transparency, pp. 1–19
45. Nguyen H (2021) TPHCM phấn đấu tăng GRDP bình quân đầu người khoảng 37.000 USD vào năm 2045," laodong.vn. https://laodong.vn/xa-hoi/tphcm-phan-dau-tang-grdp-binh-quan-dau-nguoi-khoang-37000-usd-vao-nam-2045-905641.ldo. Accessed: 29-Aug-2022
46. HCM City seeks to develop an electric bus system," VietnamPlus, 2022. https://en.vietnamplus.vn/hcm-city-seeks-to-develop-electric-bus-system/236426.vnp Accessed: 14-Sep-2022
47. European Environment Agency, Energy efficiency, 2020. https://www.eea.europa.eu/publications/ENVISSUENo12/page027.html. Accessed: 09-Sep-2022
48. United States Department of Transportation, Energy Intensity of Light Duty Vehicles and Motorcycles, Bureau of Transportation Statistics. https://www.bts.gov/content/energy-intensity-light-duty-vehicles-and-motorcycles. Accessed: 09-Sep-2022
49. Ministry of Transportation of Vietnam (2011) Circular 64/2011/TT-BGTVT on Regulations on economical and energy-efficient solutions in transportation activities, pp 2–5
50. Ministry of Industry and Trade of Vietnam (2016) Circular 36/2016/TT-BCT on Regulateting on energy labeling for vehicles and equipment using energy under the Ministry of Industry and Trade management
51. Ministry of Industry and Trade of Vietnam (2019) Draft National Action Plan on Sustainable Consumption and Production 2020–2030
52. Ministry of Transportation of Vietnam, Circular 67/2015/TT-BGTVT on national technical regulations on quality, technical safety and environmental protection for motorcycles and mopeds, vol. 13, no. 3. 2015, pp 1576–1580
53. Quan M (2020) Ho Chi Minh City begins piloting motorcycle emission testing for 5 years or more," laodong.vn. https://laodong.vn/xa-hoi/tphcm-bat-dau-thi-diem-kiem-tra-khi-thai-xe-may-tu-5-nam-tro-len-805590.ldo. Accessed: 14-Sep-2022

54. Ministry of Industry and Trade of Vietnam (2018) National Program: Vietnam-National Energy Efficiency Program 2019–2030, no. June 2018, pp 1–104
55. VietnamInsider, Development of electric vehicles in HCMC - Vietnam Insider, 2022. https://vietnaminsider.vn/development-of-electric-vehicles-in-hcmc/. Accessed: 24-Aug-2022
56. Prime Minister of Vietnam, Decision 876/QD-TTg approves the action plan on green energy transition, reducing carbon and methane emissions of the transport sector, no. 8.5.2017. 2022, pp. 2003–2005

Scalable, Low-Cost, and Versatile System Design for Air Pollution and Traffic Density Monitoring and Analysis

Thinh Gia Tran, Dat T. Vo, Long C. Tran, Hoang V. Pham, Chuong D. Le, An D. Le, Duy A. Pham, and Hien B. Vo

Abstract Vietnam requires a sustainable urbanization, for which city sensing is used in planning and decision-making. Large cities need portable, scalable, and inexpensive digital technology for this purpose. End-to-end air quality monitoring companies such as AirVisual and Plume Air have shown their reliability with portable devices outfitted with superior air sensors. They are pricey, yet homeowners use them to get local air data without evaluating the causal effect. Our air quality inspection system is scalable, reasonably priced, and flexible. The minicomputer of the system remotely monitors PMS7003 and BME280 sensor data through a microcontroller processor. The 5-megapixel camera module enables researchers to infer the causal relationship between traffic intensity and dust concentration. The design enables inexpensive, commercial-grade hardware, with Azure Blob storing air pollution data and surrounding-area imagery and preventing the system from physically expanding. In addition, by including an air channel that replenishes and distributes temperature, the design improves ventilation and safeguards electrical components. The gadget allows for the analysis of the correlation between traffic and air quality data, which might aid in the establishment of sustainable urban development plans and policies.

Keywords Air pollution monitoring · Low-cost air sensor · Traffic surveillance · IoT

T. G. Tran (✉) · H. V. Pham · C. D. Le · H. B. Vo
Vietnamese-German University, Binh Duong, Vietnam
e-mail: giathinhtran3@gmail.com

D. T. Vo
University of Windsor, Windsor, Canada

L. C. Tran
Aalto University, Helsinki, Finland

A. D. Le
University of California San Diego, San Diego, CA, USA

D. A. Pham
Computer Science Department, Hochschule Bonn-Rhein-Sieg, Sankt Augustin, Germany

V. T. Ha et al. (eds.), *Proceedings of the 8th International Conference on Sustainable Urban Development*, Advances in 21st Century Human Settlements,
https://doi.org/10.1007/978-981-99-8003-1_18

1 Introduction

The rise of the population is a serious concern that people discuss worldwide. As a consequence, the air environment in urban areas is getting more polluted. The government and researchers need air data from the city to find the root of the pollution and come up with a solution. To capture this, we come up with our instrument. The target is to capture the particulate matter (PM1, PM2.5, and PM10), temperature, humidity, and atmospheric pressure. We also aim to record the video simultaneously to study and find the relation between the number of vehicles and air quality and use that for future prediction.

At the early stage of the project, a dust measuring device that contains an STM32 microcontroller with PMS7003 for dust measuring and BME280 for temperature, humidity, and pressure capture is used with a surveillance camera for each node. This combination started to show the weakness of storing the data. There are two main problems towards it: (1) the data name format; the commercial camera fixes the name file format, and adding other information, such as location, is impossible; (2) the video coming from the surveillance camera is saved in the AVI format to save spacing storage, and to use in the visual processing model is time-consuming due to the conversion of the file format.

With the help of Raspberry Pi 3B+ as a mini-computer, all of the data files can be set to the desired output. Then the STM32 controller board with sensors is redesigned to become a module using the serial connection. A 5 Mpx camera module is used to capture visual data. We can upgrade or swap the broken modules by designing all the components in a modular form. The Raspberry Pi 3B+ connects to Microsoft Azure Blob, which provides storage for all data, and this process is automated. Azure Blob is flexible in storing the data, switching between archive and cool storage based on regularity to save unnecessary costs.

2 Related Work

Regarding the work, a research in Thailand also has a similar approach [1] in which their main goal was to record PM2.5, PM10, temperature, humidity along with vehicle flow, categorized into five groups: light-duty gasoline vehicles, light-duty liquid petroleum gas (LPG) vehicles, heavy-duty diesel vehicles, motorcycles, and light-duty diesel vehicles. To capture particulate data, a PTFE filter was used with an air pump, then weight the collected matter on an ultra-microbalance with the lowest readability of 0.1 μg. Temperature and humidity were recorded on a separate instrument for the meteorological data. Finally, the traffic flow was recorded with a closed-circuit camera. They have recognized the correlation between the traffic flow and PM is a minor factor in indoor air quality, but temperature and relative humidity seem to have a strong influence. However, they wrote that the limited monitored area could lead to inaccurate results.

In addition, Zhang et al. in the US developed a system for indoor air quality monitors using Raspberry Pi as the central controller [2]. They designed a system that can log SO_2, NO_2, TVOC, O_3, CO, CO_2, PM2.5, PM10, temperature, and humidity. They also simplified the sensor and Raspberry Pi connection using a USB connector. However, their Low-Cost Air Quality Sensor system is designed for interior monitoring and cannot thus survive outdoor conditions.

In Europe, the We-count project aims to measure mobility and air quality data using Raspberry Pi [3]. They have launched pilot tests in Madrid, Ljubljana, Dublin, Cardiff, and Leuven. From these nodes, the project can provide a data set of vehicle and air quality data for the scientists to research air pollution correlation with traffic. The We-count project includes two devices: a Raspberry Pi camera system and an IoT gadget for measuring PM. Volunteers install the visible data logging system on an interior glass window, while the dust data logging equipment is erected outdoors beneath the window roof. Two distinct gadgets might complicate the mapping network.

The TSI DustTrak [4] product line is a series of aerosol monitors designed for real-time, continuous measurement of particulate matter in the air. The monitors are widely used in a variety of settings, including industrial hygiene, environmental monitoring, and indoor air quality assessments. The TSI DustTrak 8534 is a specific model within the product line that features a built-in pump and a real-time mass concentration reading of PM1, PM2.5, and PM10 size fractions. It is capable of both manual and remote sampling, making it a versatile tool for air quality monitoring. The TSI DustTrak 8534 has been used in numerous studies to assess indoor and outdoor air quality. In our paper, we used the DustTrak 8534 as a superior reference device for comparison with our low-cost PM sensor.

3 Methodology

Our system aims to gather video traffic data and environmental parameters such as dust concentration, temperature, humidity, and atmospheric pressure. All data are stored on the cloud, through the Azure Blob Storage service. We aim to make it low-cost and scalable simultaneously.

Our system consists of three hardware components (Figs. 1 and 2): 1 Raspberry Pi mini-computer, 1 onboard camera, namely Pi Camera V2, and 1 Environmental Monitoring Module, which is custom-built by us to collect environmental parameters through a series of sensors.

3.1 Environmental Monitoring Module

This module aims to gather data on environmental parameters through a series of sensors, then process them and export the complete data through its USB port (Fig. 3).

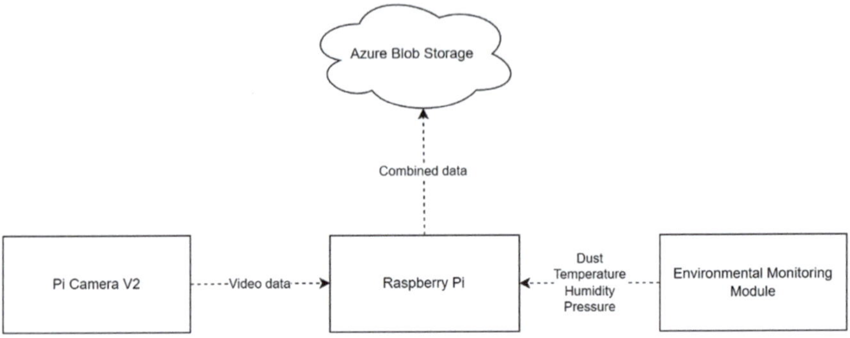

Fig. 1 Block diagram illustrating the components of the system

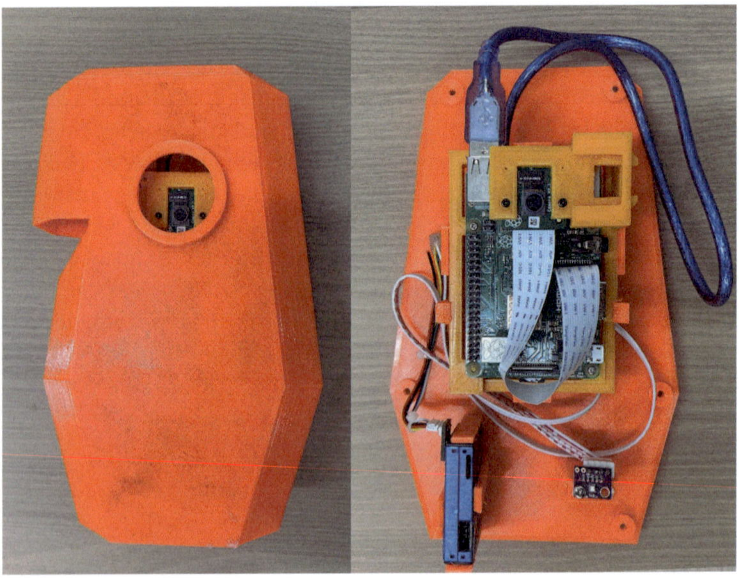

Fig. 2 The system's enclosure and internal circuitry

This module is then connected through a USB cable to the Raspberry Pi 3B+, where the data will be stored in a CSV file and uploaded daily to the cloud. Data is sampled and sent out at a 10 s interval. The whole module is mounted beneath the Raspberry Pi.

STM-32 Blue Pill development board. Manufactured by ST-Microelectronics [5], the STM-32 Blue Pill was explicitly chosen for its abundance of connection ports. It supports 37 GPIO pins, 3 UART, 2 I2C and 2 SPI, 1 USB 2.0 as well as 1 CAN 2.0 peripherals [6]. In our particular application, we only utilize 2 UART and 1 I2C interface. But if we want to make additional changes to the module in the future, such as adding a CO_2 or an insert gas sensor, we can very easily use the remaining

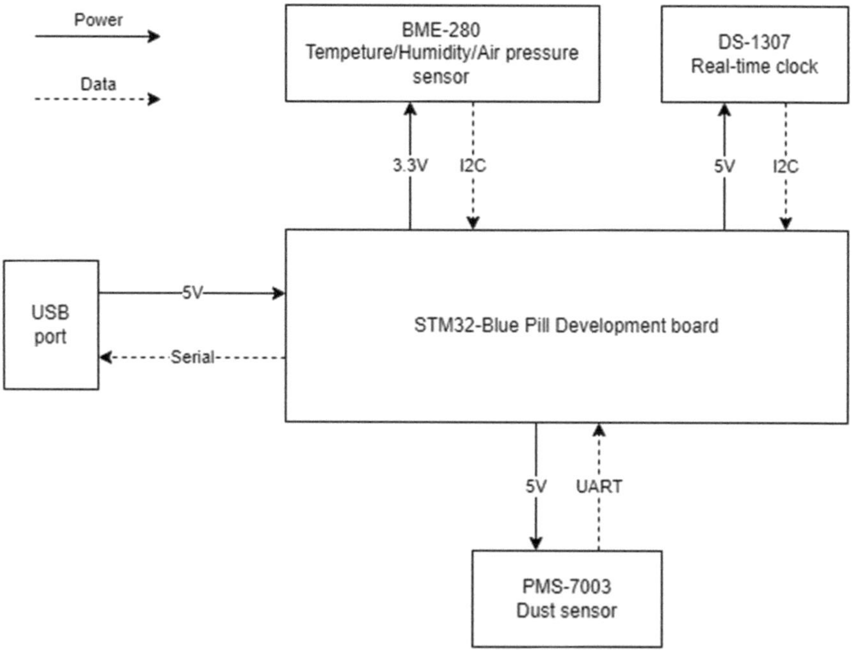

Fig. 3 The block diagram illustrating the components of the environmental monitoring module

ports. The high processing speed and large memory capacity also contribute to our choice of the particular microcontroller.

BME-280 sensor. Manufactured by Bosch Sensortec [7], the BME-280 sensor is a module used to record temperature, humidity, and atmospheric pressure data. This sensor is chosen because of its compact build and high accuracy. It communicates with the microcontroller using the standard I2C protocol [8]. An external library [9] was used in the firmware's source code to facilitate communication.

DS-1307 Real-time clock. Manufactured by Maxim Integrated [10], the clock is used to reliably keep track of the time. Raspberry Pi doesn't have an onboard real-time clock, so we need an external clock to ensure correct timekeeping. Similar to the BME-280 sensor, the DS-1307 clock also communicates using I2C protocol and an external library [11] was used to help with the communication.

PMS-7003 Dust sensor. Manufactured by Plantower [12], the sensor is used to monitor the concentration of particles of different sizes, namely, PM1.0, PM2.5, and PM10. This sensor is chosen because of its low price and high accuracy. PMS-7003 when compared to its peers shows a better general performance, with low intra-model variability, high linear fitting coefficient with superior sensor and high durability in adverse conditions [13]. It uses UART (Universal Asynchronous Receiver-Transmitter) protocol to send and receive data to and from the microcontroller. Data are handled directly by continuously requesting once every 10 s. Data will be received at the microcontroller as a 32 bytes stream. From this, a checksum is performed to eliminate error data. If the steam passes the check-sum step, bytes 12, 14, and 16 will be registered as PM1.0, PM2.5, and PM10 values respectively, with the unit of $\mu g/m^3$ [14]. Other bytes will be discarded.

3.2 Raspberry Pi and Onboard Camera

The goal of this project is to capture video data of traffic and send it to the cloud for further processing. To achieve that, we utilize a Raspberry Pi 3B+ and an onboard camera.

The camera. The camera chosen for this project is the Pi Camera V2. Manufactured by the Raspberry Pi company [15], it has an 8 Megapixel sensor, which provides 1080p resolution at 30 frames per second. It has a field of view of 62 degrees [16].

To reduce file size and save storage space, we only set the camera to operate at 10 frames per second, 1296 × 730 resolution. We determined that 10 frames per second are sufficient for our application.

The Raspberry Pi. Manufactured by the Raspberry Pi company [17], the Raspberry Pi 3B+ was chosen because it can record videos locally and upload them to cloud storage through API. Another approach was to use Microcontroller Units with Wi-Fi capabilities such as the ESP32. However, because we plan to implement this as an edge device in the future, we decided to use Raspberry Pi.

The Pi operates in headless mode and continuously runs a Python script. It will record videos and cut them into 5-min files. For air quality data, it will receive data from the Environmental Monitoring Module, parse it, and store it in a CSV file. These files will be uploaded to our Azure Blob Storage, the method of which will be discussed in later sections.

3.3 Storage Solution

In order to have high scalability, we need to implement cloud data storage. The service we chose to use is Microsoft Azure Blob Storage which provides low-cost, high-volume data storage.

We assign each node in our system to 1 blob. Each time a new node is set up, it will be automatically assigned to a blob bearing its name. All video and text data of 1 node are uploaded into the same blob. We used Azure's recommended Python library to upload the data.

First, data are collected and stored locally in a file. For video, each file is 5 min long. For text data, each file contains the data for 24 h. Second, the files are read as a byte stream, and then this byte stream is directed to the pre-specified blob. The upload process lasts around 5–7 min on a good Internet connection for a typical 5-min-long video. Due to this reason, each upload operation has to be carried out asynchronously on a separate sub-process, to avoid blocking the main loop. Lastly, after completing the upload process, the files stored in the local drive will be deleted after a period of time specified by the setting. In this sense, the local storage acts as buffer storage.

It is estimated that one node will generate approximately 30 GB worth of video data daily. The cost of storing such an amount on Azure Blob Storage will depend on the storage location, pricing plan, and other settings [18]. However, because the daily fee will increase linearly over time when data accumulates, the total cost will increase exponentially over time. Therefore, additional operations to move unused data into archive storage are highly recommended for any attempt to replicate this method. However, for the scope of this project, we didn't implement such operations.

3.4 Enclosure

The current enclosure design is for quick installation while minimizing the needed tools. The enclosure is designed to be 3D printed at the moment. The material is ABS (Acrylonitrile butadiene styrene) or PETG (Polyethylene terephthalate glycol). We highly recommend using PETG since PETG is UV resistant, high durability while staying in the solid form up to 80 °C. Thus, PETG is more 3D printing friendly in the lab due to no smell and warping without a 3D printer enclosure.

The small 1.2 mm acrylic lens (Fig. 4–3) is glued to the front cover (Fig. 4–1). Acrylic is UV resistant, low cost, and has good transparency, which makes it suitable for this design. On the back of the back cover (Fig. 4–2), there are slots (Fig. 4–6) for the zip ties and pipe clamps for the installation at the data logging location. At the bottom of the front cover (Fig. 4–1), there are passive and active air inlets (Fig. 4–4, 5).

PMS 7003, BME 280, and 4040 fan (Fig. 5—4, 5, 6) are mounted on the back cover through snapping clips (Fig. 5–1). The Raspberry Pi 3B+ and STM-32 Blue Pill development board PCB are mounted on a tray (Fig. 5–2) and then snapped to the back cover (Fig. 5–1). The tray (Fig. 5–2) holds the camera mount (Fig. 5–3) with an 8 Mpx camera.

The airflow direction shows that the fresh air travels from the inlet (yellow dashed section) to the outlet (red dashed section), although the airflow velocity direction illustrates there is some turbulence in the middle area close to the Raspberry Pi 3B+ heatsink (Fig. 6). Since the design can not achieve the laminar flow, dust inside the system will build up over time. That is why the open sensor enclosure design is better for collecting environmental data. The advantage of this closed sensor enclosure is the higher level of weather protection, especially in tropical weather with irregular rain.

According to ARM Cortex A53 (Raspberry Pi 3B+ CPU) specification, it can go up to 105 °C, although the Raspberry Pi 3B+ operating system is set to throttle at 85 °C. In Fig. 7, the hottest temperature area in the center is the CPU heatsink (around 107 °C for the worst scenario), with the airflow from the inlet (yellow dashed section) to the outlet (red dashed section), it transfers the heat to the stream and keeps the

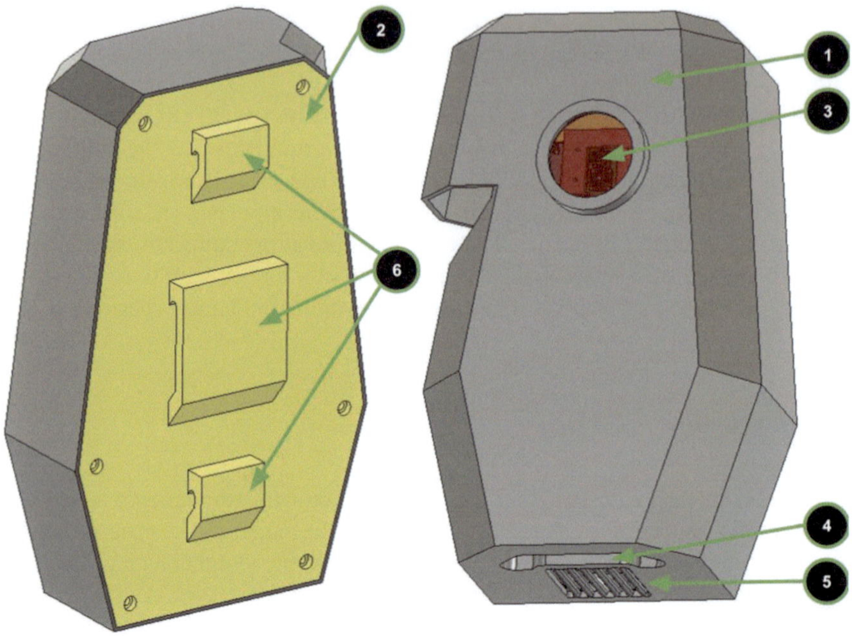

Fig. 4 Enclosure (1—Front cover, 2—Back cover, 3—Camera lens, 4—Passive air inlet, 5—Active air inlet, 6—Slots for zip ties and pipe clamps)

CPU cool. The nominal temperature in the top area is approximately 45 °C–55 °C, while the bottom area temperature (sensor area) is the environment temperature at 27 °C. It should keep the environment temperature reading value accurate.

3.5 Vehicle Counting and PM2.5 Analysis

The purpose of the following pipeline is to count the number of vehicles every 10 min over the span of 6 days, with no classification. The meaning of this is to verify the hypothesis mentioned in [19], which stated that the number of vehicles correlates positively with the PM2.5 reading of the sensor with an unknown lag.

Vehicle counting algorithm. The desire to find the correlation, the relationship between PM2.5 metric and traffic density establishes the need for a vehicle counting algorithm. In order to satisfy the urge, a simple vehicle counting method is implemented, which mainly depends on a certain background subtraction algorithm. The architecture of this method is as follows (Fig. 8):

To get a foreground object, a background subtraction algorithm called CouNT is used, as this is the fastest option with decent performance for hardware with a low spec that is available in OpenCV [20]. After subtracting the background, the

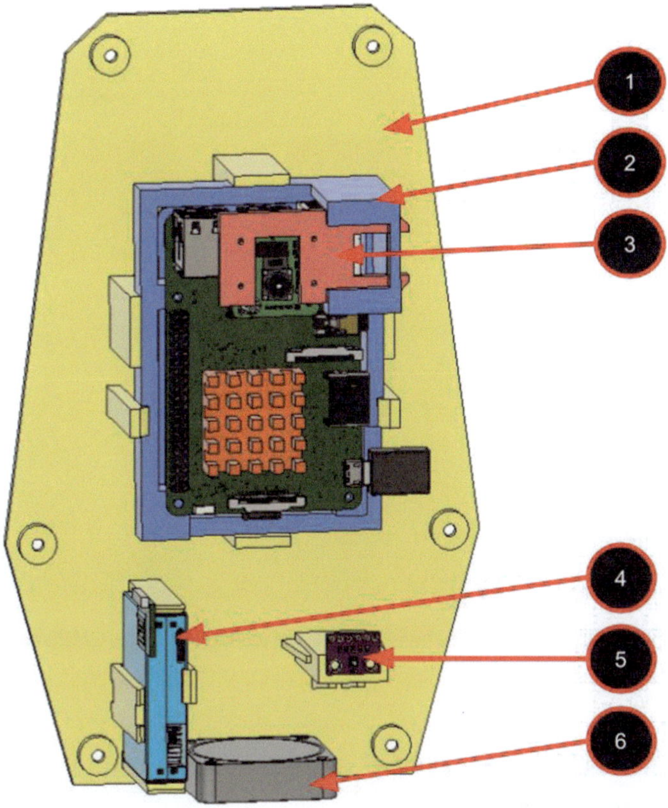

Fig. 5 Inside enclosure (1—Back cover, 2—Raspberry Pi 3B+ and STM32 sensor driver PCB mount tray, 3—Camera mount with 8 Mpx camera, 4—PMS7003 particulate matter sensor, 5—BME280 temperature and humidity sensor, 6-Active air intake 4040 fan)

foreground will be segmented into different objects through contouring, and then bounding boxes and their centers are achieved (Fig. 9).

After the previous steps, each vehicle is counted by line method, where a virtual line is placed on the frame, and when the center of the vehicle crosses this line, the number of vehicles increases by one. This method is enhanced by a tracking algorithm called SORT [21] that utilizes Kalman Filter and Hungarian algorithm for better accuracy.

PM2.5 analysis. After being collected, the PM2.5 metric is then analyzed by first being preprocessed and then being visualized via time series plot and correlation plot. The preprocessing step removes outliers and normalizes the data. It works as follows (Fig. 10):

In step 1, PM2.5 values that are larger than 20,000 (this value is hardware-specific) are filtered out, as these values are errors. In step 2, outliers are removed by using the standard deviation method. In step 3, missing data are linearly interpolated and

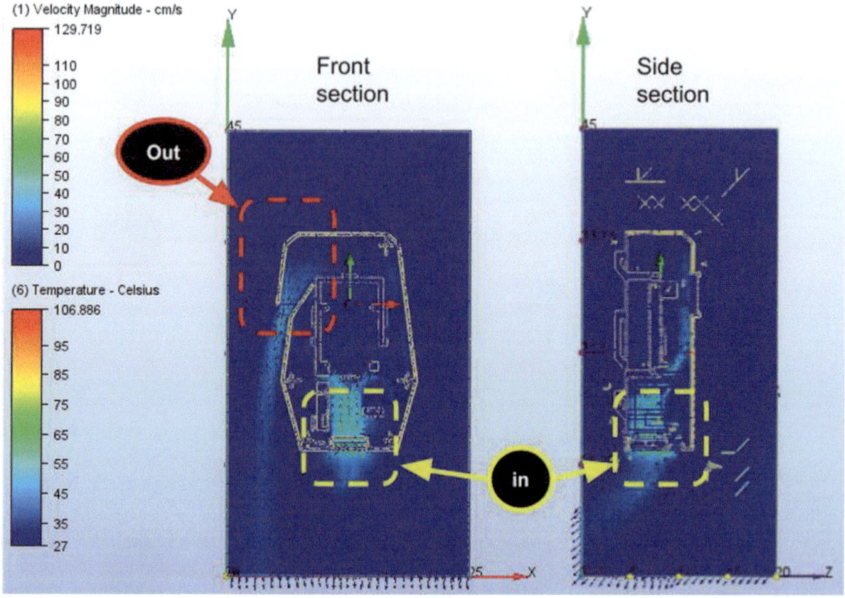

Fig. 6 CFD simulation result of heat transfer inside the model (Velocity—CMPS scale)

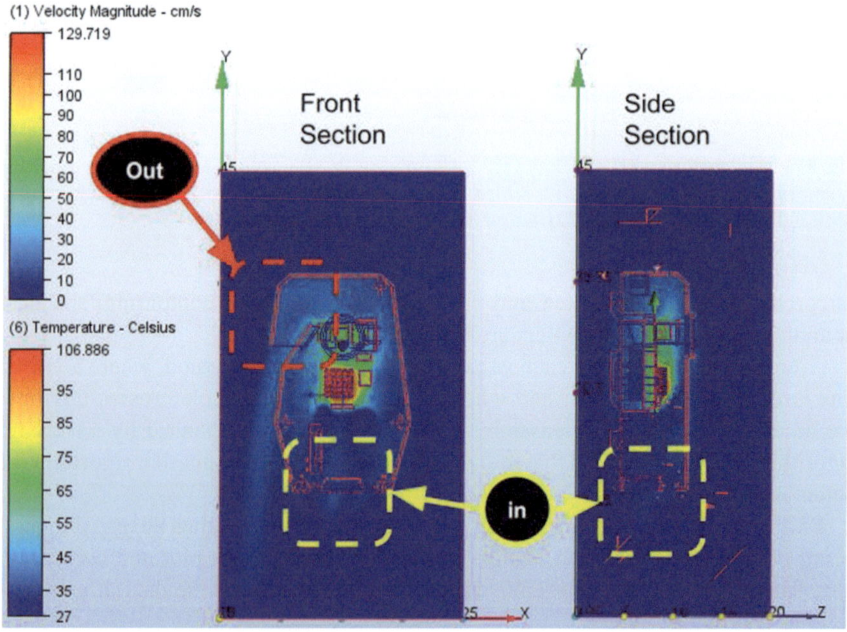

Fig. 7 CFD simulation result of heat transfer inside the model (Temperature—Celsius scale)

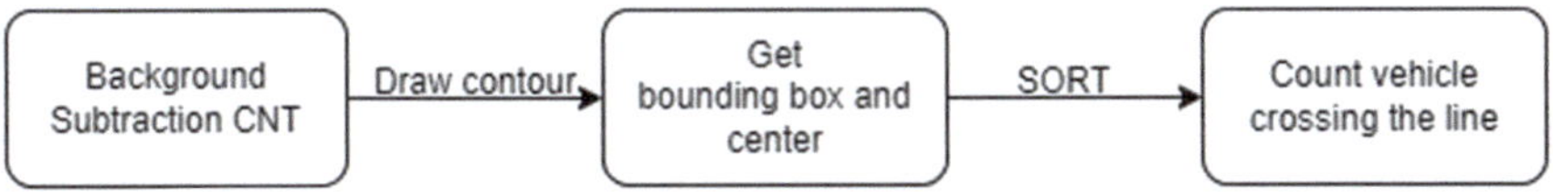

Fig. 8 Overview of counting algorithm

Fig. 9 Example of background subtraction. The left figure shows the foreground after subtracting

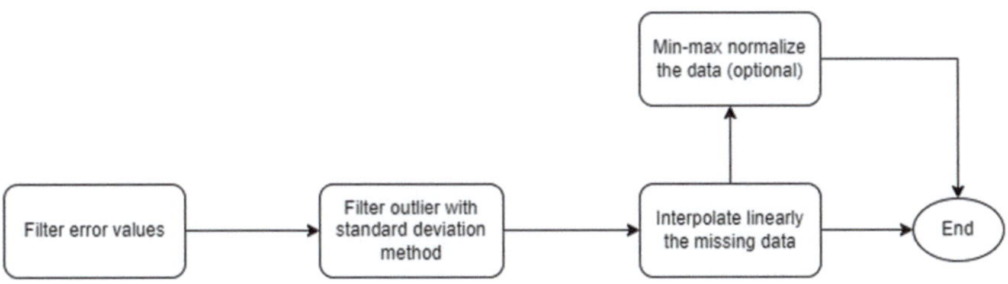

Fig. 10 Overview of PM2.5 analysis procedure

their sampling rate is altered from second to hour (Fig. 11). In step 4, the data are normalized by min–max normalization. Min–max normalization is as follows:

$$x_{scaled} = \frac{x - x_{min}}{x_{max} - x_{min}}$$

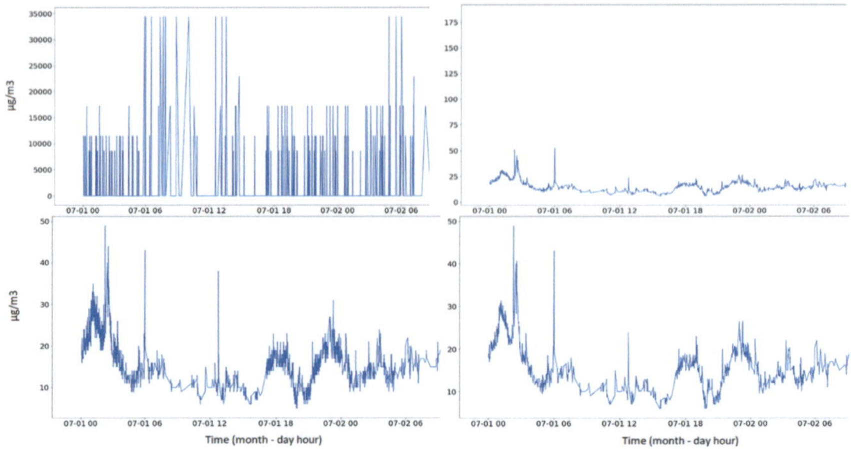

Fig. 11 Data before cleaning and after each step (steps 1–3) (plots are placed in left–right, top–bottom order)

4 Results and Discussion

4.1 PMS7003 Evaluation and Calibration

Because PMS7003 is a low-cost, off-the-shelf sensor, it is recommended that the PMS7003 be calibrated before use [13]. The method is by colocating the PMS7003 with a well-calibrated sensor, usually with superior quality compared to the PMS7003, and then the collected data is evaluated. The goal is to find an offset and scale factor that best correlates the PMS7003 data set with that of the reference. In this paper, we performed the evaluation and calibration process of the PMS7003 with an industrial-grade particle concentration sensor, namely the DustTrak DRX Aerosol Monitor 8534. Both sensors were located outside to collect PM2.5 dust data for 48 h. The result will be demonstrated below.

The time series data collected from both sensors are matched along the time axis. Then we performed DTW (Dynamic Time Warping) algorithm and then calculated the 10-min moving average. The yielded result is demonstrated below (Fig. 12).

From the data, a few statistical analysis metrics were calculated.

Root Mean Square Error (RMSE). RMSE is the square root of the average squared errors. It measures how far away data points from both sensors are, relative to each other. The result of our data set is 0.3254. Because RMSE is dependent on the scale of the data, it should be noted that the range of our data set is 0.5–33.5 μg/litter. Additionally, smaller RMSE indicates better accuracy.

Trend matching score. We used the Hodrick-Prescott filter to find the trend's cyclical value at each point for each time series. We then compared these values in a piecewise manner. If values from both time series possess the same sign, a trend match is determined. Else, it is a trend mismatch. This score allows us to quantify the

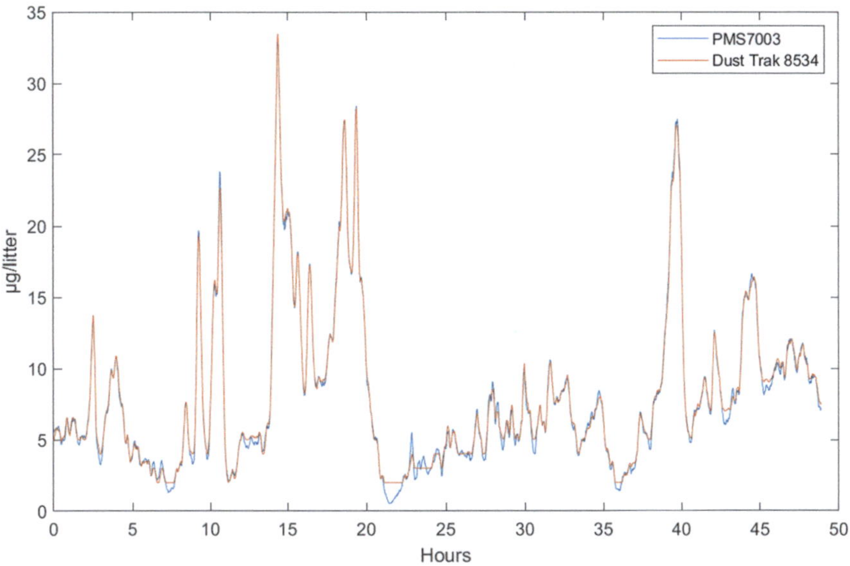

Fig. 12 10-min moving average plot of PM2.5 time series data from PMS7003 and DustTrak 8534

general trend match between the two time series. The result of our data set is 72.52%, meaning the trend from PMS7003 data matches that of DustTrak 8534 approximately 72.52% of the time.

Mean Absolute Percentage Error (MAPE). MAPE metric is used to measure the accuracy of PMS7003 relative to DustTrak 8534 as a percentage. It is calculated using the average absolute percentage error for each PMS7003 data point minus values from DustTrak 8534 divided by values from DustTrak 8534. The result of our data set is 4.95%. A MAPE value of 0% indicates perfect correlation.

Coefficient of Determination (R^2). R^2 is a statistical metric used to measure the linear correlation between two variables. An R^2 value of 1 represents perfect correlation. From our data, the R^2 value is calculated to be 0.997.

In [22], Báthory et al. conducted a similar experiment where they colocated some PMS7003 sensors in the environment for 3 months, together with a TSI DustTrak 854 301-M1 type aerosol monitor acting as a reference device. Table 1 shows the comparison of MAPE and R^2 values between our and their experiments. The difference between the results can be attributed to the difference in test duration and the reference device model used in both experiments.

Table 1 MAPE and R^2 result comparison

	Ours	Báthory et al. (2021)
MAPE	4.95%	11–14%
R^2	0.997	0.945

From the analyzed metrics and qualitative observation of the plotted data, we decided that the specific PMS7003 we used in our system correlated well, relative to the DustTrak 8534 device. Therefore, we decided that a calibration coefficient is not necessary.

4.2 *Comparing Vehicle Density and PM2.5 Metric*

We compare the PM2.5 metric with the traffic density via time-series plotting. The plot of the result can be found in Fig. 13, appendix.

The two lines have some similarities. PM2.5 rises when there are more vehicles, and decreases when there are fewer vehicles; however, there are some delays, and the pattern between these two lines is not as clear as we expected. For example, on the 1st of July at around 16 o'clock, the number of vehicles hits a local peak, but the number of PM2.5 hits its peak at around 17 o'clock. Another example would be on the 4th of July at 10 o'clock when the vehicle line hits a local peak, while the PM2.5 line hits the local peak at around 12 o'clock. This suggests two hypotheses, the first is that traffic density is not the only factor that affects the PM2.5 metric but also factors such as weather, temperature, humidity, etc.; the second is that vehicle density does not affect air quality immediately, its effect is only visible after a certain amount of delay.

This result agrees with the hypothesis mentioned in [19].

5 Limitation

Our storage system is not yet cost-optimized. All the collected data are stored in the "Cool" mode, which facilitates accessibility but generates higher accumulative costs. An algorithm could be developed to automatically move outdated data into "Archive" mode which reduces the cost.

The calibration process of the PMS7003 is still manual, which greatly slows down the building process if more nodes are to be built. An automated process could be developed to automatically analyze and calibrate new PMS7003s.

Although thorough testing has been done, the full system has not been deployed in real traffic for long periods of time. Therefore, the robustness of the system in terms of operation is not determined.

6 Conclusion and Outlook

We have integrated the visual and environmental data measurements into one system to simplify installation and management. Although the design shows some weakness in the circulation inside the enclosure, it still meets the need to protect the electronics while logging the data. Although the particular PMS7003 we used for our system did not need calibration, the analyzing process still needs to be performed for every new node built. This will slow down the building process significantly.

In the future, we aim to improve the current design to eliminate the deficiency. Plus, we plan to try another design with exposed sensors while still withstanding the outdoor environment without damaging the sensors and controller boards. A more straightforward calibration process for PMS7003 boosts the fabrication process, which is essential for placing multiple nodes around. Another goal is to optimize this gadget to become an edge device. That means each node can process the raw logged data and save the processed data to Azure Blob storage with less storage.

Appendix

See Fig. 13.

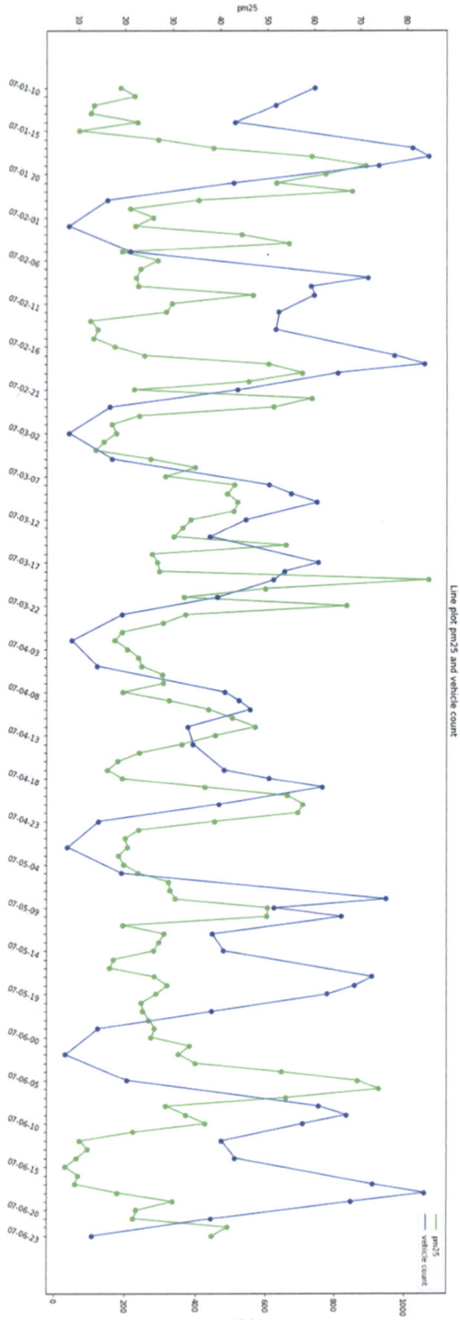

Fig. 13 Time-series plot of PM2.5 and number of vehicle crossing

References

1. Sahanavin N et al (2018) Relationship between PM10 and PM2.5 levels in high-traffic area determined using path analysis and linear regression. J Environ Sci 69: 105–114
2. Zhang H, Srinivasan R, Ganesan V (2021) Low Cost, Multi-Pollutant Sensing System Using Raspberry Pi for Indoor Air Quality Monitoring. Sustainability (Basel, Switzerland), 13(1), 370. https://doi.org/10.3390/su13010370
3. Momirski LA, Berčič T (2021) Southern Inner Ring Road in Ljubljana, Data Set from Traffic Sensors Installed as Part of the Citizen Science Project WeCount". Data Brief 41(2022):107878–107878. https://doi.org/10.1016/j.dib.2022.107878
4. TSI DustTrak product page, https://tsi.com/products/aerosol-and-dust-monitors/aerosol-and-dust-monitors/
5. ST-Microelectronics company website, https://www.st.com
6. STMicroelectronics, "Medium-density performance line Arm®-based 32-bit MCU with 64 or 128 KB Flash, USB, CAN, 7 timers, 2 ADCs, 9 com. interfaces", STM32F103x8 STM32F103xB datasheet, March 2022
7. BME-280 sensor page on Bosch Sensortec company website, https://www.bosch-sensortec.com/products/environmental-sensors/humidity-sensors-bme280/
8. BME-280 datasheet, https://www.mouser.com/datasheet/2/783/BST-BME280-DS002-1509607.pdf
9. Tyler Glenn, https://github.com/finitespace/BME280
10. Maxim Intergrated company website, https://www.maximintegrated.com/
11. Michael Miller, https://github.com/Makuna/Rtc
12. PMS7003 sensor page on Plantower company website, https://plantower.com/en/products_33/76.html
13. Marek Badura, Piotr Batog, Anetta Drzeniecka-Osiadacz, Piotr Modzel, "Evaluation of Low-Cost Sensors for Ambient PM2.5 Monitoring", Journal of Sensors, vol. 2018, Article ID 5096540, 16 pages, 2018. https://doi.org/10.1155/2018/5096540
14. PMS7003 datasheet, https://download.kamami.pl/p564008-PMS7003%20series%20data%20manua_English_V2.5.pdf
15. Raspberry Pi camera V2 module page on Raspberry Pi company website, https://www.raspberrypi.com/products/camera-module-v2/
16. Raspberry Pi camera V2 documentation page, https://www.raspberrypi.com/documentation/accessories/camera.html
17. Raspberry Pi model 3B+ page on Raspberry Pi company website, https://www.raspberrypi.com/products/raspberry-pi-3-model-b-plus/
18. Azure Blob Storage pricing page, https://azure.microsoft.com/en-us/pricing/details/storage/blobs/
19. Le CD, Pham HV, Pham DA, Le AD, Vo HB (2022) A PM2.5 concentration prediction framework with vehicle tracking system: From cause to effect," 2022 RIVF International Conference on Computing and Communication Technologies (RIVF), Ho Chi Minh City, Vietnam, pp. 714–719. https://doi.org/10.1109/RIVF55975.2022.10013864
20. Sagi Zeevi, https://github.com/sagi-z/BackgroundSubtractorCNT (2016)
21. Alex Bewley, Zongyuan Ge, Lionel Ott, Fabio Ramos, Ben Upcroft. Simple object and realtime tracking (2017)
22. Báthory C, Dobó Z, Garami A, Palotás Á, Tóth P (2022) Low-cost monitoring of atmospheric PM—development and testing. J Environ Manage **304**, 114158. Elsevier BV. https://doi.org/10.1016/j.jenvman.2021.114158

MIX
Papier aus verantwortungsvollen Quellen
Paper from responsible sources
FSC® C105338

If you have any concerns about our products,
you can contact us on
ProductSafety@springernature.com

In case Publisher is established outside the EU,
the EU authorized representative is:
Springer Nature Customer Service Center GmbH
Europaplatz 3, 69115 Heidelberg, Germany

Printed by Libri Plureos GmbH
in Hamburg, Germany